T0260938

PRACTICAL
REASONING
IN
BIOETHICS

Medical Ethics Series

DAVID H. SMITH AND ROBERT M. VEATCH, EDITORS

Norman L. Cantor. *Advance Directives and the Pursuit of Death with Dignity*

Norman L. Cantor. *Legal Frontiers of Death and Dying*

Arthur L. Caplan. *If I Were a Rich Man Could I Buy a Pancreas? And Other Essays on the Ethics of Health Care*

Cynthia B. Cohen, ed. *Casebook on the Termination of Life-Sustaining Treatment and the Care of the Dying*

Cynthia B. Cohen, ed. Commissioned by the National Advisory Board on Ethics in Reproduction. *New Ways of Making Babies: The Case of Egg Donation*

Roger B. Dworkin. *Limits: The Role of the Law in Bioethical Decision Making*

Larry Gostin, ed. *Surrogate Motherhood: Politics and Privacy*

Christine Grady. *The Search for an AIDS Vaccine: Ethical Issues in the Development and Testing of a Preventive HIV Vaccine*

A Report by the Hastings Center. *Guidelines on the Termination of Life-Sustaining Treatment and the Care of the Dying*

Paul Lauritzen. *Pursuing Parenthood: Ethical Issues in Assisted Reproduction*

Joanne Lynn, M.D., ed. *By No Extraordinary Means: The Choice to Forgo Life-Sustaining Food and Water, Expanded Edition*

William F. May. *The Patient's Ordeal*

Richard W. Momeyer. *Confronting Death*

Thomas H. Murray, Mark A. Rothstein, and Robert F. Murray, Jr., editors. *The Human Genome Project and the Future of Health Care*

S. Kay Toombs, David Barnard, and Ronald A. Carson, eds. *Chronic Illness: From Experience to Policy*

Robert M. Veatch. *The Patient as Partner: A Theory of Human-Experimentation Ethics*

Robert M. Veatch. *The Patient-Physician Relation: The Patient as Partner, Part 2*

James F. Childress

PRACTICAL
REASONING
IN
BIOETHICS

■

Indiana University Press

BLOOMINGTON AND INDIANAPOLIS

The paper used in this publication meets the minimum requirements of
American National Standard for Information Sciences—Permanence of
Paper for Printed Library Materials, ANSI Z39.48-1984.

MANUFACTURED IN THE UNITED STATES OF AMERICA

Library of Congress Cataloging-in-Publication Data

Childress, James F.
   Practical reasoning in bioethics / James F. Childress.
      p.   cm. — (Medical ethics series)
   Includes bibliographical references and index.
   ISBN 0-253-33218-4 (cl : alk. paper)
   1. Medical ethics. 2. Bioethics.   I. Title.   II. Series.
   [DNLM: 1. Ethics, Medical. 2. Bioethics.   W 50 C536pa 1997]
R724.C477   1997
174'.2—dc20
DNLM/DLC
for Library of Congress   96-25001

1 2 3 4 5   02 01 00 99 98 97

CONTENTS

PREFACE

The essays in this volume cover several topics in biomedical ethics, a field I hesitated to enter so fully in the 1970s—even after writing some well-received essays—because I thought that some new fad in practical or applied ethics would soon sweep it away. Obviously, my concern was misplaced: The field has endured and even flourished, and, as the following essays indicate, I have continued to participate in it.

These essays represent some common methodological assumptions and approaches as well as certain substantive convictions. The relevance of the term *principles* is fairly obvious: These essays continue major lines of some of my other writings, including *Principles of Biomedical Ethics,* coauthored with Tom L. Beauchamp (4th ed., Oxford University Press, 1994), in taking a principles approach—what some critics label "principlism." One essay (chapter 2) sketches varieties of principlism and develops and defends one particular version over against several types of criticism, including that of casuists. Other essays focus on the presuppositions and implications of different principles, especially (a) respect for autonomy, which receives primary attention in chapters 3–7 and 14–15, (b) justice and fairness, which are featured in chapters 10–13, where the focus is access to and rationing health care, and (c) beneficence and nonmaleficence, which receive significant attention in various chapters, especially 7–9, which are devoted to issues in withholding or withdrawing life-prolonging procedures, including medically administered nutrition and hydration. Scattered throughout these essays is an argument about the way such general principles are specified in rules and both principles and rules are balanced when conflicts cannot be eliminated. Balancing also involves a procedure of reasoning that requires attention to competing weights, to available alternatives, to effectiveness of proposed infringements, etc. (see chapters 2 and 6, among others).

Metaphors and *analogies,* which receive more comprehensive and systematic treatment in chapter 1, prior to the discussion of principles in chapter 2, are also central to my substantive analyses and arguments. In addition to the legal-like method suggested by principles and rules, practical reasoning also proceeds imaginatively, especially through various metaphors and analogies.

According to David Eerdmann, as quoted in the first chapter, imagination involves "reasoning in metaphors." I have long focused on metaphors in biomedical ethics, for instance, in my several writings on paternalism, including *Who Should Decide? Paternalism in Health Care* (Oxford University Press, 1982). Metaphors are not mere decorations; rather, they are central to our very processes of interpreting situations and appropriately responding. And they can be assessed by how well they illuminate rather than distort both what is going on and what should be done.

In the current volume, metaphors—and the models built on them—illuminate various interpretations of the relationships between clinicians and their patients. One example is my (and Mark Siegler's) analysis and assessment of several metaphors and models of relationships in health care, along with a constructive proposal that attends to respect for autonomy of professionals as well as of patients in a process of negotiation. In addition, the metaphor of warfare for medicine, health care, and research is crucial throughout these essays. It appears in debates about triage and rationing, for instance, as well as in debates about societal and professional efforts to combat the AIDS epidemic by identifying individuals who are HIV infected on the grounds that, as carriers, they are enemies who can and should be somehow incapacitated.

In recent years, casuists such as Albert R. Jonsen and Stephen Toulmin, in their influential book, *The Abuse of Casuistry* (University of California Press, 1988), have argued that analogy is central in practical reasoning. And they claim that casuistry, in contrast to a principles-based approach, starts from paradigm cases and then reasons analogically in identifying and weighting relevant similarities and differences. However, as I argue in chapters 1 and 2, analogical reasoning is indispensable in principled reasoning as well, in part because the formal principle of universalizability (or formal justice) requires that *if* we judge X as morally wrong (or right) we are committed to judging Y as also morally wrong (or right) *if* it is relevantly similar to X. Indeed, this formal requirement generates general moral norms once we judge X to be morally wrong (or right) in a particular case, because all relevantly similar Xs are also morally wrong (or right).

My first essay in biomedical ethics, "Who Shall Live When Not All Can Live?" (1970), was essentially an extended reflection on the analogy between our judgments about real and hypothetical lifeboat cases and our judgments about the allocation of scarce lifesaving medical resources (see chapter 10). A similar point can be made about Paul Ramsey's discussion of microallocation in *The Patient as Person* (Yale University Press, 1970). However, neither Ramsey nor I supposed that analogical reasoning is in any way incompatible with principled reasoning. Practical reasoning involves both.

Analogical reasoning is also central in other essays in this volume. For instance, I examine various policies of mandatory screening and testing for HIV antibodies by analogy with other actual or proposed policies of mandatory screening and testing for sexually transmitted diseases and genetic diseases, consider triage in neonatal intensive care in comparison to triage in wartime in order to determine whether social utility has a place beside medical utility in both settings, explore approaches to human fetal tissue transplantation research by analogy to the use of other cadaveric organs and tissues, and so forth.

The substantive perspective of these essays can perhaps be called "liberal communitarian." It is founded on a strong presumption in favor of respect for personal autonomy, including several rules derived from that principle, such as liberty and privacy, but it also concedes that this principle and its derivative rules are only prima facie binding, rather than absolute, and can thus sometimes be overridden for the sake of communal goods. As already noted, I defend a procedure of practical reasoning when principles such as respect for autonomy conflict with other principles and it is not possible to protect all of them in a particular case or policy (see chapters 2 and 4–7, especially chapter 6, which develops this procedure in the context of the debate about mandatory HIV screening and testing to protect other individuals and the public's health). Communal values also play a major role, even if not always a victorious one, in my arguments about access to health care, about the symbolic significance of artificial nutrition and hydration, about regulations involving nursing homes, about a market in organs for transplantation, about the ownership of donated organs, and the like.

My perspective is *liberal* communitarian in that it stresses the presumptive, though nonabsolute, priority of autonomy, liberty, and privacy, and requires that a significant, though not impossible, burden of proof be discharged to justify coercive actions or policies. It is liberal *communitarian* in its claims, for example, that social solidarity and communal beneficence, along with justice, require greater access to health care and fairer modes of allocation and distribution of health care. It stresses the importance of *expressing* rather than merely, or even mainly, *imposing* community—a distinction that becomes very important in the debate about various policies of HIV testing and screening in chapter 6—even though it also recognizes the legitimacy of some impositions of community.

Since most of the essays in this volume date mainly from the last decade—though a few are older—I had to decide how much to revise them for republication. I made several different decisions, depending on several features of each essay. A few essays were so recent that they required relatively minor revisions, often by way of expansion; examples are 1, 2, and 6. Some chapters combine earlier essays, in order to avoid duplication while making points previously covered in only one of the essays; examples are chapters 4, 14, and 16. Some I revised completely so that they are virtually new essays; chapters 12, 13, and 15 fall into this category. Some I left intact, but then added major updates to respond to critics or further develop my arguments. These include chapters 5 and 7, two of the four coauthored chapters (the other two appear as they appeared originally), and chapters 10 and 11. I was especially concerned to avoid excessive repetition, but some—though I hope only modest—repetition was unavoidable.

My debts of gratitude are many. First, I am deeply grateful to the coauthors of four chapters—Courtney Campbell, Joanne Lynn, Bettina Schoene-Seifert, and Mark Siegler—for our enjoyable and fruitful collaborations as well as for their willingness to allow me to reprint our jointly authored essays.

Second, I appreciate the criticisms and suggestions offered by many others in biomedical ethics. Even when I have not been able fully to accept their criticisms and suggestions, I hope that my work is stronger as a result of trying to address them. I note a number of these individuals in various chapters, sometimes in updates that directly respond to major criticisms or alternative positions. Of course, I have learned so much, often without even being aware of it at the time, from close to twenty years of exceedingly fruitful collaboration with Tom L. Beauchamp in preparing the various editions of *Principles of Biomedical Ethics*. Our arguments back and forth over draft after draft have provided my best education in ethics and biomedical ethics.

Third, I have also benefited greatly from feedback from students in courses and from other audiences. The questions and challenges from both undergraduate and graduate students at the University of Virginia are regularly quite stimulating. And various audiences who heard some of the ideas in this volume in lectures or papers also helped me sharpen my thoughts. My three National Endowment for the Humanities Summer Seminars for College Teachers (1981, 1983, 1986), all on the topic of "Principles and Metaphors in Biomedical Ethics," enabled me to make further progress in my reflections on metaphor and analogy, along with principles, in bioethical reasoning. Several participants in those seminars have remained good friends as well as helpful critics.

Fourth, several research assistants facilitated the preparation of these essays over the years, but I particularly want to thank Courtney Campbell and Felicia Cohn for so many years of remarkably valuable assistance.

Fifth, I owe a tremendous debt of gratitude to several colleagues and friends at the University of Virginia, who have provided indispensable support, particularly (but not only) during and following my wife's illness: John Fletcher in biomedical ethics; Don Detmer and Carolyn Engelhard in the Virginia Health Policy Center; and Diana McKenzie in the office of the Department of Religious Studies. Among so many other valued and helpful colleagues in the Department of Religious Studies and the School of Medicine, I especially want to thank Larry Bouchard, with whom I last year cotaught a graduate seminar on "Narrative in Theology and Ethics," which offered the most enjoyable and stimulating context for teaching and learning I have ever experienced.

Finally, I dedicate this book to the memory of my wonderful wife, partner, and friend, Georgia Harrell Childress, who tragically died August 17, 1994, during our thirty-sixth year of marriage, following a battle with non-Hodgkins lymphoma—words simply fail to express how much she meant and still means to me.

ACKNOWLEDGMENTS

I am grateful to the following journals and publishers for permission to use previously published materials:

Chapter 1. Much of this chapter originally appeared in "Metaphor and Analogy" in *Encyclopedia of Bioethics*, 2nd ed., ed. Warren T. Reich (New York: Simon & Schuster Macmillan, 1995), vol. 3, pp. 1765–73.

Chapter 2. "Ethical Theories, Principles, and Casuistry in Bioethics: An Interpretation and Defense of Principlism," in *Religious Methods and Resources in Bioethics,* ed. Paul Camenisch (Boston: Kluwer, 1994), pp. 181–201.

Chapter 3. With Mark Siegler, "Metaphors and Models of Doctor-Patient Relationships: Their Implications for Autonomy," *Theoretical Medicine* 5 (1984): 17–30.

Chapter 4. The first part of this chapter is drawn, in part, from "The Place of Autonomy in Bioethics," *Hastings Center Report* 20 (January/February 1990): 12–17. The second part is drawn from "If You Let Them, They'd Lie in Bed All Day . . . ," *Everyday Ethics: Resolving Dilemmas in Nursing Home Life,* ed. Rosalie Kane and Arthur L. Caplan (New York: Springer, 1990), pp. 79–89.

Chapter 5. With Bettina Schoene-Seifert, "How Much Should the Cancer Patient Know and Decide?" *Ca-A Cancer Journal for Clinicians* 36, no. 2 (March/April 1986): 85–94.

Chapter 6. "Mandatory HIV Screening and Testing," in *AIDS and Ethics,* ed. Frederic G. Reamer (New York: Columbia University Press, 1991), pp. 50–76.

Chapter 7. With Courtney Campbell, "'Who Is a Doctor to Decide Whether a Person Lives or Dies?' Reflections on Dax's Case," in *Dax's Case: Essays in Medical Ethics and Human Meaning,* ed. Lonnie Kliever (Dallas: SMU Press, 1989).

Chapter 8. With Joanne Lynn, "Must Patients Always Be Given Food and Water?" *Hastings Center Report* 13 (October 1983): 17–21.

Chapter 9. "When Is It Morally Justifiable to Discontinue Medical Nutrition and Hydration?" in *By No Extraordinary Means: The Choice to Forgo Life-Sustaining Food and Water,* ed. Joanne Lynn (Bloomington: Indiana University Press, 1986), pp. 67–83.

Chapter 10. "Who Shall Live When Not All Can Live?" *Soundings* 53, no. 4 (Winter 1970).

Chapter 11. A much briefer version of this chapter appeared as "Triage in Neonatal Intensive Care: The Limitations of a Metaphor," *Virginia Law Review* 69 (April 1983): 547–61.

Chapter 12. "Fairness in the Allocation and Delivery of Health Care: The Case of Organ Transplantation," in *A Time to Be Born and a Time to Die,* ed. Barry S. Kogan (Hawthorne, NY: Aldine de Gruyter, 1991), pp. 179–204.

Chapter 13. An earlier and very different version of this chapter appeared as "Rights to Health Care in a Democratic Society," in *Biomedical Ethics Reviews,* vol. 2, ed. Robert Almeder and James Humber (Humana Press, 1984).

Chapter 14. This chapter incorporates materials from "Ethical Criteria for Procuring and Distributing Organs for Transplantation," *Journal of Health Politics, Policy and Law* 14 (Spring 1989): 87–113, as reprinted in *Organ Transplantation Policy: Issues and Prospects,* ed. James F. Blumstein and Frank A. Sloan (Durham: Duke University Press, 1989), and from "Organ and Tissue Procurement: Ethical and Legal Issues Regarding Cadavers," *Encyclopedia of Bioethics,* 2nd ed., ed. Warren T. Reich (New York: Simon & Schuster Macmillan, 1995), vol. 4, pp. 1857–65.

Chapter 15. This chapter is a substantially revised version of "The Body as Property: Some Philosophical Reflections," *Transplantation Proceedings* 24 (October 1992): 2143–48.

Chapter 16 uses the basic structure and content of "Ethics, Public Policy, and Human Fetal Tissue Transplantation Research," *Kennedy Institute of Ethics Journal* 1 (June 1991): 93–121. It also incorporates materials from "Consensus in Ethics and Public Policy: The Deliberations of the U.S. Human Fetal Tissue Transplantation Research Panel," in *The Concept of Consensus: The Case of Technology Interventions into Human Reproduction,* ed. Kurt Bayertz (Boston: Kluwer, 1994).

Part One

■

PRINCIPLES,
METAPHORS,
AND ANALOGIES

CHAPTER ONE

□

Metaphor and Analogy in Bioethics

Many of our practices and much of our discourse in health care hinge on metaphors and analogies, whose significance is sometimes overlooked because they are considered merely decorative or escape notice altogether. Despite their relative neglect, they significantly shape our interpretations of what is going on as well as what should go on. In recent years they have received increasing attention, particularly from critics of principles-oriented approaches to bioethics who stress the role of imagination, emotion, and the like in morality and ethical reflection. I will examine metaphors before considering analogies, particularly in analogical reasoning, noting their overlap where appropriate; then I will, more briefly, consider symbols in bioethics.

All of these categories—metaphor, analogy, and symbol—capture what we might call aesthetic dimensions of moral discourse in health care and in bioethics. Even though their proponents often believe that these aesthetic dimensions are at odds with principles-oriented approaches, I disagree: They are central to moral discourse and action in health care and in bioethics, and they are too often overlooked, especially in conceptions of bioethics oriented to principles, rules, or theories. However, I will contend that they are in no way incompatible with, but actually exist in close relation with, principles, rules, and theories, and that they need and even presuppose each other. Indeed, the essays in this volume all display one or more of these aesthetic dimensions in close relation to modes of reasoning often viewed as more legal-like. Metaphor, analogy, and symbol can illuminate our moral discourse and activity, just as much as categories from law, philosophy, and theology. Our understanding and direction of actions and policies in health care, medicine, and the life sciences need the vision provided by metaphors, analogies, and symbols, as well as by principles, rules, and theories.[1]

METAPHORS IN BIOETHICS

The nature and function of metaphors. Perhaps because medicine and health care involve fundamental matters of life and death for practically everyone, frequently in mysterious ways, they are often described in metaphors. For instance, physicians may be viewed as playing God, or acting as parents, and nurses may be seen as advocates for patients; medicine itself may be interpreted as warfare against disease. Metaphors involve imagining something as something else, for example, viewing human beings as wolves or life as a journey. "The essence of metaphor," according to Lakoff and Johnson, "is understanding and experiencing one thing through another."[2] More precisely, metaphors are figurative expressions that interpret one thing in terms of something else.[3]

In the large recent philosophical literature on metaphor, critics have challenged some traditional conceptions, contending that metaphors are more than merely ornamental or affective ways to state what could be stated in a more literal or comparative way and that they can be and often are cognitively significant.[4] According to the traditional substitution view, a metaphorical expression is merely a substitute for some equivalent literal expression. For example, the metaphorical expression "John is a fox" substitutes for the literal expression "John is sly and cunning." One common version of the substitution view, what Max Black calls a comparison view (elements of which can be found in Aristotle), construes metaphor as the presentation of an underlying analogy or similarity. Hence, metaphor is "a condensed or elliptical simile,"[5] or it is a "comparison statement with parts left out."[6] "John is a fox," for example, indicates that "John is like a fox in that he is sly and cunning." According to such views, metaphors are dispensable ways to express what could be expressed differently, but they often appeal to the emotions more effectively than their equivalent literal expressions or comparisons would do.

By contrast many recent theories of metaphor stress its cognitive significance. For instance, in an early and very influential essay, philosopher Max Black defended an interaction view of metaphor, in which two juxtaposed thoughts interact to produce new meanings through the metaphor's "system of associated commonplaces" or "associated implications."[7] The metaphor—for instance, "wolf" in "man is a wolf"—serves as a "filter" for a set of associated implications that are transferred from the secondary subject ("wolf") to the principal subject ("man") in the sentence. (In a full interaction or inter-animation view of metaphor, the transfer of meaning occurs both ways, not merely from the secondary subject to the principal subject.)[8]

Metaphors highlight and hide features of the principal subject, such as the physician who is viewed as a parent or as a friend, by their systematically related implications.[9] When argument is conceived as warfare, for example, the metaphor highlights the conflict involved in argument while it hides the cooperation and collaboration, involving shared rules, that are also indispensable to argument. Our metaphors thus shape how we think, what we experience, and what we do by what they highlight and hide. They use us, just as we use them.

Metaphors are often associated with models. For instance, we have both metaphors and models of the doctor-patient relationship. The physician may be viewed through the metaphor of father and the patient through the metaphor of child, and their relationship may be interpreted through the model of paternalism. Models, for our purposes, state the network of associated commonplaces and implications in more systematic and comprehensive ways—according to Max Black, "every metaphor is the tip of a submerged model."[10]

Metaphors and models may be good or bad, living or dead. Both metaphors and models can be assessed by how well they illuminate what is going on and what should go on. We can distinguish descriptive and normative uses of metaphors and models without admitting a sharp separation between fact and value. For instance, the metaphor of physician as father (or parent), and the model of paternalism (or parentalism), may accurately describe some relationships in medicine, or they may suggest ideal relationships in the light of some important principles and values.

Medicine as war, business, etc. The metaphor of warfare illuminates much of our conception of what is, and should be, done in health care. Consider the way this metaphor emerges in the day-to-day language of medicine: The physician as the captain leads the battle against disease; orders a battery of tests; develops a plan of attack; calls on the armamentarium or arsenal of medicine; directs allied health personnel; treats aggressively; and expects compliance. Good patients are those who fight vigorously and refuse to give up. Victory is sought; defeat is feared. Sometimes there is even hope for a "magic bullet" or a "silver bullet." Only professionals who stand on the firing line or in the trenches can really appreciate the moral problems of medicine. And they frequently have "war stories" to relate. Medical organization, particularly in the hospital, resembles military hierarchy, and medical training, particularly with its long, sleepless shifts in residencies, approximates military training more than any other professional education in our society.[11]

As medicine wages war against germs that invade the body and threaten its defenses, so the society itself may also declare war on cancer, or on AIDS, under the leadership of its chief medical officer, who in the United States is the

Surgeon General. Articles and books even herald the "Medical-Industrial Complex: Our National Defense." As Susan Sontag notes, "where once it was the physician who waged bellum contra morbum, the war against disease, now it's the whole society."[12]

The military metaphor first became prominent in the 1880s when bacteria were identified as agents of disease that threaten the body and its defenses. It both illuminates and distorts health care. Its positive implications are widely recognized—for instance, in supporting a patient's courageous and hopeful struggle against disease and in galvanizing societal support to fight against disease. But the metaphor is also problematic. Susan Sontag, who was diagnosed with cancer in the late 1970s, reports that her suffering was intensified by the dominance of the metaphor of warfare against cancer. Cancer cells do not just multiply; they are "invasive." They "colonize." The body's "defenses" are rarely strong enough. But since the body is under attack ("invasion"), by "alien" invaders, counterattack is justified. Treatments are also often described in military language:

> Radiotherapy uses the metaphors of aerial warfare; patients are "bombarded" with toxic rays. And chemotherapy is chemical warfare, using poisons. Treatment aims to "kill" cancer cells (without, it is hoped, killing the patient). Unpleasant side effects of treatment are advertised, indeed overadvertised. ("The agony of chemotherapy" is a standard phrase.) It is impossible to avoid damaging or destroying healthy cells (indeed, some methods used to treat cancer can cause cancer), but it is thought that nearly any damage to the body is justified if it saves the patient's life. Often, of course, it doesn't work. (As in: "We had to destroy Ben Suc in order to save it.") There is everything but the body count.[13]

Such "military metaphors," Sontag suggests, "contribute to the stigmatizing of certain illnesses and, by extension, of those who are ill." Other ill individuals have found the military metaphor unsatisfactory for other reasons. For instance, as a teenager, Lawrence Pray originally tried to conquer his diabetes, but his struggles and battles were futile and even counterproductive. Then over time he came to view his diabetes not as an "enemy" to be "conquered" but as a "teacher." Only then did he find a personally satisfactory way of living.[14]

Still others with illness, by contrast, have found the military metaphor empowering and enabling. In her wide-ranging study of pathographies, that is, autobiographical descriptions of personal experiences of illness, treatment, and dying, Anne Hunsaker Hawkins identifies several "metaphorical paradigms" that offer themes of "an archetypal, mythic nature."[15] In addition to illness as a battle, she concentrates on illness as a game or sport (a subset of the military metaphor), illness as a journey into a distant country, illness as rebirth or regeneration, and, on a somewhat different level, healthy-mindedness as an alternative to contemporary medicine. While pathographies are individualized statements, they provide "an immensely rich reservoir of the metaphors and models that surround illness in contemporary culture." These various metaphorical paradigms structure individuals' interpretations of their experiences of illness. Patterns emerge in individuals' selection of metaphors. They vary in part according to the illness involved—for example, the military metaphor is more common in descriptions of experiences with cancer and AIDS, while the rebirth metaphor is more common in descriptions of a critical life-threatening event, such as a heart attack. Furthermore, the military metaphor is generally more prevalent than the journey metaphor because it better fits the experience of modern medicine—for instance, it is easier to construe the physician as a "general" in a war than as a "guide" on a journey. Nevertheless, these various metaphors are often mixed and complementary. And they can be evaluated, Hawkins suggests, according to their capacity to enable and empower ill persons, for instance, by restoring a sense of personal dignity and worth. While expressing larger sociocultural patterns, the individual's choice of a particular metaphor is a creative act of assigning meaning to his or her illness.

The metaphor of warfare has been further challenged in modern medicine because of its apparent support for overtreatment, particularly of terminally ill patients, where death is the ultimate enemy and trauma, disease, or illness the immediate enemy. Physicians and families under the spell of this metaphor frequently find it difficult to let patients die. "Heroic" actions, with the best available weapons, befit the military effort that must always be undertaken against the ultimate enemy. Death signals defeat and forgoing treatment signals surrender. Some clinicians even feel more comfortable withholding (i.e., not starting) a treatment for cancer, for instance, than they do withdrawing (i.e., stopping) the same treatment, in part because withdrawing treatment implies retreat. (See the discussions in chapters 8 and 9.)

According to its critics, our invocation of the military metaphor often fails to recognize moral constraints on waging war. "Modern medicine," William F. May writes, "has tended to interpret itself not only through the prism of war but through the medium of its modern practice, that is, unlimited, unconditional war," in contrast to the just-war tradition.[16] In the spirit of modern total war, "hospitals and the physician-fighter wage unconditional battle against death." One result is that many patients seek assisted suicide or active euthanasia in order to escape from this warfare's terrorist bombardment. Traditional moral limits in the conduct of war include the principle of discrimination, which authorizes direct attacks on combatants but not on non-combatants. In medical care, the opposing combatant is the disease or death, not the patient. However, the patient is regularly the battleground and sometimes even becomes the enemy. This transformation into the enemy may occur if the patient betrays the military effort by not fighting hard enough or even by surrendering before the war ends. Finally, in accord with the just-war tradition's requirement of reasonable prospect of success and proportionality, the treatment should offer the patient a reasonable chance of success, and his or her suffering and other burdens must be balanced against the probable benefits of prolongation of life.[17]

Other problematic or ambiguous implications of the military metaphor appear in the allocation of resources for and within health care. It is not surprising that the two major terms for allocation and distribution of health care under conditions of scarcity emerged from, or were decisively shaped by, military experiences. These are triage and rationing. As Richard Rettig and Kathleen Lohr note,

> Earlier, policymakers spoke of the general problem of allocating scarce medical resources, a formulation that implied hard but generally manageable choices of a largely pragmatic nature. Now the discussion increasingly is of rationing scarce medical resources, a harsher term that connotes emergency—even war-time—circumstances requiring some societal triage mechanism.[18]

A later chapter (11) will examine the military context as well as the emergency context of triage. For now I will only sketch some of the implications of the military metaphor for allocation, particularly macroallocation.

First, under the military metaphor, society's health care budget tends to be converted into a defense budget to prepare for and conduct war against disease, trauma, and death. As a consequence, the society may put more resources into health care in relation to other goods than it could justify, especially under a different metaphor, such as nursing or business (see below). Indeed, the society may overutilize health care, especially because technological care may contribute less to the national defense of health itself, through the reduction of morbidity and premature mortality, than other factors, such as the reduction of poverty.

Second, within the health care budget, the military metaphor tends to assign priority to critical care over preventive and chronic care. It tends to concentrate on critical interventions to cure disease, perhaps in part because it tends to view health as the absence of disease rather than as a positive state. It tends to neglect care when cure is impossible.

A third point is closely connected: In setting priorities for research and treatment, the military metaphor tends to assign priority to killer diseases, such as cancer and AIDS, over chronic diseases. Franz Ingelfinger once suggested that if we concentrated our research and treatment more on disabling diseases, such as arthritis, than on killer diseases, then national health expenditures would reflect the same values that individuals affirm: "It is more important to live a certain way than to die a certain way."[19] Anne R. Sommers has suggested that stroke is "a metaphor for the most difficult problems and challenges of geriatric medicine."[20] Although strokes are not limited to the elderly, they are more common among the elderly. Each year in the United States there are between 500,000 and 600,000 victims of stroke, 80 to 90 percent of them surviving their initial catastrophe, often with paralysis and aphasia, which have a terrible impact on both victims of stroke and their families. Approximately 2.5 million victims of stroke are alive today, 90 percent of them with varying degrees of incapacity and misery. Even though it has been called the single most costly disease in the United States, stroke received in 1979 only $18 million in research expenditures, in contrast to cancer, which received $937 million, and heart disease, which received $340 million. A major reason for this pattern of allocation is that, after the first acute phase, the stroke victim does not fit into the prevalent model of medical care, which emphasizes the specialist, who uses various technological weapons to fight specific problems. Hence this pattern continues.

Fourth, medicine as war concentrates on technological interventions over against nontechnological modes of care and, within technologies, it tends to concentrate on more dramatic technologies, such as intensive care units and organ transplants, rather than less dramatic technologies, such as prostheses.

In short, the military metaphor has some negative or ambiguous implications for a moral approach to health care decisions: It tends to assign priority to health care (especially medical care) over other goods and, within health care, to critical interventions over chronic care, killer over disabling diseases, technological interventions over caring, and heroic treatment of dying patients rather than allowing them to die in peace.[21]

Some of the negative or ambiguous implications of the war metaphor for health care can be avoided if, as noted earlier, the metaphor is interpreted in accord with the limits set by the just-war tradition. However, the war metaphor may require supplementation as well as limitation. It is not the only prominent metaphor for health care, and since the early 1980s its dominance has been threatened by the language of economics and business, as reflected in the language of a health care industry: Providers deliver care to consumers, seek or are forced to seek productivity in light of cost-effectiveness or cost-benefit analyses, and may be concerned with "resource management, managed care systems, and market strategies."[22]

The business metaphor also highlights and hides various features of contemporary health care. Many critics of this metaphor worry that the language of efficiency will replace the language of care and compassion for the sick and equity in distribution of health care. Nevertheless, this metaphor has become more and more pervasive and persuasive as the structure of medicine and health care has changed and concerns about costs have become more central in societal discussions. Patients now often fear undertreatment as hospitals and professionals seek to reduce costs, in contrast to their fears of overtreatment under the war metaphor.

Both military and economic metaphors illuminate contemporary health care, but they may not be adequate, even together, to guide and direct health care. Whether any particular metaphor is adequate or not will depend in part on the principles and values it highlights and hides. Others have proposed nursing, a subset of health care, as a supplementary metaphor for the whole of health care, because of its attention to caring more than curing and to hands-on rather than technological care. Even though this metaphor of nursing—in contemporary discourse it is often more than a synecdoche—is also inadequate by itself, it could direct the society to alternative priorities in the allocation of resources for and within health care, particularly for chronic care.

Another possibility is an ecologic metaphor. Focusing on failure of the

Clinton administration's health care plan and the remarkably rapid growth of managed care, George J. Annas contends that we should reframe our debate on health care reform by replacing both of our dominant metaphors—warfare, which assigns medicine the goal of a "healthy population," and the market, which assigns medicine the goal of a "healthy bottom line."[23] Annas's sketch of the systematic entailments of each metaphor fits well with the implications I identified earlier. Just as I argued that the military metaphor is often misused because warfare is distorted when the just-war or limited-war tradition is neglected, so too the market metaphor is misused when the operative picture of markets diverges so dramatically from our real markets. The market metaphor is limited because in reality, Annas stresses, our "American markets are highly regulated, major industries enjoy large public subsidies, industrial organizations tend toward oligopoly, and strong laws that protect consumers and offer them recourse through product-liability suits have become essential to prevent profits from being too relentlessly pursued."

Why not just combine the military and market metaphors? The Clinton health care plan, according to Annas, perhaps unwittingly, attempted an amalgamation, with other metaphors too, in impossible and inconsistent ways—for instance, in offering both security (military metaphor) and consumer choice (market metaphor). The solution, Annas argues, is to find a new metaphor, and his candidate is an ecologic metaphor, which involves such key terms as integrity, balance, natural, limited (resources), quality (of life), diversity, renewable, sustainable, responsibility (for future generations), community, and conservation. While both the military and market metaphors only reinforce typical but harmful American traits of wastefulness, obsession with technology, fear of death, and individualism, the ecologic metaphor could help us face and perhaps even alter those traits. Specifically in regard to medicine, "the ecologic metaphor can encourage an alternative vision of resource conservation, sustainable technology, acceptance of death as natural and necessary, responsibility for others, and at least some degree of community." However, the process of altering sociocultural metaphors is complex and uncertain, particularly when such metaphors as warfare and the market appear to be relatively accurate descriptively but problematic normatively.

The war against AIDS. Even as the military metaphor was partially displaced by business and economic metaphors in the changing structure of health care, it gained favor as a way to describe and direct society's response to the major epidemic of the acquired immunodeficiency syndrome (AIDS). Societies often resort to the metaphor of war when a serious threat to a large number of human lives requires the mobilization of vast societal resources, especially when that threat comes from biological organisms, such as viruses, that invade the human body. And AIDS activists have appealed to the military metaphor in an effort to galvanize the society and to marshal its resources for an effective counterattack against the human immunodeficiency virus (HIV), which causes AIDS. However, critics charge that the war on AIDS has diverted important resources away from other important wars, such as the war against cancer.

Other controversies have emerged. From the beginning of the war against AIDS, identification of the enemy has been a major goal. Once the virus was identified as the primary enemy, it also became possible to identify human beings who carry or harbor the virus. This technology then led to efforts to identify HIV-infected individuals, even through mandatory screening and testing, as potential enemies of the society. In social discourse and practice, the carrier tends to become an enemy as much as the virus he or she carries, especially since the society views many actions that expose individuals to the risk of HIV infection as blameworthy. Thus, the metaphor of war often co-exists with metaphors of AIDS as punishment and as otherness.[24] In the specific case of war against AIDS, just as in the general war against disease, the military metaphor would be less dangerous if the society adhered to the constraints of the just-war tradition, rather than being tempted by a crusade. (These points are further elaborated in chapter 6, "Mandatory HIV Screening and Testing.")

Culture wars. The war metaphor also appears in sharp portrayals of contemporary conflicts in American society and culture, particularly over matters relating to the human body, which are often examined in biomedical ethics. Prominent among these is abortion, but others are also important. Sociologist James Hunter published in 1994 *Before the Shooting Begins: Searching for Democracy in America's Culture War,* which was a follow-up to his earlier *Culture Wars: The Struggle to Define America,* which attempts to make "sense of the battles over the family, art, education, law, and politics."[25] In the culture war or wars "orthodox" and "progressivists" struggle over "the power to define reality," that is, to interpret society's collective myths and symbols of national identity, and this war is a proper subject of democratic debate. Through such apparently disparate issues as abortion rights and gay-lesbian rights, culture warriors struggle to define both themselves as Americans and the kind of society they want to have. While insisting that the phrase "culture war(s)" accurately describes contemporary American battles to define reality, and while displaying considerable sympathy to themes in the conservative camp, Hunter argues against a "call to arms." Rather than "drum-beating," he calls for wise and courageous leadership to direct "serious and substantive argument." This is critical, "for in a democracy, *how* we contend in public life is as important as *what* we contend for."[26] In short, Hunter uses the metaphor of war descriptively while questioning its normative use, that is, its use as a guide to action in contemporary society, other than to downplay the war itself and to concentrate on how the conflict is conducted.

However, Hunter's description makes it difficult not to take sides. For if the metaphor of war expresses what is "the inevitably dominant reality," if the stakes are as high as Hunter suggests and the "competing moral visions so non-negotiable and rational moral discussion so unlikely," then, Peter Steinfels asks, "isn't the responsible thing to choose sides and plunge in?" Although conservatives have criticized Hunter for his studied neutrality, and although he is clearly deeply dissatisfied with the polarities of the cultural conflict, it is not clear that he can avoid fanning the flames of war by his very description.

According to Steinfels, "by describing the reality, he [Hunter] wants to correct it, not perpetuate it. But can he do this without questioning the adequacy of the military metaphor itself? When culture becomes the continuation of war by other means (to paraphrase Clausewitz), something is seriously wrong."[27]

By contrast, I am not even sure how well the metaphor of warfare helps us understand what is going on in our cultural conflicts over such contested issues as abortion and the use of fetal tissue in transplantation research, but I am quite certain that it does virtually nothing to guide our actions properly in our debates. Not all efforts to define or to redefine society are conflicts, and not all conflicts are best understood as warfare. And when they are properly understood as warfare, it is important to ask whether the moral constraints from the just-war tradition are respected or whether, instead, a crusade or holy war has taken over, with a real threat to our society's fundamental commitment to ordered liberty.

In conclusion, when debating social policies and practices, including those raising bioethical issues, through the metaphor of war, we often forget the moral reality of war itself. Among other lapses, we forget important moral limits in war—both limited objectives and limited means. We largely forget the just-war tradition, with its moral conditions for resort to war and for conducting war. We are tempted by seedy realism, with its doctrine that might makes right, or by an equally dangerous illusion of a crusade or holy war, with its doctrine that right makes might of any kind acceptable. In either case, we neglect important moral constraints, such as right intention, discrimination, and proportionality, which protect the humanity of all parties in conflict. And in general the loose use of the metaphor of war both trivializes real wars and exaggerates other conflicts and problems our society faces.

Relationships between health care professionals and recipients of care. Relationships between physicians and other health care professionals, on the one hand, and patients, on the other, have been described and directed by a wide variety of metaphors and models.[28] For example, May has identified images of the physician as fighter, technician, parent, covenanter, and teacher, and Robert M. Veatch has identified several major competing models of physician-patient relationships: engineering, priestly (which includes the paternalistic model), collegial, and contractual models.[29] Other metaphors such as friend and captain of the ship have also been used.[30]

Some critics contend that such models are "whimsical gestalts," that many other arbitrary models could be invented—for example, bus driver or back-seat driver—and that moral points can and should be made more directly.[31] Such criticisms overlook how metaphors and models function in the interpretation and evaluation of interactions between physicians and patients. They miss the role of imagination, which can be defined as "reasoning in metaphors."[32] For example, opponents of paternalistic medical relationships usually do not eschew all use of metaphor; instead they offer alternative metaphors, such as partnership or contracts. And these various metaphors may be more or less adequate to describe what occurs and to direct what should occur in health care.

Metaphors and models highlight and hide features of the roles of physicians and other health care professionals by their various associated implications. For example, viewing the physician as a parent, or specifically as a father, based on the nineteenth-century model of the family, highlights some features of medical relationships, such as care and control, while hiding others, such as the payment of fees. The use of such metaphors to describe, interpret, and explain relationships is subject to criticism if they distort more than they illuminate. And when they are offered to guide relationships and actions, they are subject to criticism if they highlight only one moral consideration, such as the physician's duty to benefit the patient or to respect patient autonomy, while hiding or obscuring other relevant moral considerations. It is also appropriate to consider the feasibility of various ideal relationships in light of significant personal, professional, and institutional constraints.

Several metaphors may be necessary to interpret health care as it is currently structured and to guide and direct actions, practices, and policies in health care. Some metaphors may fit some relationships better than others; for example, relations in clinical research, family practice, and surgery may be illuminated respectively by the metaphors of partner, teacher-student, and technician-consumer. Furthermore, not all of these metaphors conflict with each other; some may even be mutually supportive as well as compatible, for example, contractor and technician. However, conflicts can be expected if a physician interprets and directs relationships paternalistically, while his or her patient interprets and directs the interaction through metaphors of negotiation and accommodation. (See, in general, chapter 3, "Metaphors and Models of Doctor-Patient Relationships," as well as the discussion of paternalism in other chapters, including chapter 7 on Dax's case.)

Nursing as advocacy. Major changes in the conception of nursing correlate with alterations in its primary metaphors. Whether situated within the military effort against disease or viewed as physicians' handmaidens and servants, nurses have traditionally been expected to cultivate passive virtues, such as loyalty and obedience. Their moral responsibility was primarily directed toward physicians and institutions, such as hospitals, and only secondarily toward patients. This interpretation of responsibility was shaped in part by nursing's military origins in the nineteenth century as well as by societal conceptions of gender.[33] Then in the 1970s, nursing was reconceived through the metaphor of advocacy. Nurses became advocates for "clients" and "consumers" (the term "patient" was often rejected as too passive). This legal metaphor, drawn from the advocate as one who pleads another's cause, especially before a tribunal of justice, highlights active virtues, such as courage, persistence, perseverance, and courage, and views the nurse as primarily responsible to the patient or client. This metaphor is explicit or implicit in formal nursing codes, and it is also featured in a large number of nurses' stories of advocacy and conflict in health care.[34]

Critics, such as Ellen Bernal, note that the metaphor of advocacy reduces the range of services traditionally offered by nurses; it is thus insufficiently comprehensive. In addition to distorting the human experience of illness, it distorts nursing by focusing almost exclusively on patients' or clients' rights, construed mainly in terms of autonomy, and it neglects positive social relationships in health care.[35] It highlights conflict among health care professionals because it implies that some of them do not adequately protect the rights of patients. Thus, the metaphor frequently supports a call for increased nursing autonomy as a way to protect patient autonomy. Because of its adversarial nature, many question whether the metaphor of advocacy can adequately guide relationships among health care professionals in the long run, even if it is useful in the short run. The metaphor may also assume that the nurse's responsibility to the patient/client is always clear-cut and overriding, even though nurses may face serious conflicts of responsibility involving patients, other individuals, associates, and institutions.[36] At the very least, sympathetic commentators call for further clarification of the metaphor of advocacy, while critics seek alternative metaphors and models, such as covenant, partnership, teamwork, or collegiality, which appear to offer more inclusive, cooperative ideals.

Playing God and other metaphors of limits. "Playing God" has been a common metaphor for both describing and (re)directing the activities of scientists, physicians, and other health care professionals. They have been criticized for usurping God's power—for instance, the power over life and death—by letting patients die or by using new reproductive technologies.

There are theological warrants for playing God in the Jewish and Christian traditions, which affirm the creation of human beings in God's image and likeness. Philosopher David Heyd builds on this idea of the image of God: "If indeed the capacity to invest the world with value *is* God's image, it elevates human beings to a unique (godly) status, which is not shared by any other creature in the world. This is playing God in a creative, 'human-specific' way."[37] And Paul Ramsey calls on those who allocate health care to play God in a fitting way: "Men should then 'play God' in the correct way: he makes his sun rise upon the good and the evil and sends rain upon the just and the unjust alike." We should emulate God's indiscriminate care by distributing scarce lifesaving medical technologies randomly or by a lottery rather than on the basis of judgments of social worth.[38] (See the discussion in chapter 10.)

Despite a few such positive uses of the metaphor of "playing God," the metaphor is generally used to identify two aspects of divine activity that should not be imitated by humans: God's unlimited power to decide and unlimited power to act. On the one hand, users of this metaphor demand scientific and medical accountability over against unilateral decision making, for example, regarding life and death. On the other hand, they call for respect for substantive limits, for example, not creating new forms of life in violation of divinely ordained natural limits.[39] Critics frequently focus on human arrogance and rebellion in daring to "play God." In a more typical statement (in contrast to his positive use of the metaphor cited earlier), Ramsey writes: "Men ought not to play God before they learn to be men, and after they have learned to be men they will not play God."[40] Thus, critics of "playing God" usually demand scientific and medical accountability along with respect for substantive limits, such as not creating new forms of life. Objectors to negative uses of the metaphor of "playing God" often challenge the rationale for holding that a particular course of action, such as human genetic engineering, is wrong.

Edmund L. Erde contends that statements such as "doctors should not play god" are so unclear that they cannot function as commands and do not articulate a principle; thus, they cannot be followed because agents do not know how to conform their actions to them. Nor do they explain why certain actions should not be undertaken. Such phrases are, Erde argues, "metaphoric in that they tuck powerful feelings and images into descriptive language that cannot be understood literally." Any activity, such as mercy killing, that is "labeled 'playing god' carries the implication that it is clearly wrong." These phrases are used for situations in which agents face choices but one option is considered immoral and is rejected as arrogantly and presumptuously playing God. The background of intelligibility of this metaphor, according to Erde, is found in the Western idea of the great chain of being, which identifies appropriate responsibilities at each level and opposes the usurpation of power and the failure to respect limits.[41]

Other important and widespread metaphors of limits include the thin edge of the wedge and the slippery slope, both of which warn against undertaking certain actions because other unacceptable actions will inevitably follow. Examples regularly appear in debates about euthanasia. Such arguments about limits may take at least two different forms: (1) conceptual, and (2) psychological-sociological. The first focuses on the logic of moral reasoning, the hammer back of the wedge, and the other on what the wedge is driven into. Or, to shift to the slippery slope metaphor, the slope may be slippery for two different reasons. According to the first reason, the slope is slippery because the concepts and distinctions are vague, inadequately drawn, or ultimately indefensible. This first version derives its power from the principle of universalizability (discussed below), which commits us to treating relevantly similar cases in a similar way. Because of this principle of universalizability, Paul Ramsey argues, ethical (and legal) mistakes tend to replicate themselves: "It is quite clear that at the point of medical, legal, and ethical intersections at the edges of life . . . , the so-called wedge argument is an excellent one. This is true because legal principles and precedents are systematically designed to apply to other cases as well. This is the way the law 'works,' and . . . also the way moral reasoning works from case to similar case."[42]

By contrast, the second version, which is more plausible in arguments against the legalization of physician-assisted suicide and euthanasia, considers the personal, social, institutional, and cultural context in order to determine the possible impact of changing rules or making exceptions. Even if certain distinctions between various acts are, in principle, clear and defensible, agents may not be able over time to draw them and act on them because of various psychological, social, institutional, and cultural forces, such as racism, sexism, or ageism.[43]

Even though such metaphors of limits are often misused, they are appropriate in some contexts. In each use of these metaphors of limits, important moral questions require attention—the evaluation of the first action and subsequent actions—and important conceptual and empirical questions must be addressed in order to determine whether the putatively bad consequences will inevitably follow what might be innocuous first steps. (Similar points hold for analogies invoking moral limits, as will be evident when I examine the Nazi analogy.)

Metaphors for bioethics and bioethicists. The role and function of the bioethicist have often been construed in metaphorical terms. The common language of "applied ethics" invokes the metaphor of engineering as an application of basic science that does not contribute to basic science. The expertise of applied ethicists resides in their ability to apply general theories and principles to specific arenas of human activity. The metaphor of application has been widely challenged on the grounds that it is too narrow and distorts much that is important in bioethics. The term "applied" suggests that ethicists are problem solvers rather than problem setters, that they solve puzzles rather than provide perspectives, that they answer rather than raise questions, and that they begin from theory rather than from lived experience. It implies a limited technical or mechanical model of ethics.

The term "applied" distorts the numerous theoretical controversies in bioethics and neglects the way bioethics may help to resolve or recast some theoretical controversies. At the very least, the metaphor of application needs to be supplemented by various other metaphors for the task of practical ethics and the role of the practical ethicist: "Theoretician, diagnostician, educator, coach, conceptual policeman, and skeptic are also supplemental or alternative roles to that of the technician."[44] Some other metaphors are drawn from ancient religious roles, such as prophet or scribe. Yet another metaphor is "conversation," which is prominent in approaches to bioethics that emphasize interpretation, hermeneutics, and narrative. And the "stranger" has been proposed as the best metaphor for the ethicist in professional education because his or her outside perspective can challenge ordinary assumptions.[45]

No doubt several such metaphors are needed to interpret and direct the activities of ethicists. Which ones appear to fit best will depend in part on our operative conception of ethics and, in particular, on its breadth and richness. Part of my task in this volume is to show through various concrete studies that (at least some) principle-oriented approaches to biomedical ethics are broader and richer than is often supposed, particularly in their attention to dimensions of experience and discourse reflected in metaphors (as well as analogies, symbols, and narrative).

Generative metaphors. Suggestions emerge at various times to retire all metaphors, not merely some metaphors, in some realm of discourse—for instance, Susan Sontag proposes retiring all metaphors for illness.[46] However, it is not possible to strip our discourse in science, medicine, and health care, or in biomedical ethics, of all metaphors. Instead, we must use metaphors with care and carefully assess their adequacy in their descriptive and normative functions.

For each use of metaphor, we have to ask whether, through highlighting and hiding features of subjects, it generates insights about what is or about what ought to be. A simple example of what Donald Schoen calls a generative metaphor, that is, one that generates insights, occurred when researchers were trying to improve the performance of a new paintbrush made with synthetic bristles. The new brush applied the paint to the surface in a "gloppy way." Nothing the researchers tried made the artificial bristles work as well as natural bristles. Then one day someone observed, "You know, a paintbrush is a kind of pump!" That was a generative metaphor: Pressing a paintbrush against a surface forces paint through the spaces or "channels" between the bristles, and painters sometimes even vibrate brushes to increase the flow. Once the researchers began to view the paintbrush as a kind of pump, they were able to improve the brush with synthetic bristles.[47] That is the kind of insight, both descriptive and directive, that we seek. Rarely, however, will it be so dramatic. In most issues confronted in biomedical ethics, the tests of adequacy of particular metaphors will be more complex and subtle. At the very least, tests of metaphors that function normatively to guide being and doing need to incorporate general moral considerations.

ANALOGIES IN BIOETHICS

Analogies and analogical reasoning. Often metaphors and analogies are presented in ways that indicate their substantial overlap. Indeed, in the comparison view of metaphor, there is little difference between them, because metaphors are compressed analogies. Some recent theories of metaphor have stressed, by contrast, that metaphors create similarities rather than merely expressing previously established and recognized similarities or analogies. According to Max Black, comparison views of metaphor fail because they reduce the ground for shifts of meaning (from the secondary subject to the primary subject) to similarity or analogy.[48] Nevertheless, there is a strong consensus that metaphorical statements presuppose some resemblance, even when they also create resemblance.[49] Black himself later conceded that metaphors "mediate an analogy or structural correspondence." Metaphor is, roughly speaking, "an instrument for drawing implications grounded in perceived analogies of structure between two subjects belonging to different domains."[50] And yet metaphor does not merely compare two things that are similar, but rather enables us to see similarities in what would be regarded as dissimilar.

Metaphors and analogies are thus closely related, with metaphors both expressing and creating similarities. In general, good metaphors function cognitively to generate new meaning and insight by providing new perspectives, while good analogies extend our knowledge by moving from the familiar to the unfamiliar, the established to the novel. In stretching language, concepts, and so forth for new situations, analogy does not involve the imaginative strain often evident in the use of metaphors.[51] Nevertheless, the differences in function between metaphors and analogies should not be exaggerated.

The term analogy derives from the Greek *analogia,* which referred to mathematical proportion. "An analogy in its original root meaning," Dorothy Emmet observes, "is a proportion, and primarily a mathematical ratio, e.g., 2:4::4:X. In such a ratio, given knowledge of three terms, and the nature of the proportionate relation, the value of the fourth term can be determined. Thus analogy is the repetition of the same fundamental pattern in two different contexts."[52]

Analogical reasoning thus proceeds inductively, moving from the known to the unknown. It appears prominently in problem solving and thus is featured in research in cognitive science and artificial intelligence.[53] For instance, computer problem-solving programs must search for analogous problems that have been successfully solved to generate solutions to new problems whether in highly structured domains such as law or in less structured domains.

Analogical reasoning has an important place in moral discourse, not only because of its importance in problem solving, but also because of the widely recognized moral requirement to treat similar cases in a similar way. Often stated as a principle of universalizability or of formal justice or formal equality, and dating back at least to Aristotle, the requirement to treat similar cases in a similar way also appears in the common law's doctrine of precedent. The basic idea is that one does not make an acceptable moral or legal judgment—perhaps not even a moral or legal judgment at all—if one judges that X is wrong but that a similar X is right, without adducing any relevant moral or legal difference between them. In general, analogical reasoning illuminates features of morally or legally problematic cases by appealing to relevantly similar cases that reflect a moral or legal consensus (precedent). Of course, much of the moral (or legal) debate hinges on determining which similarities and differences are both relevant and significant. A good example is the debate about the appropriateness of the Nazi analogy in contemporary bioethics.

The Nazi analogy. The Nazi analogy is widely invoked to oppose such practices as mercy killing, withdrawing artificial nutrition and hydration from patients in a persistent vegetative state, such as Nancy Cruzan, and using neural tissue from aborted fetuses in transplantation research. (For the last, see chapter 16, "Ethics, Public Policy, and Human Fetal Tissue Transplantation Research.") Such appeals to this analogy assert a "moral equivalence" between what Nazi physicians and researchers did, for example, in the euthanasia program or in human experimentation, and what contemporary physicians and researchers are doing. However, as Caplan argues, more than "similarities in overt conduct" or in outcomes of such conduct are required to establish "moral equivalence."[54] Reasons for the action and the motives of those participating must also be analyzed and assessed: "Merely noting similarities in conduct, rhetoric, or context does not prove that two actions are the same. Two actions can appear to be substantively the same, but unless some insight is available about motives, intentions, capacities, and goals, it is impossible to tell whether they are morally equivalent." Hence, Caplan argues, it is important to examine the rationale and justification Nazi physicians used at the Nuremberg Trials and elsewhere for their acts before the analogy can be properly constructed. This examination establishes major and decisive dissimilarities between Nazi practices and most contemporary practices, which are criticized by the Nazi analogy. While such contemporary practices may still be judged wrong, they cannot properly be opposed on grounds of their morally relevant similarities to Nazi practices. Furthermore, as Caplan contends, invoking the Nazi analogy in inappropriate contexts to claim moral equivalence—for example, between murderous Nazi experiments on living subjects and the use of aborted fetal tissue—dishonors those who died in the Holocaust.

Analogy and casuistry. Over the last decade or so analogical reasoning has received renewed attention from philosophers and theologians focusing on case-oriented or casuistical judgments in bioethics and elsewhere, particularly in the wake of perplexities often associated with new technologies that appear, at least at first glance, to create or occasion unprecedented problems. In *The Abuse of Casuistry,* Albert R. Jonsen and Stephen Toulmin identify "the first feature of the casuistic method" in its classical formulations as "the ordering of cases under a principle by paradigm and analogy."[55] For instance, the rule prohibiting killing is set out in "paradigm cases" that illustrate its most manifest breaches according to its most obvious meaning. Moving from simple and clear cases to complex and uncertain ones, casuists examine various alternative circumstances and motives to determine whether those other cases violate the rule against killing. They seek analogies that permit the comparison of "problematic new cases and circumstances with earlier exemplary ones," that is, the similar type cases that constitute presumptions.[56]

Despite the claims of some modern casuists, it is not clear that analogical reasoning distinguishes casuistical from principlist approaches. For instance, in analyzing the novel microallocation problems of modern medicine, Ramsey, a strongly rule-oriented ethicist, appealed to the analogous "lifeboat" cases— when some passengers have to be thrown overboard in order to prevent the lifeboat from sinking—as a way to interpret the requirements of the principle of equality of opportunity in distributing such scarce lifesaving medical technologies as kidney dialysis.[57] Because principles and rules are indeterminate and because they sometimes conflict, analogical reasoning can be expected in case judgments—mere application cannot be sufficient in ethical frameworks that appeal to principles and rules. (See the fuller discussion in chapter 2.)

Analogies are often divided into two main types, analogies of attribution and analogies of proportion.[58] The analogy of attribution involves a comparison of two terms or analogates, both of which have a common property, the analogon, that appears primarily in one and secondarily in the other. As Thomas Aquinas noted, "healthy" is used primarily for a person in a state of health (a "healthy" person) and secondarily for those medicines and practices that help to maintain or restore health (e.g., a "healthy" diet) or specimens that provide evidence of the body's health (e.g., "healthy" blood). By contrast, in the analogy of proportion, the analogates lack a direct relationship, but each of them involves a relationship that can be compared to a relationship in the other.[59] This second type is most common in analogical reasoning in biomedical ethics, as is evident in debates about maternal-fetal relations and abortion, where analogies of attribution also appear, particularly with reference to the fetus.

Analogical reasoning in debates about maternal-fetal relations. Debates about maternal-fetal relations, including pregnant women's decisions to abort and to decline caesarean sections, illustrate the pervasiveness and importance of analogical reasoning. Traditionally, abortion has been construed as directly killing the fetus, an innocent human being, in violation of the duty of nonmaleficence. Hence, in traditional Roman Catholic moral theology, direct abortions are tantamount to homicide. Sometimes the analogy of the "unjust aggressor" appears in situations where the pregnancy threatens the pregnant woman's life or health, but it has not been accepted in official Catholic thought the way the similar analogy of the "pursuer" has been accepted in some Jewish thought to justify abortions when the pregnant woman's life or health is threatened.

Some feminists and others have attempted to recast the debate about abortion to focus on the basis and extent of the pregnant woman's obligation to provide bodily life support to the fetus. Often accepting, at least for purposes of argument, the premise that the fetus is a human being from the moment of conception (or at some time during the pregnancy), they argue that this premise does not entail that the pregnant woman always had a duty to sustain the fetus's life, regardless of the circumstances of pregnancy, the risks and inconveniences to the pregnant woman, and so forth. Their arguments often proceed through analogies to other hypothetical or real practices or cases, on the assumption that a judgment about those practices or cases will entail a similar judgment about abortion.

The fantastic abortion analogies introduced by Judith Jarvis Thomson have been particularly influential and controversial.[60] In one of her artificial cases, an individual with a rare blood type is kidnapped by the Society of Music Lovers and attached to a famous violinist who needs to purify his system because of his renal failure. Part of the debate is whether relevant analogies can be found in such fantastic, artificial cases, in contrast to actual real cases. For example, over against Thomson, John Noonan opposes abortion in part by appeal to a U.S. tort law case, in which the court held liable the hosts who had invited a guest for dinner but then put him out of their house into the cold night even though he had become sick and fainted and requested permission to stay.[61]

Some feminists and others contend that other analogous real-life legal and moral cases support the pregnant woman's free decision to continue or to discontinue her pregnancy. For many the relevant analogous cases concern living organ and tissue donation. Such donations are conceived as voluntary, altruistic acts that should not be forced by others even to save the potential recipient's life. They are "gifts of life." Requiring a pregnant woman to continue the pregnancy until birth imposes on her a heavier burden than others are expected to bear in analogous circumstances, such as a parent who could save a child's life by donating a kidney. Thus, the provision of bodily life support, whether through donating an organ or allowing the fetus to use the uterus, has been conceived as a gift of life that should not be legally enforced.[62]

According to Lisa Sowle Cahill, much analogical reasoning about pregnancy overlooks what is unique about maternal-fetal relations and thus obscures the morally relevant features of pregnancy or makes some relevant features more significant than they are. Many analogies problematically narrow our moral perspective on abortion by portraying the inception of pregnancy as accidental and the fetus as strange, alien, and even hostile. Furthermore, they often rely on the connotative meanings of their terms, particularly as embedded in a story, such as Thomson's case of kidnapping the unconscious violinist. Examples also appear in the rhetoric of abortion opponents who, for instance, speak of the fetus as a "child" and thereby distort the unique dependence of the fetus on the pregnant woman. Finally, Cahill contends, justifications of abortion based on analogy often rest on liberal convictions that special responsibilities derive only from free choice.[63]

For all these reasons, Cahill holds that analogical reasoning needs supplementation through direct examination of the unique features of maternal-fetal relations, particularly total fetal dependence, and of the ways these unique features qualify maternal, professional, and societal obligations. She argues that, as a category or class of moral relations, pregnancy "is unique among human relations at least because in it one individual is totally and exclusively dependent on a particular other within a relation which represents in its physical and social aspects what is *prima facie* to be valued positively."[64] Hence, she argues, most analogies hide what is distinctive and unique about pregnancy, even though they identify some morally relevant features of maternal-fetal relations.

With the emergence of other maternal-fetal conflicts, particularly regarding caesarean sections to benefit the fetus, similar debates have emerged about the appropriateness of the analogy with living organ and tissue donation. For instance, in the case of *In re* A.C. (No. 87–609, District of Columbia Court of Appeals, April 26, 1990), the majority of the court held that just as courts do not compel people to "donate" organs or tissue to benefit others, so they should not compel caesarean sections against the will of pregnant women to benefit potentially viable fetuses. The dissenting opinion rejected the analogy with organ and tissue donation, insisting that the pregnant woman "has undertaken to bear another human being, and has carried an unborn child to viability," that the "unborn child's" dependence upon the mother is unique and singular, and that the "viable unborn child is literally captive within the mother's body."

Even though analogies with organ and tissue donation are now widely invoked to oppose state control of pregnant women's decisions regarding both abortion and caesarean sections, there are important differences between these two sorts of issues in maternal-fetal relations. In the abortion debate, pregnancy is viewed as the provision of bodily life support and is itself analogous to the donated organ. In the debate about caesarean sections, the surgical procedure is analogous to organ donation—the potentially viable fetus is removed for its own benefit rather than to benefit some other party as in organ or tissue donation. In the abortion debate, the pregnancy is viewed as invasive; in the debate about caesarean sections, the surgical procedure is invasive. However, the central issue is whether state coercion to benefit the fetus is morally and legally acceptable in these cases, and the debate hinges in part on the appropriateness of living organ and tissue donation as an analogy. Even the critics of the analogy engage in analogical reasoning, but they deny that the similarities are more morally or legally relevant and significant than the dissimilarities. Defenders of governmental coercion could also hold that the moral or legal precedent is mistaken and that organs and tissues should sometimes be conscripted or expropriated from living persons.

Similar disputes appear in other areas of contemporary bioethics, for instance, in debates about whether mandatory testing or screening for antibodies to the human immunodeficiency virus, which causes AIDS, can be justified by analogy to accepted practices of mandatory testing or screening (and what precedents it would create for additional testing and screening, for example, for various genetic conditions), and in debates about whether transplantation experiments using human fetal tissue, following deliberate abortions, are analogous to the complicitous use of materials or data from the morally heinous Nazi experiments. In these cases, as in many others, the debates focus to a great extent on the relevance and significance of the proposed analogies. (For these disputes, see chapters 6 and 16, "Mandatory HIV Screening and Testing" and "Ethics, Public Policy, and Human Fetal Tissue Transplantation Research.")

SYMBOLS AND SYMBOLIC RATIONALITY

Less prominent than metaphor and analogy in the essays that follow in this volume is another aspect of human experience and discourse: symbol. However, they are all often connected rather than being merely discrete aspects of our discourse and experience. In considering symbols I will concentrate on actions, policies, and practices that stand for or suggest something else, especially principles, values, attitudes, commitments, and the like. (The word "symbol" comes from Greek roots which mean "throwing together" and "comparing".)

In this volume the role of symbols is particularly prominent in my discussions of artificial nutrition and hydration and human fetal tissue transplantation research, where there is a particular interest in what is symbolized through providing or withdrawing artificial nutrition and hydration, and through the use of human fetal tissue, following deliberate abortions, in transplantation research. (See especially chapters 8, 9, and 16.) In addition, symbolic rationality is featured in several essays that attend to the values symbolized or expressed, rather than simply realized, through certain actions or policies.

Following Max Weber, it is possible to distinguish between "goal-rational" (*zweckrational*) and "value-rational" (*wertrational*) conduct.[65] Conduct that is goal-rational involves instrumental rationality—reasoning about means in relation to ends. Conduct that is value-rational involves matters of value, virtue, character, and identity that are not easily reduced to ends, effects, or even rules of right conduct. There is thus an important distinction between *realizing* a goal and *expressing* a value, attitude, or virtue. In the context of debates about the allocation of resources for and within health care, with particular attention to the relative merits of prevention versus rescue intervention, the goal-rational approach concentrates on effectiveness and efficiency in statistical terms, while the value-rational approach focuses on the values, attitudes, and virtues that policies express.

This distinction between value-rational and goal-rational conduct may illuminate the 1972 congressional decision to make funds available for almost everyone who needs renal dialysis or renal transplantation. This decision followed widespread publicity in the media about particular individuals who were dying of renal failure. One patient was even dialyzed before the House Ways and Means Committee. The result was a program with unexpectedly massive expenditures (over three billion dollars each year). Some argue that this decision was an attempt to preserve society's cherished myth that it will not sacrifice individual lives in order to save money. Of course, we make those sacrifices all the time (e.g., when we fail to pass and enforce some safety measures). However, society's myth is not as threatened when the sacrificed lives are statistical rather than identified—"statistical lives," a phrase from Thomas Schelling, refers to unknown persons in possible future peril. Hence, efforts to rescue identified individuals have symbolic value. They are gestures, not merely tasks.

It has been said that Universalists believe that God is too good to damn human beings, while Unitarians believe that human beings are too good to be damned. In a similar vein, the symbolic-value argument suggests that rescue attempts show both that individuals are "priceless" and that society is "too good" to let them die without vigorous efforts to save them. This is society's myth. And, so the argument goes, when Congress acted to cover the costs of end-stage renal failure, it acted in part to preserve this myth, for the "specific individuals who would have died in the absence of the government program were known."[66] They were identified lives. The End-Stage Renal Disease Program is thus value-rational, if not goal-rational—it symbolizes the value-commitment of the agent (our society) and the victims whose lives the program seeks to save.

Yet serious problems arise for either feature of value-rational allocation between preventive and rescue efforts. On the one hand, within a set budget, to concentrate on a symbolically valuable but less effective approach would, as

Charles Fried notes, "symbolize our concern for human life by actually doing less than we might to save it," which is, at best, an odd strategy. On the other hand, if the life-protection budget, covering both prevention and rescue, is increased by taking resources from other areas of life, such as the arts, then "we symbolize our concern for human life by spending more on human life than in fact it is worth."[67] Symbolic actions and policies may shift over time in order to express the significance of some principles or values that were earlier neglected, compromised, or even sacrificed in order to express the importance of some other principles or values.[68]

Often symbolic rationality is set over against instrumental rationality. While hard-core utilitarians would be inclined to accept only the latter, there is also what might be called "symbol utilitarianism," which is evident in my discussion of artificial nutrition and hydration, where one of the main concerns is what medically administered nutrition and hydration symbolizes about our relation to patients who are terminally ill or in a persistent vegetative state. (See chapters 8 and 9, which are devoted to issues surrounding artificial nutrition and hydration.)

CONCLUSIONS

Debates in biomedical ethics are often debates about which metaphors and analogies illuminate more than they distort. Far from being merely decorative or affective, metaphors and analogies are central to both discourse and practice, for framing our problems as well as for shaping our responses to them. They cannot be evaluated in general but rather must be evaluated specifically according to how well they function to describe and/or direct actions, relationships, and the like. Even though in recent bioethics they have sometimes been offered as ways to circumvent or transcend principles and rules, particularly through attention to cases, narratives, and aesthetic dimensions of experience, they are not necessarily incompatible with principles and rules. For instance, analogical reasoning is important within frameworks of principles and rules, as well as in casuistry, and metaphors and models often succeed or fail depending on how well they express the full range of relevant moral considerations. Finally, regarding symbols and symbolic actions, what is symbolized is often an individual's, a profession's, or a society's commitment to certain values, including principles and rules, such as respect for persons and equal opportunity. Then the debate frequently concerns how much significance a particular symbol has relative to other values and goals.

CHAPTER TWO

□

Ethical Theories, Principles, and Casuistry in Bioethics

AN INTERPRETATION AND DEFENSE OF PRINCIPLISM

INTRODUCTION

For the last several years a debate has raged about methods in general ethics, whether theological or philosophical, and in bioethics as one area of applied or practical ethics. It has often focused on what has been characterized as the dominant paradigm in bioethics—the appeal to principles. This approach has recently been disparagingly labeled "principlism."[1] As a defender of principles in bioethics, I will nevertheless use this label of "principlism" as a shorthand expression for a position that I have presented in several works, especially in *Principles of Biomedical Ethics*—hereafter *PBE*—which I coauthored with Tom L. Beauchamp.[2]

In brief, *PBE* offers four core principles—respect for autonomy, beneficence (including both positive beneficence and utility), nonmaleficence, and justice—with several derivative rules—especially truthfulness, privacy, confidentiality, and fidelity—that "are intended to provide a framework of moral theory for the identification, analysis, and resolution of moral problems in medicine" (*PBE*, 3rd ed., p. 16; 4th ed., pp. 37–38). These principles and rules are viewed as prima facie binding, rather than absolute requirements or mere maxims, with debates about their meaning and weights occurring in situations. Principles and rules are used in judgments in cases through application and balancing, structured by a modest decision procedure (and, I will add below, specification). Rather than being supported by only one theory, these principles and rules find support in several converging or overlapping theories.

In view of the widespread controversy about the best or most defensible approach(es) to bioethics, I have sometimes been puzzled about claims that "principlism" is (still) the dominant approach, if it ever was. When I asked Daniel Callahan about the warrant for his claim that "principlism" (about which he has serious reservations) is the dominant paradigm in bioethics, he responded by noting the preponderance of principlist articles submitted to the *Hastings Center Report*. Nevertheless, as a defender of (one form of) principlism, I am impressed by the number and the vigor of the criticisms, even when I fail to be convinced of their cogency. I shall argue that many of the criticisms do not challenge all forms of principlism, that much of the debate is misplaced, and that some forms of principlism can either avoid or accommodate the most serious criticisms. I concede that the form of principlism that I defend has sometimes been packaged in metaphors that distort the argument (e.g., metaphors regarding bases, grounds, and founda-tions of various judgments) and accompanied by diagrams that mislead (e.g., the arrows connecting the different levels of moral discourse point only one way).

In defending principlism, I will attempt to clear away some misunder-standings, but at points I will also reinterpret, modify, and attempt to improve *PBE*'s version of principlism, particularly in response to the direct and indirect challenges that have emerged over the last decade. These challenges come from a motley crew of theorists, casuists, feminists, narrativists, virtuists, etc. In so describing several opponents of principlism, I only want to suggest the various perspectives they take and by implication the various targets they find in principlism. I will obviously be unable to address all the important criticisms here. Instead I want to concentrate on two major types of criticism—one from the standpoint of a stronger conception of theory (and also of rules) than principlism offers, and the other from the standpoint of a stronger conception of case-judgments (casuistry) than principlism provides. In terms of the levels of moral discourse noted in *PBE* (3rd ed., p. 6; 4th ed., p. 15)—(1) ethical theories, (2) principles, (3) rules, and (4) particular judgments—some strong theorists reject (2), in favor of a combination of (1) and (3) (the latter derived from the former), and some strong casuists give primacy to (4), reject (1) and assign only a modest role to (2) and (3). After addressing these theory-based and casuistry-based criticisms, I will close with a few observations about some other major criticisms and about these three methods or approaches in religious ethics.

THEORY, PRINCIPLES, AND CASUISTRY IN PROPOSALS TO USE ABORTED HUMAN FETAL TISSUE IN TRANSPLANTATION RESEARCH

How would ethicists using these different approaches participate in the debate about the legitimacy of using (human) fetal tissue in transplantation research (particularly fetal neural tissue for patients with Parkinson's disease) following deliberate abortions? In the United States, this debate has focused not only on the rightness or wrongness of acts of using such tissue following deliberate abortions, but especially on whether federal funds should be used to support such research. In March 1988, the Department of Health and Human Services (DHHS) declared a moratorium on the use of federal funds in such research until an independent advisory committee (or committees) could address several questions. In response, the overwhelming majority of the National Institutes of Health Human Fetal Tissue Transplantation Research Panel supported the use of federal funds in such research as "acceptable public policy" and proposed guidelines and safeguards to reduce the chance that such research would lead to additional abortions.[3] However, in fall 1989 DHHS indefinitely extended its moratorium on the use of federal funds in such research.[4] President Clinton lifted the moratorium upon taking office in January 1993. (The debate about human fetal tissue transplantation research is discussed in detail in chapter 16 below.)

In order to show how the three approaches might work in practice, I will take my examples from three defenders of human fetal tissue transplantation research: Albert R. Jonsen, a casuist; K. Danner Clouser, who stresses a unified theory with derivative rules; and myself, a principlist. By choosing three representatives who reach roughly the same conclusion about what ought to be done, and who have also addressed principlism, I should be able to focus attention on the methodological issues without background noise from substantive disagreements.

Jonsen offers a casuistical argument in defense of the use of human fetal tissue in transplantation research following deliberate abortions.[5] He notes that the National Commission for the Protection of Human Subjects of Biomedical and Behavioral Research, on which he served, used casuistical analysis when it considered research on the living fetus. In his terms, casuistical analysis identifies particular cases that represent unquestionably immoral acts and distinguishes them from morally permissible and obligatory acts. While such distinctions are difficult to draw in the abstract, they can be drawn with less difficulty in the concrete through the use of paradigm cases accompanied by analogical reasoning.

A paradigmatic case of unethical research is the ancient Roman practice of vivisection of condemned criminals, while a paradigmatic case of the ethical use of human tissue is the "removal of tissue from a cadaver, either for pathologic analysis, research, or, in recent years, for transplantation." Jonsen concludes that the proposed transplantation of fetal tissue "will be done by way of a procedure analogous to autopsy, rather than vivisection." From his standpoint, the argument that fetal tissue should not be used when the fetus's death is brought about by abortion is not convincing. The only relevant difference between this situation and that of the adult cadaver is the pregnant woman's decision to abort. However, according to Jonsen, objection to a woman's donation of fetal tissue because of her decision to abort would be cogent "if consent were intrinsically necessary for the moral acceptability of using cadaver tissue"; but the *primary* ethical purpose of consent, that is, the manifestation and protection of the moral autonomy of persons, plays no role in dealing with the cadaver:

> This ethicist's—or casuist's—perspective, then, views the use of fetal tissue for therapy as morally acceptable in a specific case, namely when the fetus is dead. In this case, salvaging of fetal tissue is no different than the salvaging of an organ from a cadaver. The suspect consent of the woman about to abort, does not invalidate its moral acceptability, because consent is incidental rather than essential to the morality of salvaging cadaver tissue.

In considering "the features and conditions" of the different practices of Roman vivisection of criminal and modern autopsy, Jonsen notes that

> from our modern viewpoint, the features and conditions associated with vivisection satisfy our sense of a 'cruel and unusual punishment'; with autopsy, in which no pain is inflicted, no coercion is exercised, and no unfairness is imposed, our sense of moral propriety is satisfied. In particular vivisection visits great harm on certain unfortunate individuals, with the prospect of some vague benefit to an unspecified future population. In the autopsy, no harm comes to the individual, who is dead; even if the benefit is unclear or relatively unimportant, it can be justified easily in the absence of any harm or coercion.[6]

I will argue later that these "features and conditions" of cases are materially very close to rules and principles.

By contrast, when the Human Fetal Tissue Transplantation Research Panel presented its report to the NIH Director's Advisory Committee, Clouser, a member of the Panel, offered testimony that focused less on the Panel's recommendations and more "on the moral framework within which such discussions . . . take place."[7] This moral framework is the one defended in Bernard Gert, *Morality: A New Justification of the Moral Rules:*[8]

> [It] is based on rationality, is applicable to all rational persons, and serves the mutual self-interest of all by deriving its moral rules from rationality. These rules proscribe us from causing specified harms to each other, and thus comprise a moral code which would have universal agreement, since all rational persons would avoid harm unless they had a reason

not to. . . . It is a basic morality, universal and public, that all rational persons by virtue of their rationality alone would espouse.[9]

Even though there may be individual or personal reasons for wanting to grant certain rights to those outside the community, such as fetuses, "there are no universally compelling reasons as there are for our moral rules which pertain to all rational persons within the moral community." There is thus a struggle for consensus and for compromise. "Within the community of rational persons it is clearly immoral to cause each other harm—such as depriving them of life or liberty, or causing them pain, or deceiving them. And that is why analogies between what has happened to persons in the past and what is happening to fetuses now will not work."[10] From this theoretical perspective, Clouser supports the "guidelines" or "safeguards" recommended by the Panel—for example, separating as much as possible the pregnant woman's decision to abort and her decision to donate fetal tissue.

Finally, as a member of the same Human Fetal Tissue Transplantation Research Panel, I also testified before the NIH Director's Advisory Committee in support of the Panel's recommendations.[11] My testimony focused on one of the most controversial recommendations of the Panel—the necessity and sufficiency of maternal consent to transfer fetal tissue after an elective abortion (except where the father's objection is known—unless the pregnancy resulted from rape or incest). I agreed with the Panel that among the several modes of acquisition of fetal tissue (express donation, presumed donation, abandonment, sale, or expropriation), express donation is (ethically) preferable. And I argued that the pregnant woman's legal decision to abort does not and should not disqualify her from donating fetal tissue. As the Panel noted, the pregnant woman who chooses to abort "still has a special connection with her fetus, and she has a legitimate interest in its disposition and use. Furthermore, the dead fetus has no interests that the pregnant woman's donation would violate."[12] I closed by paraphrasing Winston Churchill to the effect that "the alternatives to express maternal donation have even worse features. Of the possible ways to transfer fetal tissue, maternal donation is the most congruent with our society's traditions, laws, policies, and practices, including the UAGA [The Uniform Anatomical Gift Act]."[13] My brief argument presupposed several moral principles, embedded in our society's traditions, laws, policies, and practices, in its recognition of the potential benefit of such research, the autonomy of pregnant women, the unfairness of alternative decision makers, and the lack of harm to dead fetuses.

What differences emerge as defenders of three different approaches to bioethics offer support for (at least) some forms of human fetal tissue transplantation research? They provided respectively a casuistical analysis, involving paradigm cases and analogical reasoning; a single unified ethical theory with definite rules; and an (implicit) appeal to embedded principles.[14] Although there may be some substantive differences (e.g., regarding maternal consent), they came to much the same conclusions about acceptable public policy—the permissibility of human fetal tissue transplantation research using tissue from deliberately aborted fetuses within certain guidelines or safeguards. And each appealed at least indirectly to similar moral considerations, whether couched in the language of "features" of the case, or rules (supported by moral theory), or principles. In assessing different critiques of principlism, I will return to this example to show both the overlap and the distinctiveness of the approaches. I will concentrate on what principlism can account for or offer that they miss or downplay. I will show that at least some forms of principlism can accommodate the concerns of casuistry, but I will argue that strong theorists overlook much that is crucial in the moral life and moral discourse. In their concern for a tight, unified, rationalistic theory, they neglect or distort much of our moral experience.

THEORY-BASED CRITIQUE OF PRINCIPLISM

As articulated by Clouser and Gert, one critique of principlism rests on a view of moral theory that establishes a set of normative rules (rather than principles), derived from a conception of rationality and governing all rational creatures, and that provides a procedure to resolve practical problems.[15] As noted above, this conception of moral theory is reflected in Clouser's testimony about human fetal tissue transplantation research, and it appears in its full-blown form in Bernard Gert, *Morality: A New Justification of the Moral Rules*.[16] It will not be possible to do justice to all their criticisms, much less to their conception of moral theory, but I will examine some of the major criticisms.

As suggested earlier, Clouser and Gert argue that principlism, as represented especially by William Frankena's *Ethics*[17] and Beauchamp and Childress's *Principles of Biomedical Ethics,* uses principles "to replace both moral theory and particular moral rules and ideals in dealing with the moral problems that arise in medical practice" (p. 219).[18] In general, these (strong) theorists charge, the principles do not function as claimed and are misleading both theoretically and practically.

More specifically, *PBE*'s principles are not *usable or meaningful guides* to action but "merely names [other labels are chapter headings, check lists, and hooks] for a collection of sometimes superficially related matters for consideration when dealing with a moral problem" (p. 219). By contrast, they believe principles should provide "a specific directive for action" (p. 222). Thus, they believe, agents appealing to *PBE*'s principles as firmly established and secure may not recognize their real grounds for a moral decision because these principles fail to direct action (pp. 222, 226). Since, as I will stress later, Clouser and Gert recognize a requirement of nonmaleficence, specified through their several moral rules, their fundamental problems are with the principles of respect for autonomy, beneficence, and justice. We are subject to the charge of not fully determining action through these principles, because we view principles as general, as prima facie binding, as needing specification through rules, and as sometimes coming into conflict that cannot always be clearly resolved through a moral theory. Principles are necessary though not sufficient or exhaustive, and they can guide without fully determining action. This is the most we can expect of principles or rules. And this perspective is, I believe, warranted by the nature of morality as captured in our moral experience and by theoretical reflections on morality.

A related Clouser and Gert charge focuses on the alleged lack of *relationship* of the various principles with each other. In addition to failing to offer useful guidance, our principles do not offer a "unified guide" to action because our principles often conflict, and these conflicts are unresolvable because there is no unified moral theory from which they are derived (p. 227). I also gladly plead guilty to this charge. Clouser and Gert are, I believe, mistaken about the nature of morality and about the nature of ethics because of their conception of morality as a unified system. By contrast, I contend that there are genuine dilemmas in the moral life and that some conflicts are not resolvable by theory. Again the Clouser and Gert theory oversimplifies the richness and complexity of the moral life (which is not reducible to morality as a system).

Clouser and Gert also claim that "the value of using a single unified moral theory to deal with the ethical issues that arise in medicine and all other fields, is that it provides *a single clear, coherent, and comprehensive decision procedure for arriving at answers"* (p. 233, my emphasis). Because of our lack of a decision procedure, we allegedly "can pick and choose whatever combination we like" or "see fit" (pp. 222–23). Some of the issues involved in relating principles and cases will emerge below in my discussion of casuistry-based critiques of principlism. To focus only on situations in which principles come into conflict, we do not deny that there is a ineliminable role for intuitive judgments in actual situations; but particularly in the third edition, which they do not consider, we do propose a decision procedure to try to reduce the reliance upon intuition. Although our conception of principles as prima facie binding may appear to be too intuitive and flexible, we contend that "the logic of prima facie duties does contain moral conditions that prevent just any judgment based on a grounding principle from being acceptable in a moral conflict" (*PBE*, 3rd ed., p. 53). And we propose several "requirements for justified infringements of a prima facie principle or rule"—there must be a realistic prospect of realizing the moral objective that appears to justify the infringement; the infringement of a prima facie principle must be necessary in the circumstances; the infringement should be the least possible, commensurate with achieving the primary goal of the action; and the agent must seek to minimize the negative effects of the infringement (*PBE*, 3rd ed., p. 53; 4th ed., p. 34). We believe that moral experience and moral theory do not warrant a stronger decision procedure for resolution of conflicts between prima facie principles whose weights cannot be established a priori.[19] While our decision procedure would probably not satisfy Clouser and Gert, it is hard to see how their notion of public advocacy of violations when rules conflict is concretely operational or what it adds to the indispensable principle of universalizability.

Clouser and Gert also charge that our approach is *relativistic*. This charge appears to have several aspects. In particular, they criticize our "anthology approach" for recognizing ethical theor*ies*, rather than ethical theory, and for using principles as "surrogates" for major types of ethical theories (e.g., beneficence for utilitarianism). However, it is our view that no single theory, including Gert's, is ultimately defensible (or at least has been adequately defended), and that a theory is not needed to establish these principles as binding (even though several theories do converge to support these principles). Our principles are not derived from a single unified theory, based on rationality, but rather can be ferreted out of the practices, policies, laws, and institutions in our society. Ethics involves critical reflection on those practices, including stepping back without ever reaching a universal perspective. Hence, Clouser and Gert are right to see in our perspective the seeds of relativism, but we believe we are able to thwart their growth into a vicious relativism; and as I will argue in the discussion of casuistry, these embedded principles can provide a basis for criticism of the practices, policies, laws, and institutions themselves.

Whether our principlism is ultimately antitheoretical depends to a great extent on what is meant by theory. On the one hand, in its rationalistic formulation, according to Stanley G. Clarke and Evan Simpson, ethical theory "requires a set of normative principles governing all rational beings and providing a dependable procedure for reaching definite moral judgments and decisions." On the other hand, ethical theory may be interpreted in terms of the method of reflective equilibrium, which "construes ethical inquiry as a matter of reflective testing of ethical beliefs against others or against particular ethical conceptions presented for exploration."[20] The first version tends to be deductivist, the second tends to be coherentist. And instead of deriving principles or rules from moral theory, our approach (perhaps more clearly now than earlier) involves a dialectic between theory, principles, rules, and particular judgments (see the discussion of casuistry below), with a reflective equilibrium operating at each level as well as between the levels. Over against the Clouser and Gert charge of lack of "rigor" (p. 228), we believe that we have gained as much theoretical rigor as morality, which is not a rigid system, permits. Furthermore, as Beauchamp contends, the distinction between theoretical and applied or practical ethics needs rethinking and perhaps even abandoning.[21] Some of the Clouser and Gert charges that we lack a "comprehensive moral theory" (e.g., regarding principles and ideals) neglect the theoretical distinctions we do draw about such matters (e.g., the distinction between obligatory and ideal beneficence).

As part of their attack on principlism, Clouser and Gert contend that "there is neither room nor need for principles between the theory and the rules or ideals which are applied to particular cases" (pp. 234–35). In *PBE*'s construal of principles, "for all practical and theoretical purposes there are no moral principles" (p. 235). They continue: "Our quarrel is not so much with the content of the various 'principles' as it is with the use of 'principles' at all"

(p. 220). However, I do not believe that form and content can be as easily separated here as they suppose. Part of the debate is about substantive morality, even when it appears to be about the language or perspective of theory and rules over against principles. Viewing morality as primarily concerned about the avoidance of harm (nonmaleficence), and construing the moral rules as ways to avoid harm, Clouser and Gert are not at all inclined to recognize principles of beneficence, justice, or respect for autonomy that appear to require positive actions.

Can't what is important in these moral principles be captured in the ten rules Gert offers?[22] The rules are: don't kill; don't cause pain; don't disable; don't deprive of freedom; don't deprive of pleasure; don't deceive; keep your promise; don't cheat; obey the law; and do your duty. These rules specify the harms that are to be avoided under the requirement of nonmaleficence. All are negative except for promise-keeping, obedience of law, and performance of duty (mainly role related). One possibility would be to include what is omitted under the heading *ideals,* as Clouser and Gert do with such positive actions as relieving pain, preventing death, and helping the needy (pp. 229, 234).[23] The main difference is that the rules can be obeyed impartially all the time, but the ideals of such positive actions cannot be followed all the time and impartially.

Clouser and Gert note about the positive duty of beneficence: "It is impossible for us to do good toward everyone, impartially, all the time" (p. 228). That we recognize, and we draw some relevant distinctions, which are clearer in the third edition (and further developed in the fourth edition) than in the second edition, between general, specific, role-related, and ideal beneficence. However, we do insist that there is a positive duty or obligation to benefit others beyond our promises or roles. Here again Clouser and Gert succumb to the temptation to excise from morality what is messy and complicates their neat model, rather than challenging the model itself. Furthermore, Gert argues, we can use force against some violations of the moral rules but never against failures to realize the ideals.[24] However, strong moral sanctions of blame, short of physical force, may be warranted for some failures to realize beneficence or justice.

Elsewhere I have argued that the neglect of respect for autonomy as a moral requirement leads Gert and his colleagues to bizarre interpretations of particular cases, especially in the discussion of paternalism.[25] Consider this case: Following a serious accident, a patient while still competent refuses a blood transfusion on religious grounds; he then falls unconscious and the physicians believe he will die unless he is given a transfusion. Gert and Charles Culver argue that the physician's provision of a blood transfusion under these circumstances would be paternalistic because, after the patient regains consciousness, it would *lead to* a violation of the moral rule against deception or the moral rule against causing pain.[26] By neglecting the principle of respect for autonomy, Gert and Culver are forced to take a circuitous and misleading route of moral analysis and argument. They fail to see that the transfusion is paternalistic and prima facie wrong because it violates the patient's autonomously expressed wishes and choices in order to provide medical benefit. Paradoxically, the Gert and Culver analysis implies that if the patient dies without regaining consciousness, no paternalistic act has been committed because the moral rules have not been violated. The content of the Gert and Culver moral rules does not adequately express the principle of respect for persons and their autonomy; such moral rules as "don't deprive of freedom" will not cover all the critical cases involving self-determination.

In *PBE* Beauchamp and I recognize tests of ethical theories, perspectives, or frameworks that are both formal (clarity, consistency, coherence, and simplicity) and substantive (congruence with moral experience). The kind of guidance offered is also relevant. And within a tradition, such as a religious tradition, fidelity to that tradition is also important. My main response to the Clouser and Gert critique is that it oversimplifies our moral experience in viewing morality as a system when in fact it consists of fragments; in believing that conflict can be overcome through a unified moral theory based on rationality as such; and in seeking specific guides to action that will yield concrete judgments. I should note what is probably obvious: I have not dealt with, much less dealt with adequately, all the important criticisms raised by Clouser and Gert and, in particular, with the moral theory on which their critique is based. As an effort to illuminate and guide action, their moral theory offers concrete directives through its moral rules; but when these rules are not directly involved, as in the case of paternalism or, for that matter, in the case of human fetal tissue transplantation research, the illumination and direction are limited.

CASUISTRY-BASED CRITIQUE OF PRINCIPLISM

Whereas some theory-based critiques challenge principlism for its failure to develop a rationalistic, universalistic theory with concrete directives for action (in the form of rules), a casuistry-based critique holds that principlism fails to give independent and sufficient attention to particular judgments about cases. It might appear that my decision to start with a case study of the policy debate about human fetal tissue transplantation research plays into the hands of the casuists, particularly because the representatives of these three approaches reached general agreement about the policy despite their different approaches. In fact, when casuist Jonsen distinguishes ethically wrong, permissible, and obligatory actions, he focuses on relevant similarities and differences and, in the process, implicitly or indirectly appeals to what may be viewed as principles. For example, as noted earlier, in identifying the features and conditions that distinguish autopsy from vivisection of criminals, Jonsen focuses on cases of autopsy that do not involve harm or pain (because the subject is dead), or coercion, or unfairness, even if the prospect of benefit is limited.[27] All four of the major principles that Beauchamp and I defend are implicit in Jonsen's argument: the noninfliction of pain expresses the principle of nonmaleficence; avoidance of coercion expresses the principle of respect for autonomy; and avoidance of unfairness expresses the principle of justice. In addition, the interest in autopsy stems from the principle of beneficence: producing good. A similar point can be made about the casuistical categories of medical indications, patient preferences, quality of life, and socioeconomic factors, which are featured in the book *Clinical Ethics,* which Jonsen coauthored with Mark Siegler and William Winslade.[28]

And yet Jonsen and Stephen Toulmin's excellent and important book, *The Abuse of Casuistry,*[29] vigorously opposes "the tyranny of principles" (a formulation that appeared earlier in Toulmin's article by the same title).[30] However, a careful reading of their book reveals that their target is not the inevitable tyranny of principles as such, but the tyranny of some conceptions of principles, particularly "eternal, invariable principles, the practical implications of which can be free of exceptions or qualifications" (p. 2). Such principles lead to problems, particularly deadlocks and fruitless standoffs.

Central to Jonsen's and Toulmin's argument for casuistry over against (absolutist) principlism is their interpretation of the work of the National Commission for the Protection of Human Subjects of Biomedical and Behavioral Research. (Jonsen was a commissioner, and Toulmin a staff philosopher.) According to Jonsen and Toulmin, "the locus of certitude in the commissioner's discussions did not lie in an agreed set of intrinsically convincing general rules or principles, as they shared no commitment to any such body of agreed principles. Rather, it lay in a shared perception of what was specifically at stake in particular kinds of human situations" (p. 18). Nevertheless, the commissioners did issue the *Belmont Report,*[31] which articulates three fundamental principles: beneficence (which includes what Beauchamp and I call nonmaleficence), respect for persons (which includes what Beauchamp and I call respect for autonomy), and justice. To be sure, this report came late in the National Commission's deliberations, after much of the work on problem areas had been completed (p. 356, n. 14), and it may have been motivated in part by the legislative mandate to identify the ethical principles that should govern research involving human subjects. But another plausible interpretation is that these general principles were already embedded in the Commissioners' agreements about problem areas, such as research involving children and prisoners, and that the *Belmont Report* simply articulated these principles with greater clarity. A clear articulation of these underlying principles could both illuminate the particular recommendations and also provide a basis for testing their consistency and their adequacy.

Furthermore, the National Commission's analysis of types of cases in research involving human subjects occurred within a general consensus about a paradigm case (the Nazi experiments on unconsenting human subjects) and also about the relevant moral principles and rules (the Nuremberg Code, which embodies virtually all the principles and rules that would now be taken to govern research involving human subjects).[32] For purposes of moral education and enforcement, it was important to formulate the Nuremberg Code. Even though it is true that the Nuremberg Code itself emerged after moral revulsion to the Nazi experiments (a case judgment), it would be a gross oversimplification to suppose that historically it was only after a negative judgment about the Nazi experiments that moral principles and rules emerged for judging such experiments. The relations of particular judgments and principles and rules are more complex. Because of consensus about the principles and rules embodied in the Nuremberg Code, it was easier for the National Commission to deal with new or ambiguous cases not captured by the paradigm case.

In fact, Jonsen and Toulmin do recognize principles. At one point, they state that their aim is to argue for "good casuistry," that is, casuistry "which applies general rules to particular cases with discernment [in contrast to] bad casuistry, which does the same thing sloppily" (p. 16). And elsewhere Toulmin concedes that principles have special relevance in relations between strangers rather than intimates.[33] Thus, one important question is how we can best characterize relations in research and also in medicine and health care—are they relations between strangers or intimates? In short, this casuistry-based attack on tyrannical principlism focuses on its absolutist versions rather than on versions that view principles and rules as prima facie binding and require balancing and other modes of interpretation in situations of decision.

The authors go on to claim, among other points, (1) that "casuistry is unavoidable," (2) that "moral knowledge is essentially particular, so that sound resolutions of moral problems must always be rooted in a concrete understanding of specific cases and circumstances," and that (3) moral reasoning proceeds by paradigm cases and analogies (p. 330). I largely agree with all three claims but would offer a more complex interpretation of the essential particularity of moral knowledge.

What does it mean to affirm the primacy or priority of particular judgments? First, it might mean that particular judgments chronologically come first. But then the question is, for whom? Each individual participates in communities, many of which involve traditions of moral reflection. These traditions include both general principles and judgments about cases. Even if the particular judgments came first historically (a claim that is difficult to establish), we do not encounter the moral wisdom of our communities and their traditions only through their particular judgments, for these communities and traditions usually present general principles and rules, too. Indeed, it is hard to imagine moral education proceeding without general principles and rules as well as paradigm cases. Hence, the Jonsen and Toulmin view of principles as "top down" holds only if we are thinking in terms of theoretical derivation; it does not fit with an historical, communal interpretation of morality.

Second, the primacy of particular judgments might mean logical priority. However, it is not clear that logical priority can be assigned to either principles or case-judgments; the relation between them is better viewed as dialectical, with neither one fully and completely derived from the other but each potentially modifying the other.

Third, the primacy of particular judgments might suggest normative priority. Here again the language of dialectic is appropriate. Where particular judgments appear to conflict with general principles, adjustments are required, but it is not possible to say that either should always take priority.

If we make a judgment that an action in a particular set of circumstances is wrong, what is implied? R. M. Hare argues that "if we, as a result of reflection on something that has happened, have made a certain moral judgment, we have acquired a precept or principle which has application in all similar cases. We have, in some sense of that word, learnt something." If we learn something useful from reflection on a particular case, Hare claims, the principle we gain must be somewhat general rather than having unlimited specificity. Since no two real cases are exactly alike, we have results of reflection that can be useful in the future only if we "have isolated certain broad features of the cases we were thinking about—features which may recur in other cases."[34] Hare's argument presupposes the principle of universalizability, which Beauchamp and I also take to be a condition or safeguard of moral judgments (*PBE*, 3rd ed., pp. 18–20; 4th ed., p. 26). Acceptance of the principle of universalizability does not distinguish principlism from casuistry, for this formal principle is also presupposed by the Jonsen and Toulmin approach to casuistry, which after all depends on identifying relevant similarities and differences between cases. This principle of universalizability also appears in the common law doctrine of precedent, which serves as a model for casuistry.

Now, the approach to principlism that I have defended with Beauchamp has not always been stated in such a way as to make clear our (current) view of the relation of principles and cases. (I include the qualifier "current" because our position may have become clearer to us as we worked on it over time; but I would argue that even the first edition of *PBE* in 1979 was richer on this point than is sometimes recognized.) In particular, as I have already noted, problems in interpretation of our principlism could easily arise because of our continued use of the metaphor of "application," along with such metaphors as hierarchical tiers, foundation, basis, and ground, and the use of diagrams that sketch the relations between theories, principles, rules, and particular judgments in ways that seem to deny or diminish the certitude in casuistical judgments other than by way of application of general principles and rules. However, in the text we stress that particular moral judgments have relative independence and can lead to a modification or reformulation of general principles; for example, we note that it is a mistake to say that ethical theory and principles are "not *drawn* from cases but only *applied* to cases" (*PBE*, 3rd ed., p. 16; 4th ed., p. 23). And, throughout our editions, we have viewed the relation between principles and rules, on the one hand, and particular judgments about cases, on the other, as *dialectical*, with each potentially modifying the other.[35]

A principle-oriented morality itself requires interpretation. It is often unclear whether a case falls under a principle or rule; a single principle or rule may point in two different directions in the same situation; and there may be apparent or real conflicts among principles and rules—perhaps even dilemmas as well as interpersonal conflicts. Henry S. Richardson identifies three models of connection between principles and cases: (1) application, which involves the deductive application of principles and rules, (2) balancing, which depends on intuitive weighing, and (3) specification, which proceeds by "qualitatively tailoring our norms to cases."[36]

Even though Beauchamp and I use metaphors both of applying and of balancing, it is clear that what we most often do in the text is balancing, especially if application is construed as deduction. The balancing approach is the most consistent with our conception of prima facie principles. However, we have attempted to reduce the reliance on intuition by developing a rough decision procedure (sketched above). And what Richardson calls "specification" is also involved, for as Hare notes, "any attempt to give content to a principle involves specifying the cases that are to fall under it. . . . Any principle, then, which has content goes some way down the path of specificity."[37] Even when we did not explicitly identify and develop specification as an approach to principles in *PBE,* specification was involved in determining the meaning, scope, and range of principles such as respect for autonomy, and in indicating how principles can take shape in rules (such as rules requiring voluntary informed consent). It also appears in our discussion of the meaning of lying (*PBE,* 3rd ed., pp. 49–50; 4th ed., pp. 29–30).

Richardson's model of specification is promising, and David DeGrazia, whose interpretation and critique of principlism have been very stimulating and helpful, has developed what he calls "specified principlism."[38] And the fourth edition of *PBE* explicitly develops specification along with balancing to connect principles to particular judgments. However, I am not yet convinced that specification can be developed as the exclusive model for relating principles and particular judgments, in part because of my sense of the inevitability of moral conflict, not only between people but also within the moral universe. My view about conflict and harmony is itself rooted in theological and philosophical convictions about tragedy and the like. Instead of viewing application, balancing, and specification as three alternative models, it is better, I believe, to recognize that all three are important in parts of morality and for different situations or aspects of situations. Sometimes there is the application of a principle or rule and sometimes there is specification, but at times there is conflict that can only be resolved by balancing, within the constraints and direction of the decision procedure I identified. It is a substantive, and not simply a formal, moral debate as to which approach fits where.

Perhaps the most adequate general category is that of interpretation as long as interpretation includes both the meaning (scope, range, etc.) and the weights of different principles as well as attention to situations.[39] If determinations of meaning and weight are not possible in total abstraction from situations, or at least types of situations, as we argue, then it is not clear that there are terribly important differences between our principlism and casuistry. In actual practice, Jonsen and Toulmin do not reject principles (but only those with certain characteristics or qualities, such as absoluteness, invariability, and exceptionlessness) and concede that casuistry involves the discerning application of principles; and Beauchamp and I recognize a more complex relation of principles and cases than some of our language and diagrams of principlism suggest. Thus, our approaches may share more than is sometimes recognized. Beauchamp and I do not claim that principles are indispensable for every particular judgment and moral decision, but we do claim that principles play an important role in many moral judgments and decisions, in moral education, and in moral justification in a communal setting (e.g., regarding public policy). At the very least, there is an important similarity in spirit between Jonsen and Toulmin's casuistry and our principlism.

Cases play a prominent role in *PBE*. Not only was there a large number (thirty-eight) of cases in the appendix of the third edition, but these cases also function significantly in the text. And the fourth edition integrates most of these cases into the text itself. While we often refer to these cases as examples, we also attend to their complexity and richness and, in particular, to the relevance of several different principles and rules for various cases. John D. Arras has argued that, within a casuistical pedagogy, cases should be used in the following ways: (1) they should be real as opposed to hypothetical; (2) there should be "lengthy and richly detailed case studies . . . [with] the degree of complexity, uncertainty, and ambiguity encountered" in the real world; (3) there should be "whole sequences of cases bearing on a related principle or theme" rather than merely isolated case studies; (4) finally, more attention should be devoted to the "problem of 'moral diagnosis.'"[40]

Regarding (1), virtually all of our cases in *PBE* are real cases drawn from practice and literature in health care and from legal decisions; however, as Hare suggests, the main contrast should be between realistic cases, whether actual or hypothetical, and fantastic ones.[41] But there may even be an occasional role for such fantastic cases to stimulate our reflection and test our consistency as well as our principles; a good example is Judith Jarvis Thomson's famous case of the unconscious violinist.[42]

Regarding (2), many of the cases in *PBE* meet this test, though we also recognize a legitimate role for briefer illustrative cases. And regarding (3), we include and sometimes build our discussions in *PBE* around a sequence of cases, especially in the discussion of death and dying.

Finally regarding (4), we try to avoid what Arras criticizes by not treating cases only under one principle or heading—for example, we resisted supplying labels for our thirty-eight cases in the appendix to the third edition, preferring to number them instead, but we arranged them in an order roughly to reflect certain major themes. "Moral diagnosis" remains a problem not only for pedagogy but also for substantive analysis in principlism and in other approaches.

In my formulation of principlism, principles are discerned in practices, etc., rather than established by a unified theory; but they can serve a critical function, perhaps more readily than the taxonomic approach of a pure casuistry that attempts to operate without principles. Even if one could locate the historical origins of principles in case judgments, principles operate in communities' and traditions' moral discourse and education. And general principles may offer ways to criticize practices, etc., that are not available in case-to-case analysis. Building on some suggestions by Arras, I would note that in moving from actual case to actual case, the ethicist (casuist) may be limited to what is presented by practitioners and others for analysis and assessment as *felt* problems or dilemmas. General principles may help identify other cases that should be on the moral agenda, and they may direct our attention to *real* problems and dilemmas, which have not yet been experienced as such.

Appealing to Rawlsian "reflective equilibrium," Arras concludes, rightly in my judgment,

> the casuistical interpretation of cases, on the one hand, and moral theories, principles and maxims, on the other, exist in a symbiotic relationship. Our intuitions on cases will thus be guided, and perhaps criticized, by theory; while our theories and moral principles will themselves be shaped, and perhaps reformulated, by our responses to paradigmatic moral situations. Whether we attempt to flesh out this method of reflective equilibrium or further develop the casuistical program, it should be clear by now that the methodological issue between theory and cases is not a dichotomous 'either/or' but rather an encompassing 'both-and.'[43]

This is the perspective of principlism in *PBE*.

How can various principles be justified? Here Beauchamp and I (*PBE*, 3rd ed., p. 24, n. 20; 4th ed., chap. 1) have more clearly taken an historicist turn in contrast to strong conceptions of theory (of the sort defended by Clouser and Gert). Our first edition in 1979 was probably still under the sway of a strong conception of ethical theory, even when many of our arguments took a somewhat different direction. And we still discuss various theories, particularly deontological and utilitarian theories, at least schematically. However, we do not really press justification to ultimate foundations of our principles; instead, we rely on convergence between different theories on a set of prima facie principles and rules. Now the Rawlsian language of "overlapping consensus" is attractive. Furthermore, I would now emphasize that ethical reflection usually occurs within a community (or communities), and within a tradition (or traditions), and that this reflection does not involve an ahistorical theory as much as the discernment of principles that are embedded in laws, policies, and practices. This mode of reflection is not inconsistent with the language of principles. While principles are sometimes viewed as ultimately foundational, rationalistically established, ahistorical norms, from which other norms or judgments are deduced, that is not our conception of principles.

SOME OTHER CRITICISMS OF PRINCIPLISM

I will only identify and briefly respond to a few other criticisms of principlism. One criticism is shared by both strong theorists and strong casuists: Many works in biomedical ethics simply invoke the "mantra" of principles and then apply them ritualistically. Critics lament the "tiresome invocation of the applied ethics mantra (i.e., the principles of respect for autonomy, beneficence, and justice)."[44] Clouser and Gert start their critique with this graphic description:

> Throughout the land, arising from the throngs of converts to bioethics awareness, there can be heard a mantra ' . . . beneficence . . . autonomy . . . justice. . . .' It is this ritual incantation in the face of biomedical dilemmas that beckons our inquiry. . . . Brandishing these several principles, adherents to the 'principle approach' go forth to confront the quandaries of biomedical ethics.[45]

This criticism sometimes suggests that principlism has distorted bioethics, and scholarship in bioethics, by offering some categories in relatively abstract, philosophical language (especially "autonomy," "beneficence," and "nonmaleficence") that has misled some newcomers to philosophical or theological ethics to suppose that they have been initiated into a mysterious realm and thus have become experts in ethics. Then these newcomers mechanically apply the "mantra." My response is that there is good and bad scholarship in bioethics, whether from the standpoint of principles, or theory, or casuistry, or virtues.[46] After all, as Jonsen and Toulmin note, casuistry fell into disfavor because it developed into a form of self-serving logic-chopping (or so, at least, the critics charged). Various works in bioethics present careless casuistry or sermonic renditions of virtues and offer unsophisticated interpretations and mechanical applications of principles. Thus, it is important to compare the best scholarly presentations of each approach, as well as to note deficient versions.

Principlism is often attacked from feminist and feminine perspectives. At a bioethics conference where Beauchamp and I debated our position with Clouser, none of the women present (women constituted over a third of the group) spoke for over an hour, and the first comment made by a woman was, "What difference does it make?" Another woman remarked after the session that she did not intervene in the discussion because, in effect, "it was important for the men to play their game." Often the care perspective, as sketched by Carol Gilligan, among others, is set over against principlism.[47] Gilligan's own approach depends on distinguishing care and justice perspectives, with the latter often formulated in terms of principles. It is important to distinguish a feminine approach from a feminist approach, and some feminists, who take seriously the situation of oppression of women under a system of patriarchy, are suspicious of the care perspective because it appears to be a feminine perspective that originates under oppression (and thus may lack independent standing) and that may actually perpetuate oppression through the liberal division of private and public (with care seen as appropriate for the private sphere of life but not for the public sphere and with women as bearers of caring often being confined to the private sphere).[48] Obviously, these matters are very complicated, in part because feminists often appeal to principles of respect for autonomy and justice, particularly in the form of equality, to argue for women's rights.

At any rate, principlism must be carefully (re)formulated in light of the whole range of human moral experience (one of our tests of ethical theories, perspectives, or frameworks), and this includes women's experiences both as carers and as oppressed. In addition, more conceptual and normative work is needed on the distinction and relations between care (including caring for, caring about, and taking care of) and the principle of beneficence, on the distinction and relations between care and justice, and more broadly, on the relations between care and principles. In particular, it will be important to determine which metaphors—such as complementarity and transformation—most adequately express the relations between care and principles in general and justice in particular. Many of these issues are similar (but not identical) to those raised in Christian theological discussions of the relations of love *(agape)* and justice.[49]

Methodologically, as Alisa L. Carse notes, "'care' reasoning is concrete and contextual rather than abstract; it is sometimes principle-guided, rather than always principle-driven, and it involves sympathy and compassion rather than dispassion."[50] Insofar as feminine and feminist moral reasoning is contextual, it overlaps with some important themes of casuistry. Insofar as it focuses on relationships, it raises themes that appear in several forms of communitarianism. But a move away from principles to context, including relationships, still leaves open some difficult questions about priorities when responsibilities to concrete others in various relationships come into conflict, and also about whether caring is too parochial, without pressures generated by a more principled approach. Whatever the outcome of such reflections and debates, feminism is certainly one of the most important and potentially fruitful areas for further reflection about norms and context, and future versions of principlism, including the one in *PBE,* must be appropriately revised to take account of its charges and claims.

Feminism also overlaps with ethical perspectives that stress virtue and virtues. In particular, care is often best understood as a virtue category, including emotions so often overlooked or downplayed in rationalistic approaches usually associated with principlism. As sharp critics of principlism, proponents of the virtue perspective—though by no means a monolithic group—tend to stress that (1) a virtuous agent can *discern* the right course of action in a situation without reliance on principles and rules, and/or (2) a virtuous person will *desire* to do what is right and avoid what is wrong without the necessity of such ulterior motives as avoiding sanctions. Without developing the necessary argument here, it is my conviction that sufficient attention can be paid to the virtues within the context of principlism without restructuring ethical reflection around these categories (*PBE,* chap. 8). Even though the virtues cannot, I believe, replace principles and casuistry, they are indispensable to moral being and doing and to moral development and education. To focus again on discernment, Carse is right when she argues that "recognizing that a general principle or rule is relevant to the situation at hand, and knowing how it is fittingly to be acted upon requires a capacity for discernment that is *distinct from,* and *presupposed by,* the application of principles themselves."[51] And she notes further that discerning responses are not always principle-driven or the result of principled deliberation; they may involve sensitivity to other people through a sympathetic attunement to their needs and concerns. Principlism need not and should not deny these points: principles do not exhaust the moral life and its decisions, and when principles are relevant they must be used discerningly in the situation.

CONCLUDING REFLECTIONS

I have focused on three approaches to bioethics—theory, principles, and casuistry—generally without regard to their philosophical or theological supports, because of my belief that all three can be developed from either philosophical or theological perspectives. Theology as such or religion as such does not necessarily dictate one of these approaches. In fact, we do not have religion as such or theology as such, but only particular religions and theologies within particular traditions. Even within the major traditions that we embody and study in the West, a strong emphasis on law often coexists with an emphasis on virtues, and there have been major efforts to develop principles and rules (e.g., under natural law) that are universalistic, as well as to develop casuistry. Implicit in my analysis and assessment of the three main approaches considered in this essay is a recognition of the role of larger theological and philosophical convictions, for example, about human moral agency; about self and community; about the locus and power of human evil; as well as about God's relation to the world.[52] Apart from the points I have stressed in this essay, theological and philosophical convictions will help render plausible or implausible these different approaches.[53]

CHAPTER THREE

□

Metaphors and Models of Doctor-Patient Relationships

THEIR IMPLICATIONS FOR AUTONOMY

with Mark Siegler

INTRODUCTION

Many metaphors and models have been applied to relationships between patients and physicians. One example is an interpretation of physician-patient relationships as paternalistic. In this case, the physician is regarded as a parent and the patient is regarded as a child. Opponents of such a paternalistic view of medicine rarely reject the use of metaphors to interpret medical relationships; rather, they simply offer alternative metaphors, for example, the physician as partner or the patient as rational contractor. Metaphors may operate even when patients and physicians are unaware of them. Physician-patient conflicts may arise if each party brings to their encounter a different image of medicine, as, for example, when the physician regards a paternalistic model of medicine as appropriate but the patient prefers a contractual model.

As these examples suggest, metaphors involve seeing something as something else, for example, seeing a lover as a red rose, human beings as wolves, or medical therapy as warfare. Metaphors highlight some features and hide other features of their principal subject.[1] Thus, thinking about a physician as a parent highlights the physician's care for dependent others and his or her control over them, but it conceals the patient's payment of fees to the physician. Metaphors and models may be used to describe relationships as they exist, or to indicate what those relationships ought to be. In either the descriptive or the prescriptive use of metaphors, this highlighting and hiding occurs, and it must be considered in determining the adequacy of various metaphors. When metaphors are used to describe roles, they can be criticized if they distort more features than they illuminate. And when they are used to direct roles, they can be criticized if they highlight one moral consideration, such as care, while neglecting others, such as autonomy.

Since there is no single physician-patient relationship, it is probable that no single metaphor can adequately describe or direct the whole range of relationships in health care, such as open heart surgery, clinical research, and psychoanalysis. Some of the most important metaphors that have shaped health care in recent years include: parent-child, partners, rational contractors, friends, and technician-client. We want to determine the adequacy of these metaphors to describe and to direct doctor-patient relationships in the real world. In particular, we will assess them in relation to patient and physician autonomy.

METAPHORS AND MODELS OF RELATIONSHIPS IN HEALTH CARE

(1) The first metaphor is *paternal* or *parental,* and the model is paternalism. For this model, the locus of decision-making is the health care professional, particularly the physician, who has "moral authority" within an asymmetrical and hierarchical relationship. (A variation of these themes appears in a model that was especially significant earlier—the priest-penitent relationship.)

Following Thomas S. Szasz and Marc H. Hollender, we can distinguish two different versions of paternalism, based on two different prototypes.[2] If we take the *parent-infant relationship* as the prototype, the physician's role is active, while the patient's role is passive. The patient, like the infant, is primarily a dependent recipient of care. This model is applied easily to such clinical situations as anesthesia and to the care of patients with acute trauma, coma, or delirium. A second version takes the *parent-adolescent child* relationship as the prototype. Within this version, the physician guides the patient by telling him or her what to expect and what to do, and the patient cooperates to the extent of obeying. This model applies to such clinical situations as the outpatient treatment of acute infectious diseases. The physician instructs the patient on a course of treatment (such as antibiotics and rest), but the patient can either obey or refuse to comply.

The paternalistic model assigns moral authority and discretion to the physician because good health is assumed to be a value shared by the patient and the physician and because the physician's competence, skills, and ability place him or her in a position to help the patient regain good health. Even if it was once the dominant model in health care and even if many patients and physicians still prefer it, the paternalist model is no longer adequate to describe or to direct all relationships in health care. Too many changes have occurred. In a pluralistic society such as ours, the assumption that the physician and patient have common values about health may be mistaken. They may disagree about the *meaning* of health and disease (e.g., when the physician insists that cigarette smoking is a disease, but the patient claims that it is merely a nasty habit) or about the *value* of health relative to other values (e.g., when the physician wants to administer a blood transfusion to save the life of a Jehovah's Witness, but the patient rejects the blood in order to have a chance of heavenly salvation). As a normative model, paternalism tends to concentrate on care rather than respect, patients' needs rather than their rights, and physicians' discretion rather than patients' autonomy or self-determination. Even though paternalistic actions can sometimes be justified, for example, when a patient is not competent to make a decision and is at risk of harm, not all paternalistic actions can be justified.[3]

(2) A second model is one of *partnership,* which can be seen in Eric J. Cassell's statement: "Autonomy for the sick patient cannot exist outside of a good and properly functioning doctor-patient relation. And the relation between them is inherently a *partnership.*"[4] The language of collegiality, collaboration, association, co-adventureship, and covenant is also used. This model stresses that health care professionals and their patients are partners or colleagues in the pursuit of the shared value of health. It is similar to the paternalist model in that it emphasizes the shared general values of the enterprise in which the participants are involved. But what makes this model distinctive and significant is its emphasis on the equality of the participants' interpretations of shared values such as health, along with respect for the personal autonomy of all the participants.[5] The theme of equality does not, however, cancel a division of competence and responsibility along functional lines within the relationship.

Szasz and Hollender suggest that the prototype of the model of "mutual participation" or partnership is the adult-adult relationship. Within this model physicians help patients help themselves, while patients use expert help to realize their own and their physicians' ends. Some clinical applications of this model appear in the care of chronic diseases and psychoanalysis. It presupposes that "the participants (1) have approximately equal power, (2) [are] mutually interdependent (i.e., need each other), and (3) engage in activity that will be in some ways satisfying to both." Furthermore, "the physician does not know what is best for the patient. The search for this becomes the essence of the therapeutic interaction. The patient's own experiences furnish indispensable information for eventual agreement, under otherwise favorable circumstances, as to what 'health' might be for him."[6]

Although this model describes a few practices, it is most often offered as a normative model, indicating the morally desirable and even obligatory direction of practice and research.[7] As a normative model, it stresses the equality of value contributions and the autonomy of both professionals and other participants, whether sick persons or volunteers for research.

(3) A third model is that of *rational contractors*. Health care professionals and their patients are related or should be related to each other by a series of specific contracts. The prototype of this model is the specific contract by which individuals agree to exchange goods and services, and the enforcement of such contracts by governmental sanctions. According to Robert Veatch, one of the strongest proponents of the contractual model in health care, this model is the best compromise between the ideal of partnership, with its emphasis on both equality and autonomy, and the reality of medical care, where mutual trust cannot be presupposed. If we could realize mutual trust, we could develop partnerships. In the light of a realistic assessment of our situation, however, we can only hope for contracts. The model of rational contracts, according to Veatch, is the only realistic way to share responsibility, to preserve both equality and autonomy under less than ideal circumstances, and to protect the integrity of various parties in health care (e.g., physicians are free not to enter contracts that would violate their consciences and to withdraw from them when they give proper notice).[8]

Such a model is valuable but problematic both descriptively and normatively. It neglects the fact that sick persons do not view health care needs as comparable to other wants and desires, that they do not have sufficient information to make rational contracts with the best providers of health services, and that current structure of medicine obstructs the free operation of the marketplace and of contracts.[9] This model may also neglect the virtues of benevolence, care, and compassion that are stressed in other models such as paternalism and friendship.

(4) A fourth attempt to understand and direct the relationships between health care professionals and patients stresses *friendship*. According to P. Lain Entralgo,

> Insofar as man is a part of nature, and health an aspect of this nature and therefore a natural and objective good, the *medical relation* develops into comradeship, or association for the purpose of securing this good by technical means. Insofar as man is an individual and his illness a state affecting his personality, the medical relation ought to be more than mere comradeship—in fact it should be a friendship. All dogma apart, a good doctor has always been a friend to his patient, to all his patients.[10]

For this version of "medical philia," the patient expresses trust and confidence in the physician while the doctor's "friendship for the patient should consist above all in a desire to give effective technical help—benevolence conceived and realized in technical terms."[11] Technical help and generalized benevolence are "made friendly" by explicit reference to the patient's personality.

Charles Fried's version of "medical philia" holds that physicians are *limited, special-purpose friends* in relation to their patients. In medicine as well as in other professional activities such as law, the client may have a relationship with the professional that is analogous to friendship. In friendship and in these relationships, one person assumes the interests of another. Claims in both sets of relationships are intense and demanding, but medical friendship is more limited in scope.[12]

Of course, this friendship analogy is somewhat strained, as Fried recognizes, because needs (real and felt) give rise to medical relationships, even if professionals are free not to meet them unless they are emergencies, because patients pay professionals for their "personal care," and because patients do not have reciprocal loyalties. Nevertheless, Fried's analysis of the medical relationship highlights the equality, the autonomy, and the rights of both parties—the "friend" and the "befriended." Because friendship, as Kant suggested, is "the union of two persons through equal and mutual love and respect," the model of friendship has some ingredients of both paternalism (love or care) and antipaternalism (equality and respect).[13] It applies especially well to the same medical relationships that fit partnership; indeed, medical friendship is very close to medical partnership, except that the former stresses the intensity of the relationship, while the latter stresses the emotional reserve as well as the limited scope of the relationship.

(5) A fifth and final model views the health care professional as a *technician.* Some commentators have referred to this model as plumber, others as engineer; for example, it has been suggested that with the rise of scientific medicine the physician was viewed as "the expert engineer of the body as a machine."[14] Within this model, the physician "provides" or "delivers" technical service to patients who are "consumers." Exchange relations provide images for this interpretation of medical relations.

This model does not appear to be possible or even desirable. It is difficult to imagine that the health care professional as technician can simply present the "facts" unadorned by values, in part because the major terms such as health and disease are not value-free and objective. Whether the "technician" is in an organization or in direct relation to clients, he or she serves some values. Thus, this model may collapse into the contractual model or a bureaucratic model (which will not be discussed in this essay). The professional may be thought to have only technical authority, not moral authority. But he or she remains a moral agent and thus should choose to participate or not in terms of his or her own commitments, loyalties, and integrity. One shortcoming of the paternalist and priestly models, as Robert Veatch notes, is the patient's "moral abdication," while one shortcoming of the technician models is the physician's "moral abdication."[15] The technician model offers autonomy to the patient, whose values dominate (at least in some settings) at the expense of the professional's moral agency and integrity. In other models such as contract, partnership, and friendship, moral responsibility is shared by all the parties in part because they are recognized, in some sense, as equals.

RELATIONS BETWEEN INTIMATES AND
BETWEEN STRANGERS

The above models of relationships between physicians and patients move between two poles: intimates and strangers.[16] In relations of intimacy, all the parties know each other very well and often share values, or at least know which values they do not share. In such relations, formal rules and procedures, backed by sanctions, may not be necessary; they may even be detrimental to the relationships. In relations of intimacy, trust rather than control is dominant. Examples include the relationships between parents and children and between friends. Partnerships also share some features of such relationships, but their intimacy and shared values may be limited to a specific set of activities.

By contrast, in relations among strangers, rules and procedures become very important, and control rather than trust is dominant.[17] Of course, in most relations there are mixtures of trust and control. Each is present to some degree. Nevertheless, it is proper to speak about relations between strangers as structured by rules and procedures because the parties do not know each other well enough to have mutual trust. Trust means confidence in and reliance upon the other to act in accord with moral principles and rules or at least in accord with his or her publicly manifested principles and rules, whatever they might be. But if the other is a stranger, we do not know whether he or she accepts what we count as moral principles and rules. We do not know whether he or she is worthy of trust. In the absence of intimate knowledge, or of shared values, strangers resort to rules and procedures in order to establish some control. Contracts between strangers, for example, to supply certain goods, represent instances of attempted control. But contractual relations do not depend only on legal sanctions; they also presuppose confidence in a shared structure of rules and procedures. As Talcott Parsons has noted, "transactions are actually entered into in accordance with a body of binding rules which are not part of the ad hoc agreement of the parties."[18]

Whether or not medicine is now only a series of encounters between strangers rather than intimates, medicine is increasingly regarded by patients and doctors, and by analysts of the profession—such as philosophers, lawyers, and sociologists—as a practice that is best understood and regulated *as if it were* a practice among strangers rather than among intimates. Numerous causes can be identified: first, the pluralistic nature of our society; second, the decline of close, intimate contact over time among professionals and patients and their families; third, the decline of contact with the "whole person," who is now parceled out to various specialists; fourth, the growth of large, impersonal, bureaucratically structured institutions of care, in which there is discontinuity of care (the patient may not see the same professionals on subsequent visits).[19]

In this situation, Alasdair MacIntyre contends, the modern patient "usually approaches the physician as stranger to stranger; and the very proper fear and suspicion that we have of strangers extends equally properly to our encounters with physicians. We do not and cannot know what to expect of them. . . ."[20] He suggests that one possible response to this situation is to develop a rule-based bureaucracy in which "we can confront *any* individual who fills a given role with exactly the same expectations of exactly the same outcomes. . . ." Our encounters with physicians and other health care professionals are encounters between strangers precisely because of our pluralistic society: Several value systems are in operation, and we do not know whether the physicians we encounter share our value systems. In such a situation, patient autonomy is "a solution of last resort" rather than "a central moral good." In the end, patients have to decide for themselves what will be done to them or simply delegate such decisions to others, such as physicians.

Just as MacIntyre recognizes the value of patient autonomy in our pluralistic society, so John Ladd recognizes the value of the concept of rights among strangers. He notes that a legalistic, rights-based approach to medicine has several important advantages because rules and rights "serve to define our relationships with strangers as well as with people we know. . . . In the medical context . . . we may find ourselves in a hospital bed in a strange place, with strange company, and confronted by a strange physician and staff. The strangeness of the situation makes the concept of rights, both legal and moral, a very useful tool for defining our relationship to those with whom we have to deal."[21]

Rules and rights that can be enforced obviously serve as ways to control the conduct of others when we do not know them well enough to be able to trust them. But all the models of health care relationships identified above depend on some degree of trust. It is too simplistic to suppose that contracts, which can be legally enforced, do away with trust totally. Indeed, as we have argued, a society based on contracts depends to a very great extent on trust, precisely because not everything is enforceable at manageable cost. Thus, the issue is not simply whether trust or control is dominant but, in part, the basis and extent of trust. Trust, at least limited trust, may be possible even among strangers. There may be a presumption of trust, unless a society is in turmoil. And there may be an intermediate notion of "friendly strangers." People may be strangers because of differences regarding values or uncertainty regarding the other's values; they may be friendly because they accept certain rules and procedures, which may ensure that different values are respected. If consensus exists in a pluralistic society, it is primarily about rules and procedures, some of which protect the autonomy of agents, their freedom to negotiate their own relationships.

PHYSICIAN-PATIENT INTERACTIONS AS NEGOTIATIONS

It is illuminating, both descriptively and prescriptively, to view some encounters and interactions between physicians and patients as negotiations. The metaphor of negotiation has its home in discussions to settle matters by mutual agreement of the concerned parties. While it frequently appears in disputes between management and labor and between nations, it does not necessarily presuppose a conflict of interests between the parties. The metaphor of negotiation may also illuminate processes of reaching agreement regarding the terms of continuing interaction even when the issue is mainly the determination of one party's interests and the means to realize those interests. This metaphor captures two important characteristics of medical relationships: (1) it accents the autonomy of both patient and physician, and (2) it suggests a process that occurs over time rather than an event that occurs at a particular moment.

The model of negotiation can both explain what frequently occurs and identify what ought to occur in physician-patient interactions. An example can make this point: A twenty-eight-year old ballet dancer suffered from moderately severe asthma. When she moved from New York to Chicago, she changed physicians and placed herself in the hands of a famed asthma specialist. He initiated aggressive steroid therapy to control her asthma, and within several months he had managed to control her wheezing. But she was distressed because her dancing had deteriorated. She suspected that she was experiencing muscle weakness and fluid accumulation because of the steroid treatment. When she attempted to discuss her concerns with the physician, he maintained that "bringing the disease under complete control—achieving a complete remission of wheezes—will be the best thing for you in the long run." After several months of unhappiness and failure to convince the physician of the importance of her personal goals as well as her medical goals, she sought another physician, insisting that she didn't live just to breathe, but breathed so that she could dance.[22]

As in this case—and despite the claims of several commentators—people with medical needs generally do not confront physicians as strangers and as adversaries in contemporary health care. As we suggested earlier, even if they can be viewed as strangers in that they often do not know each other prior to the encounter, both parties may well proceed with a presumption of trust. Patients may approach physicians with some trust and confidence in the medical profession, even though they do not know the physicians before them. Indeed, codes of medical ethics have been designed in part to foster this trust by indicating where the medical profession stands and by creating a climate of trust. Thus, even if patients approach individual physicians as strangers, they may have some confidence in these physicians as members of the profession as they negotiate the particular terms of their relationship. At the other extreme, some patients may approach physicians as adversaries or opponents. But for negotiation to proceed, some trust must be present, even if it is combined with some degree of control, for example, through legal requirements and the threat of legal sanctions.

The general public trust in the medical profession's values and skills provides the presumptive basis for trust in particular physicians and can facilitate the process of negotiation. But as we noted earlier, in a pluralistic society even people who are strangers, that is, who share very few substantive values, may be "friendly" if they share procedural values. Certain procedural values may provide the most important basis for the trust that is necessary for negotiation; indeed, procedural principles and rules should structure the negotiation in order to ensure equal respect for the autonomy of all the parties.

First, the negotiation should involve adequate disclosure by both parties. In this process of communication—much broader and richer than most doctrines of informed consent recognize—both parties should indicate their values as well as other matters of relevance. Without this information, the negotiation cannot be open and fair. Second, the negotiation should be voluntary, that is, uncoerced. Insofar as critical illness can be viewed as "coercing" individuals through the creation of fear, etc., it may be difficult to realize this condition for patients with certain problems. However, for the majority of patients this condition is achievable. Third, the accommodation reached through the negotiation should be mutually acceptable.[23]

What can we say about the case of the ballet dancer in light of these procedural requirements for negotiation? It appears that the relationship foundered not because of inadequate disclosure at the outset, or along the way, but because of the patient's change in or clarification of her values and the physician's inability to accommodate these other values. The accommodation reached at the outset was mutually acceptable for a time. Initially their values and their metaphors for their relationship were the same. The physician regarded himself as a masterful scientist who was capable technically of controlling a patient's symptoms of wheezing. In fact, he remarked on several occasions: "I have never met a case of asthma I couldn't help." The patient, for her part, selected the physician for the same reasons. She was unhappy that her wheezing persisted, and she was becoming discouraged by her chronic health problem. Because she wanted a therapeutic success, she selected an expert who would help her achieve that goal. Both the patient and the physician made several voluntary choices. The patient chose to see *this* physician and to see him for several months, and the physician chose to treat her asthma aggressively with steroids.

In a short time, the patient reconsidered or clarified her values, discovering that her dancing was even more important to her than the complete remission of wheezing, and she wanted to renegotiate her relationship so that it could be more mutual and participatory. But her new metaphor for the relationship was incompatible with the physician's nonnegotiable commitment to his metaphor—which the patient had also accepted at the outset. Thus, the relationship collapsed. This case illustrates both the possibilities and the limitations of the model of negotiation. Even when the procedural requirements are met, the negotiation may not result in a satisfactory accommodation over time, and the negotiation itself may proceed in terms of the physician's and the patient's metaphors and models of the relationships, as well as the values they affirm.

Autonomy constrains and limits the negotiations and the activities of both parties: Neither party may violate the autonomy of the other or use the other merely as a means to an end. But respecting autonomy as a constraint and a limit does not imply seeking it as a *goal* or praising it as an *ideal*.[24] This point has several implications. It means, for example, that patients may exercise their autonomy to turn their medical affairs completely over to physicians. A patient may instruct the physician to do whatever he or she deems appropriate: "You're the doctor; whatever you decide is fine." This relationship has been characterized as "paternalism with permission,"[25] and it is not ruled out by autonomy as a constraint or a limit. It might, however, be ruled out by a commitment to autonomy as an ideal. Indeed, commitment to autonomy as an ideal can even be paternalistic in a negative sense; it can lead the health care professional to try to force the patient to be free and to live up to the ideal of autonomy. But our conception of autonomy as a constraint and a limit prevents such actions toward competent patients who are choosing and acting voluntarily. Likewise, maintenance, restoration, or promotion of the patient's autonomy may be, and usually is, one important goal of medical relationships. But its importance can only be determined by negotiation between the physician and the patient. The patient may even subordinate the goal of autonomy to various other goals, just as the ballet dancer subordinated freedom from wheezing to the power to dance.

This view of autonomy as a limit or a constraint, rather than an ideal or a goal, permits individuals to define the terms of their relationship. Just as it permits the patient to acquiesce in the physician's recommendations, it permits the physician to enter a contract as a mere technician to provide certain medical services, as requested by the patient. In such an arrangement, the physician does *not* become a mere means or a mere instrument to the patient's ends. Rather, the physician exercises his or her autonomy to enter into the relationship to provide technical services. Such actions are an expression of autonomy, not a denial of autonomy. If, however, the physician believes that an action requested by the patient—for example, a specific mode of therapy for cancer or a sterilization procedure—is not medically indicated, or professionally acceptable, or in the patient's best interests, he or she is not obligated to sacrifice autonomy and comply. In such a case, the professional refuses to be an instrument of or to carry out the patient's wishes. When the physician cannot morally or professionally perform an action (not legally prohibited by the society) he or she may have a duty to inform the patient of other physicians who might be willing to carry out the patient's wishes. A refusal to be an instrument of another's wishes is very different from trying to prevent another from realizing his or her goals.

Negotiation is not always possible or desirable. It is impossible, or possible only to a limited extent, in certain clinical settings in which the conditions for a fair, informed, and voluntary negotiation are severely limited, often because one party lacks some of the conditions for autonomous choices. First, negotiation may be difficult if not impossible with some types of patients, such as the mentally incompetent. Sometimes paternalism may be morally legitimate or even morally obligatory when the patient is not competent to negotiate and is at risk. In such cases, parents, family members, or others may undertake negotiation with the physician, for example, regarding defective newborns or comatose adults. But health care professionals and the state may have to intervene in order to protect the interests of the patient who cannot negotiate directly. Second, the model of negotiation does not fit situations in which patients are forced by law to accept medical interventions such as compulsory vaccination, involuntary commitment, and involuntary psychiatric treatment. In such situations, the state authorizes or requires treatment against the wishes of the patient; the patient and the physician do not negotiate their relationship. Third, in some situations physicians have dual or multiple allegiances, some of which may take priority over loyalty to the patient. Examples include military medicine, industrial medicine, prison medicine, and university health service. The physician is not free in such settings to negotiate in good faith with the patient, and the patient's interests and rights may have to be protected by other substantive and procedural standards and by external control. Fourth, negotiation may not be possible in some emergencies in which people desperately need medical treatment because of the risk of death or serious bodily harm. In such cases, the physician may *presume* consent, apart from a process of negotiation, if the patient is unable to consent because of his/her conditions or if the process of disclosing information and securing consent would consume too much time and thus endanger the patient. Finally, procedural standards are important for certain types of patients, such as the poor, the uneducated, or those with "unattractive medical problems" (e.g., drug addiction, obesity, and hypochondriasis). In such cases, there is a tendency—surely not universal—to limit the degree of negotiation with the patient because of social stigmatization. A patient advocate may even be appropriate.

In addition to the procedural requirements identified earlier, there are societal constraints and limits on negotiation. Some actions may not be negotiable. For example, the society may prohibit "mercy killing," even when the patient requests it and the physician is willing to carry it out.[26] Such societal rules clearly limit the autonomy of both physicians and patients, but some of these rules may be necessary in order to protect important societal values. However, despite such notable exceptions as "mercy killing," current societal rules provide physicians and patients with considerable latitude to negotiate their own relationship and actions within that relationship.

If negotiation is a process, its accommodations at various points can often be characterized in terms of the above models—parent-child, friends, partners, contractors, and technician-consumer. An accommodation reached through the process of negotiation is not final or irrevocable. Like other human interactions, medical relationships change over time. They are always developing or dissolving. For example, when a patient experiencing anginal chest pain negotiates a relationship with a cardiologist, he may not have given or even implied consent to undergo coronary angiography or cardiac surgery if the cardiologist subsequently believes that it is necessary. Medical conditions change, and people change, often clarifying or modifying their values over time. In medical relationships either the physician or the patient may reopen the negotiation as the relationship evolves over time and may even terminate the relationship. For example, the ballet dancer in the case discussed above elected to terminate the relationship with the specialist. That particular relationship had not been fully negotiated in the first place. But even if it had been fully negotiated, she could have changed her mind and terminated it. Such an option is essential if the autonomy of the patient is to be protected over time. Likewise, the physician should have the option to renegotiate or to withdraw from the relationship (except in emergencies), as long as he or she gives adequate notice so that the patient can find another physician.

Part Two

■

RESPECT FOR

AUTONOMY

ITS IMPLICATIONS
AND LIMITATIONS

CHAPTER FOUR

□

If You Let Them, They'd Stay
in Bed All Morning

THE PRINCIPLE OF RESPECT FOR AUTONOMY AND THE

TYRANNY OF REGULATION IN NURSING HOME LIFE

PART I: THE PRINCIPLE OF RESPECT FOR
AUTONOMY: A MATTER OF LIMITS

The general theme of the fifteenth anniversary of the founding of the Hastings Center was "Autonomy—Paternalism—Community." Hearing several sharp criticisms—indeed, virtual rejections—of autonomy, I stressed in my oral remarks and later in my published paper that we "need several independent moral principles, such as individual and communal beneficence and respect for personal autonomy." It is "unfortunate and even pernicious," I continued, to suggest that "biomedical ethics is allegedly moving beyond autonomy to community and paternalism," for such an approach would reduce "ethical reflection to a mere mirror of societal concerns at a particular time, when in fact the task for serious ethical reflection is to indicate the importance and relative weight of several moral considerations that should be maintained in some tension or balance."[1]

In reaffirming that statement, I want to sketch briefly the principle of respect for personal autonomy (PRA) as one among several important moral principles in biomedical ethics, since it will play such a prominent role in the remainder of the book, particularly later in this chapter when I consider the tyranny of regulation in nursing home life; in the next chapter, which considers "how much the cancer patient should know and decide"; in the succeeding chapter, which considers when mandatory HIV screening and testing can be justified; and in the first chapter in the next section, which extends the discussion of the PRA to a severely burned patient's request to stop treatment. The PRA also has a prominent role in my discussion of organ and tissue donation. In the first part of this chapter, I will point the direction for much that follows in this volume.

My brief defense will proceed by sketching and clarifying some presuppositions and implications of the PRA in light of several major criticisms. I will argue that many of those criticisms are misplaced because they are (perhaps deliberately) directed at some of the least defensible conceptions of the PRA. I will contend that an adequate conception of the PRA can meet the main criticisms leveled by various critics, whether communitarians, narrativists, virtue theorists, traditionalists, or religionists. My main argument focuses on the *principle of respect for autonomy as an important moral limit but also as limited.* As a moral limit, the PRA constrains actions; but it is also limited in scope and in weight, in addition to being complex in its application. Both critics and defenders tend to neglect these senses of limit in their focus on an oversimplified, overextended, overweighted PRA.

Some Clarifications in Response to Misdirected Criticisms of the PRA

In several respects the PRA has been misunderstood and misinterpreted, in part as a result of flawed formulations and defenses by its supporters. As a result critics have found easier targets than they should have in particular formulations of the PRA.

First, as I argue at greater length in the next chapter, it has been a mistake to use the term "autonomy" or even the phrase "principle of autonomy" as a shorthand expression for "the principle of respect for autonomy."[2] This point is important because many critics seem to suppose that proponents of this principle have an *ideal* of personal autonomy and believe that we *ought* to be autonomous persons and make autonomous choices. However, the ideal of personal autonomy is neither a presupposition nor an implication of the principle of respect for personal autonomy, which obligates us to respect the autonomous choices and actions of others. (In the postscript to chapter 5, I further develop these points in relation to recent empirical studies that confuse ideals and duties of [patient] autonomy with respect for [patient] autonomy.)

The *conditions* for autonomous choice must be distinguished from the ideal of autonomy. It is important for the moral life that people be competent, be informed, and act voluntarily. However, they may choose to surrender their first-order decisions (i.e., their decisions about the rightness and wrongness of particular modes of conduct). For example, they may yield to their physicians when medical treatment is proposed or to their religious community or tradition in matters of sexual ethics. Surrender of first-order autonomy appears to involve heteronomy, that is, rule by others. However, if a person autonomously chooses to yield first-order decision making to a professional or to a religious institution, that person has exercised what may be called second-order autonomy.[3] People who are subservient to a professional or to a religious institution may lack first-order autonomy, that is, self-determination, regarding the content of their first-order decisions and choices because they have exercised and continue to exercise second-order autonomy, that is, selection of the professional or institution to which they choose to be subordinate. Hence, in those cases, respect for their second-order autonomy is central, even though their first-order choices are heteronomous. This point is important because of the common supposition that the principle of respect for autonomy is at odds with all forms of heteronomy, authority, tradition, and the like.

The term "respect" also requires amplification. One meaning of respect is to have reference to or regard for or to consider. For example, a boxer may respect his opponent's right hand. A second meaning is more relevant—to consider worthy of high regard, to esteem, or to value. This meaning reflects the attitude that is proper in relation to autonomous choices. Although this attitude does not depend on the content of those choices, it is not inconsistent with criticism of those choices. In a third sense, respect is more than an attitude. It is an action of refraining from interfering with, or attempting to interfere with, the autonomous choices and actions of others, through subjecting them to controlling influence, usually coercion or manipulation of information.[4]

The PRA can be stated negatively as "it is [prima facie] wrong to subject the actions (including choices) of others to controlling influence." This principle provides the justificatory basis for the right to make autonomous decisions. This right in turn takes the form of specific autonomy-related (if not autonomy-based) rights, such as liberty and privacy. This negative formulation focuses on avoidance of controlling influences, including coercion and deception. However, the PRA also has clear positive implications in contexts, including health care relationships. In research, medicine, and health care, for example, it engenders a positive or affirmative obligation to disclose information and to foster autonomous decision making. Nevertheless, it is important to distinguish negative and positive rights based on or related to the PRA, and the limits on positive rights may be greater than the limits on negative rights. Not surprisingly, the positive right to request a particular treatment may be severely limited by research protocols and by just allocation schemes.

Finally, the PRA is ambiguous because it focuses on only one aspect of personhood, namely self-determination, and defenders often neglect several other aspects, including our embodiment. A strong case can be made for a principle of "respect for persons," with respect for their autonomous choices being simply one aspect—though perhaps the main aspect—of respect for them as persons. But even then we would have to stress that persons are embodied, social, historical, etc. Some of these issues emerge when we try to explicate what the principle of the PRA requires.

Complexity of Respect for Personal Autonomy

In determining what the PRA requires, it is important to recognize its complexity, which is widely neglected by both defenders and critics. Some of my earlier points highlighted aspects of this complexity—for example, the distinction between first-order and second-order choices. Because of the complexity of persons, judgment is required, rather than the mechanical application of a clear-cut moral principle.

One difficulty in respecting people's choices is determining what they are choosing, what preferences they are expressing, etc. This complexity is magnified because people communicate not only through written statements (such as signed consent forms) or through words. The process of communication requires attention to various verbal and nonverbal signs.[5] Furthermore, people may be ambivalent and may even express contradictory preferences. In the maze of signals, the professional may have to make a judgment about whether a patient really wants full or only partial disclosure, whether a patient really wants to undergo a test to determine whether he could donate a kidney to a sibling, and the like.

A major difficulty in respecting personal autonomy stems from the fact that people exist in and through time and their choices and actions occur over time. Consent itself is given and withdrawn over time, and a patient's present statements should not always be taken at face value. Hence in discharging our obligations under the PRA, we not only have to determine whether a patient is autonomous and what he or she chooses but we also have to put that patient's present consents and dissents in a broad temporal context including the past and the future. As temporal beings through and through, people may express different preferences at different times. Often the discussion of the PRA focuses on the present moment—for example, is there an informed consent or refusal at this time? However, respecting persons becomes very complex when their temporality is properly included. Which choices and actions should we respect? In particular, is it justifiable to override a patient's present autonomous choices and actions in the light of his/her past or (anticipated) future choices and actions? And is a decision to do so an instance of respect for personal autonomy or a paternalistic breach of the PRA?[6]

Past or prior consent/refusal poses no problem if the patient cannot currently autonomously express his or her wishes. As in the case of advance directives, we respect personal autonomy by acting on that past or prior statement. Matters become more problematic, however, when a person's present choices appear to contradict those previous choices, which may even have been made with a view to preventing future change. For example, in one case a twenty-eight-year-old man decided to terminate chronic renal dialysis because of his restricted lifestyle and the burdens on his family—he had diabetes, was legally blind, and could not walk because of progressive neuropathy. His wife and physician agreed to provide him medication to relieve his pain while he died and agreed not to put him back on dialysis even if he requested it under the influence of uremia, morphine sulfate, and ketoacidosis (the last resulting from the cessation of insulin). While dying in the hospital, the patient awoke complaining of pain and asked to be put back on dialysis. The patient's wife and physician decided to act on the patient's earlier request that he be allowed to die, and he died four hours later.[7] In my judgment, the spouse and physician should have put the patient back on dialysis in view of his current request and the irreversibility of the decision to let him die in accord with his earlier statements. After putting him back on dialysis, they could have determined if he had autonomously revoked his prior choice. If he then persisted in his prior decision, they could have proceeded again with more confidence.

A critical question in this case and others is whether people have autonomously revoked their previous consents/refusals. Thus, it is necessary to continue to assess a person's level of autonomy over time in order to determine whether he or she is autonomously revoking previous consents or dissents. The PRA requires that we attend to a person's prior consent/refusal *and* present revocation, but the present revocation takes priority if it is autonomous.

Judgments about the authenticity or inauthenticity of a person's decision may be important in our decisions about how to respond—as in the case featured in the second part of this chapter. The consistency or inconsistency of a present choice or action with a person's life plan and risk budget over time may help us determine, for example, whether his or her revocation is genuine. For Bruce Miller, authenticity means that "an action is consistent with the attitudes, values, dispositions and life plans of the person."[8] Its intuitive idea is "acting in character." We wonder whether actions are autonomous if they are out of character (e.g., a sudden and unexpected decision to discontinue dialysis by a woman who has displayed considerable courage and zest for life despite years of disability). Similarly, we are less likely to challenge actions as nonautonomous if they are in character (e.g., refusal of a blood transfusion by a Jehovah's Witness). Nevertheless, as important as the idea of character is, it would be a mistake to make authenticity a criterion of autonomy. At most, actions apparently out of character and inauthentic can be caution flags that warn others to press for explanations and justifications in order to determine whether the actions are autonomous. It is important, however, not to rule out in advance the possibility of a change or even a conversion in basic values.

In some situations the health care professional may have good reason to believe that if a patient is kept alive, for example, by a particular treatment, which she is now refusing, then she will eventually ratify the coercive or deceptive treatment on her behalf, perhaps even thanking the professional. Such a ratification does occur in some cases. One question is whether anticipation of future consent justifies present actions against a patient's express choices, in part on the grounds that the present actions respect what the person will be rather than what she now is. My response is that actual or predicted future consent is neither necessary nor sufficient to justify interventions against current choices. At most a patient's probable future consent may provide evidence that the criteria for justified paternalistic interventions have been met.[9] (For further discussion, see chapter 7, devoted to Dax's case.)

Finally, respecting personal autonomy is complex because there are several varieties of consent and refusal. Although express consent (or refusal) is the primary model, consent (or refusal) may also be implicit, tacit, or presumed. To take one example, solid organ procurement in the United States is structured around express consent or donation, whether by the individual while alive or by the family after the individual's death. But there is also presumed consent in the donation of corneas in several states. Presumed donation is not necessarily a breach of the PRA. In some circumstances, silence or a failure to refuse donation could appropriately be construed as donation. For presumed donation—perhaps better viewed as tacit donation—to be autonomous and valid, society needs to make sure that the conditions of understanding and voluntariness have been met. Otherwise, the appeal to presumed donation may only be expropriation. (For further discussion, see chapter 14 below.)

Respect for Autonomy as Limited in Scope and Range

In explicating the PRA as limited, I want to focus on its limited scope or range and on its limited weight or strength. If these limits are not recognized, it is too easy to dismiss the principle as extending too far or as outweighing or overriding too much. Deflation of claims for and about the PRA are essential to its preservation.

Respect for persons who are autonomous may legitimately differ from respect for persons who are not autonomous. The presence, absence, or degree of autonomy is a morally relevant characteristic (though hardly the only morally relevant characteristic). When people are autonomous, respect for them requires (or prohibits) certain actions that may not be required (or prohibited) in relation to nonautonomous persons. Several principles may establish minimum standards of conduct, such as noninfliction of harm in relation to all persons, whatever their degree of autonomy. However, what the principle of respect requires (and prohibits) in relation to autonomous persons and in relation to nonautonomous persons will differ. Hence, Immanuel Kant excluded children and the insane from his discussion of the principle of respect for persons, and John Stuart Mill applied his discussion of liberty only to those in the "maturity of their faculties."

Nevertheless, it is appropriate to operate with a presumption in favor of adults' autonomy, unless and until they are determined to be nonautonomous. Several failures of autonomy are relevant; these include incompetence, that is, an inability to perform certain tasks, lack of understanding, and lack of voluntariness (both internal and external). All of these failures require more explication than I can provide here. At any rate, when these failures occur, and people are at risk of harm or loss of benefits to themselves, interventions based on beneficence can be justified, and they are not opposed by the PRA even if the person refuses. This is limited or constrained beneficence or even limited or constrained paternalism.[10]

There are several examples of the overextension of the PRA's scope and range in ways that are misleading and even dangerous. One simple but risky overextension is to refer to the cadaveric *source* of organs for transplantation as a *donor* even if he or she never donated because he/she never had autonomy or never chose to donate. The donor is one who autonomously decides to donate, whether an individual while alive or a family member after the individual's death. If the decedent never made a decision to donate while alive, the family is the donor. The significance of this language—as well as the language of presumed donation—will be developed in the three chapters devoted to organ and tissue procurement in part 4 of this volume).

Another troubling example is the appeal to substituted judgment in some circumstances where it does not plausibly apply. If a person has previously (and competently) expressed preferences with sufficient clarity, that person's autonomous preferences can and should be extended to periods of lack of autonomy. The metaphor "conveyed" is better than "substituted" for the kinds of cases I have in mind. However, for never-autonomous patients or for previously autonomous patients whose prior preferences and values cannot be reliably traced, it is more defensible to rely on a best-interests standard based on nonmaleficence and beneficence, rather than on a substituted judgment standard based on autonomy. The standard of substituted judgment should be rejected in such cases as an illegitimate fiction.

A final point needs to be made about scope and range. The principle of (respect for) autonomy has been criticized as minimalist and perhaps even egoistic in nature or at least in application in a sociocultural context.[11] This criticism focuses on a person's claim to have his or her autonomy respected rather than on a person's obligation to respect the autonomy of others. The PRA, however, involves correlative obligations and rights. And it is thus a principle of obligation, rather than liberation from obligation. Here again the confusion may stem in part from the misleading language of "principle of autonomy," which should be replaced by the "principle of respect for autonomy."

Even as a principle of obligation, this principle does not stand alone or exhaust the moral life. Other principles are important, not only where autonomy reaches its limits. For example, focusing narrowly on the PRA can foster indifference; thus care and beneficence are necessary. However, without the limits set by the PRA, these other principles may support arrogant enforcement of the good for others. Nevertheless, these and other principles sometimes outweigh or override the PRA.

The requirements of the PRA are often specified in various rules, both moral rules and institutional rules. For instance, rules involving informed consent and informed refusal specify some of the requirements of the PRA, as do rules of liberty, privacy, and confidentiality, which will figure significantly in the discussion of mandatory HIV screening and testing in chapter 6.[12]

Respect for Autonomy as Limited in Weight or Strength

The principle of PRA is more than a maxim. Yet it is not absolutely binding and does not outweigh all other principles all the time. Two major alternatives remain. It could be viewed as lexically or serially ordered, for example, taking absolute priority over some other principles; or it could be viewed as prima facie binding, competing equally with other prima facie principles in particular circumstances. I take the latter approach throughout these chapters.[13] Even though this approach obviously avoids a priori rankings and is thus case-oriented or situational, it is different from some perspectives on casuistry because the logic of prima facie principles dictates a procedure of reasoning or justification for infringements of principles in particular circumstances. For example, the prima facie PRA can be overridden or justifiably infringed when the following "justificatory conditions" are met:

1. When in the circumstances there are *stronger* competing principle(s) [proportionality].
2. When infringing the PRA would *probably protect* the competing principle(s) [effectiveness].
3. When infringing the PRA is *necessary* to protect the competing principle(s) [last resort or necessity].
4. When the infringement of the PRA is *least possible* in the circumstances, consistent with protecting the competing principle(s) [least infringement].
5. And, finally, when explanations and justifications for the infringement of the PRA are made, wherever possible and appropriate, to the autonomous agents whose autonomy is infringed.[14]

These justificatory conditions will be explicated later in this volume, for instance, in the chapter devoted to mandatory screening or testing for HIV infection. As the first public health crisis in an era of firmly established civil rights and liberties, AIDS poses some important questions for further reflection about the place and significance of the PRA, especially in relation to the community as well as to other individuals. The needs of the community in public health may well override the PRA-related rights, such as liberty, privacy, and confidentiality, of some individuals under some circumstances in order to reduce the spread of HIV infection.

Consider, for example, the principles or rules of liberty, privacy, and confidentiality. These may be derived from the PRA, but even if they have independent standing, they are nevertheless closely related to the PRA, for individuals may exercise or waive their rights to liberty, privacy, and confidentiality and thereby remove the constraints in particular cases. But even when individuals do not waive their rights, their rights and their autonomous choices regarding those rights may sometimes be overridden.

Even in actions to protect the community, it is important to start with a presumption in favor of the PRA as expressed in liberty and privacy, and then to determine whether that presumption can be rebutted by arguments for mandatory screening or testing. Critics sometimes doubt whether it is appropriate for the community to have to bear the burden of proof, but in view of the community's power and temptations, along with the importance of the PRA, this is not an inappropriate burden, and it can sometimes be met. For example, if we apply the conditions, identified above, for overriding prima facie obligations, it is necessary to consider the effectiveness and proportionality of any proposed mandatory screening or testing; the absence of an alternative; the least infringement of autonomy and privacy (the least restrictive and intrusive options) consistent with achieving the end; and finally explanation and justification to those whose autonomy and liberty are infringed on behalf of a communal good. In view of what we now know about HIV and its transmission, very few types of mandatory screening and testing actually meet these conditions—certainly donations of blood, semen, and organs, and perhaps a few others. (See chapter 6 below.)

This pattern of justification holds in efforts to protect the community or other individuals, including health care professionals, within the community. Whatever the target, it is important to recognize when the PRA—and associated rules—are being overridden, rather than camouflaging the justification as one of respect for autonomy through certain fictions of consent such as "presumed consent" in some circumstances and "deemed consent" in others, which will be examined in later chapters.

The danger of both overextending and overweighting the PRA is evident in this move to "deemed consent." It is an inappropriate fiction to construe testing and release of information about HIV infection as based on the PRA in situations where individuals did not consent and perhaps even explicitly refused to consent. It is more defensible to face directly the conflict between the PRA and other principles rather than to reinterpret the PRA by extending it to circumstances where it does not apply. Then we can determine more clearly whether the PRA can be outweighed by competing principles in the circumstances.

The PRA: One Principle among Several

The PRA is very important in the firmament of moral principles guiding science, medicine, and health care, as are the rules (whether stated as rights or obligations) that specify some of its requirements. However, it is not the only principle, and it cannot be assigned unqualified preeminence. A clear example of overconcentration on the PRA and its implications can be seen in research involving human subjects, where for years the subject's voluntary, informed consent tended to overshadow all other ethical issues. As a consequence other important moral principles were neglected, even though they must be satisfied prior to soliciting the potential subject's consent to participate—for example, research design, probability of success, risk-benefit ratio, and selection of subjects.[15] To be sure, if researchers do not receive a potential subject's voluntary, informed consent, they may not enlist that subject in research. However, the right of the potential subject to refuse to participate in research became for many the only moral constraint worthy of attention, even though this issue should not be reached until other important ethical issues have been resolved.

In addition, this concentration on the PRA invited inadequate reasons for rejecting or redirecting some research on some populations. Critics of research involving prisoners, for example, tended to argue that the PRA cannot be met in an inherently coercive environment. However, a more defensible ethical criticism emerges from the principles of justice—the unfair imposition of the burdens of research on a captive and vulnerable population many of whom have already suffered serious deprivations in the society.

We should go beyond the principle of respect for autonomy—in the sense of going beyond its misconceptions and distortions and in the sense of incorporating other relevant moral principles. Going beyond, however, should not mean abandoning. Despite its complexity in application, despite its limits in scope and range and in weight and strength, and despite various social changes, the principle of respect for personal autonomy will be critically important in biomedical ethics in the decades to come. But that role requires a sense of limits; we must not overextend or overburden respect for autonomy.

Furthermore, as this discussion suggests, interpreting the PRA's requirements, both its direct requirements and its specified requirements, as expressed in various rules, involves sensitivity to numerous contextual factors, as well as to competing moral principles and rules. Some of the difficulties built into such interpretations can be identified through an analysis of a somewhat routine case in the setting of a nursing home, where there may be a tyranny of regulations, some of them even aimed at protecting autonomy.

PART II: RESPECTING RESIDENT AUTONOMY IN THE NURSING HOME

Case: If You Let Them, They'd Stay in Bed All Morning

At Mansion Manor the nurse's aides pretty well all agree that residents need to get up for breakfast, which is served at 7:30. Some of the aides feel bad waking the residents up, especially knowing that many residents have little in particular to do all day. Mildred, one of the senior aides, says she sometimes makes an exception and lets a resident sleep through breakfast, especially if the resident was feeling poorly the night before. But "if we left the residents alone, some would just stay in bed all morning." Not only is it bad for residents to get "lazy," but also the facility must get everyone up so that the baths and morning routines can be done after breakfast.

Battle of the breakfast. Mrs. Hollinger, aged seventy-six, is one resident who really is not a breakfast fan. She has lived in the home for two years, ever since her first stroke, when she also started using a walker. Recently she had a second stroke, and she now uses a wheelchair for mobility. She finds that making the effort for breakfast is almost intolerable—she has been feeling so tired lately. All of her life she has skipped breakfast, preferring just a cup of coffee. The aides insist on having her up, washed, and dressed for breakfast by 7:20.

Partly because of her disregard for breakfast, and partly because she is slow in navigating the wheelchair, Mrs. Hollinger has been late for breakfast every morning for the last two weeks. The aides say that her late arrival disrupts the feeding of other residents. Also, Mrs. Hollinger is a slow eater and doesn't finish if she comes late. The floor nurse warned Mrs. Hollinger that if she continues to be late for breakfast, they will have to put her with the "feeders" (those unable to feed themselves) in a separate dining room. Mrs. Hollinger now has trouble sleeping at night because she worries about being late for breakfast. This stress tires her even more and makes the morning more difficult.

Other battles. This is not the first time Mrs. Hollinger has run afoul of nursing home policy. When she first came to the facility, she thought she could keep a bottle of sherry in her room to use in the evening and to entertain guests. The nurse put a stop to that. She said the sherry could be kept in the medications area, and if the doctor agreed, an evening drink could be prescribed.

Mrs. Hollinger has also objected to the way staff "poke around" in her room. Just before the inspectors came six months ago, a staff member actually looked in the drawers to see if anything was unsafe. Mrs. Hollinger's shoe polish and her bottle of aspirins were taken from her night table drawer, and the decorative candle she was given for Christmas was also removed. When she protested, the staff explained that although Mrs. Hollinger could handle these things safely, someone who was confused might wander in and harm themselves with these poisonous substances. Regulations stipulated what could be kept in a resident's room. All of these matters were upsetting, but the breakfast situation was the worst thing that had happened to Mrs. Hollinger since her admission.

The regulations did it! This week when Mrs. Hollinger's son James came to visit, he found his mother nervous and teary. When he learned about the breakfast problem and the threat to put his mother with the feeders, he furiously confronted the floor nurse and director of nursing (DON). He said that his mother was not one of their "seniles." Moreover, Mr. Hollinger questioned the reasoning behind the breakfast policies. His mother lived very well for years without having breakfast, and he himself never ate in the morning.

Mr. Hollinger exerted so much pressure that the DON took him to her office to show him the regulations about feeding. She explained that federal requirements state that no more than fourteen hours can elapse between a "substantial evening meal" and the first meal of the next day. Dinner gets served at 4:30 for some residents and at 5:15 for a second group. Therefore, to be in compliance with the regulations, the residents must have an early breakfast.[16]

Case Commentary: Time in Ethical Analyses of Long-Term Care

Several features of long-term care in nursing homes generate ethical problems and dilemmas that often differ from those encountered in critical care, the primary focus and locus of biomedical ethics. These problems and dilemmas frequently involve routine care in which no course of action is life-threatening, at least in the short run. The situations often evolve over time and do not require immediate, definitive answers. Both of these features point to the significance of time in long-term care and to the lack of urgency of most decisions in that setting.

In long-term care there is often time for trial and error and for rational resolutions, negotiated settlements, or compromises.[17] In addition to the significance of the present and the future, the past is also important in various decisions in long-term care. For example, what has happened in the recent past in the care of Mrs. Hollinger is important for interpreting and assessing the actions of the different parties. Because of a series of conflicts, or "battles," as the case report calls them—over the bottle of sherry and over the staff's "poking around" in her room, looking in the drawers, and removing shoe polish, a bottle of aspirins, and a decorative candle received as a gift—the staff appears to view Mrs. Hollinger as a difficult and uncooperative resident. The very language—"battles"—indicates the perceived intensity of the conflict. In turn, it is not unreasonable that Mrs. Hollinger would view the staff as hostile intruders and meddlers, bent on breaking her independence and establishing control over every aspect of her life.

Agents make their moral judgments not only in light of various moral principles but also in relation to their interpretations of situations in which they have to act.[18] In this case it is probable that neither party's interpretation of the current situation is fully adequate; but their interpretations are surely shaped by their memories and images of past actions and conflicts, and their interpretations include assessments of the motives, intentions, attitudes, and traits of character of the other party.

The staff's view that Mrs. Hollinger is a difficult and uncooperative resident may stem from their sense that she is asserting rights without accepting her obligations and responsibilities within the institution. Her responsibilities include respecting the rights of other residents as well as the legitimate demands of institutional order.[19]

One major source of uncertainty in analyzing this case is whether Mrs. Hollinger's actions really infringe on the rights of others and disrupt institutional order. Even if Mrs. Hollinger's actions cannot be characterized in these terms, the staff may nonetheless attempt to persuade her to act in ways they deem appropriate. However, for forcible actions (such as removing her valuables) and coercive threats (such as threatening to put her in a separate dining room with the "feeders," who are unable to feed themselves), it is essential that the staff meet a heavy burden of proof in overriding her autonomous choices, privacy, and liberty.

The aides could respond that they are not overriding Mrs. Hollinger's autonomy but respecting it because, after all, she accepted the rules and regulations that restrict privacy and liberty when she voluntarily entered the nursing home. Thus, she has a duty of compliance, not only because of the need for institutional order but because of her consent.

Before such a claim can be sustained, we need to know exactly what Mrs.

Hollinger (and her son) were told at the outset about the rules and regulations. If they were not adequately informed, they would have additional grounds to complain about some of the interventions. However, even if they were adequately informed and accepted admission on those terms, they may still legitimately complain about those terms and seek to have them changed. In this case, their complaints could be directed at the nursing home's excessively legalistic and narrow interpretation of governmental and institutional rules and regulations.

State and federal regulations also set the context for what the long-term care facility may require of admittees. It is important to determine whether some of the institution's demands themselves violate federal and state regulations, which serve as the background and justification for several of the nursing home's restrictions of Mrs. Hollinger's liberty and privacy. As Ambrogi and Leonard note, federal regulations and some state regulations contain a "patients' Bill of Rights" designed to protect nursing home residents by "preserving the dignity and individual liberties that are central to residents' autonomy."[20] Even though the law usually does not require the potential residents or their proxies receive notice of those rights when executing the admission agreement, the acceptance of the proffered terms of admission cannot be construed as *ethically* valid without clear notification of those rights. In short, then, appealing to Mrs. Hollinger's acceptance of the rules in the past, upon entry, will not necessarily dispose of her current complaints.

The nature, value, and limits of resident autonomy. The principle of respect for autonomy (PRA) implies *individualized care* in the nursing home setting; it suggests that the provision of care, within limits, should follow the tracks set by the resident's autonomous choices, and it establishes both negative rights (e.g., liberty and privacy) and positive rights (e.g., to information and assistance).

Neither the PRA nor the rights (and correlative duties) that express it can be construed as absolute, that is, as outweighing or overriding all other ethical principles and values. At most the principle and the rights that specify it—particularly, for our purposes, liberty and privacy—are only prima facie binding. Insofar as respect for autonomy, liberty, or privacy is a feature of an action, that action is right or obligatory, all things being equal. However, in a given situation it may be outweighed or overridden by competing prima facie principles, such as beneficence, nonmaleficence, and justice. A judgment that a prima facie principle can be overridden or outweighed in particular circumstances requires justification. And that justification should meet what I referred to above as "justificatory conditions": proportionality, effectiveness, last resort or necessity, least infringement, minimization of negative effects, and due explanation.

Even when the PRA (or liberty or privacy) is justifiably overridden, it still exerts pressure. It is not canceled, as the above conditions make clear. Consider, for example, the PRA-based rights of liberty and privacy. When an agent justifiably overrides another's liberty, he or she should choose the least restrictive alternative; and when an agent justifiably overrides another's privacy, he or she should choose the least intrusive alternative. Furthermore, respect for autonomy requires an explanation and justification to the person whose autonomy, liberty, and privacy are justifiably overridden. The justificatory conditions are not rigid, but they identify relevant considerations in the justification of the infringement of a principle such as respect for autonomy in situations of conflict.

Two sorts of limits are involved in interpretations of the PRA, one internal and one external. Determining when the PRA actually applies raises a question of *internal limits*. If residents lack autonomy—for example, the capacity for effective deliberation—then it is inappropriate to talk about respecting their autonomous choices even though they are due other forms of respect. If an intervention is designed to protect the nonautonomous agent, it is a form of weak or limited paternalism and is more easily justified than strong or extended paternalism that overrides an autonomous patient's choices in order to protect him or her. In this case, there is no evidence to suggest that Mrs. Hollinger is not autonomous in the sense of having the capacity to deliberate rationally and to choose freely.

Even though, as noted earlier, I do not accept authenticity or "acting in character" as a criterion of autonomy, I think it important to consider whether a person's actions are in accord with that person's character.[21] Actions apparently out of character alert us to press for justifications and explanations in order to determine whether they are actually autonomous—for example, whether the agent has undergone a change in settled preferences, basic values, and life plans. In this case, Mrs. Hollinger's actions regarding breakfast are in character; she has skipped breakfast all of her life, preferring instead a cup of coffee each morning. Thus, there is no reason on grounds of inauthenticity to suspect that her choices are not autonomous.

External limits may justify overriding the PRA. In this case, as in real life, different reasons are mixed together. First, in the breakfast battle, one senior aide notes that she sometimes makes an exception to the rule about awakening residents in time for breakfast but that residents, if permitted to do so, would stay in bed and get lazy (a paternalistic justification), and the staff needs to take care of baths and morning routines after breakfast (institutional and staff needs). Furthermore, Mrs. Hollinger's late arrival allegedly disturbs other residents at breakfast. Finally, there is the institution's stated concern for compliance with the federal regulations (which will be discussed later). Other external limits include costs to the institution and to the society of respecting autonomy, as well as the society's interest in preventing exploitation of vulnerable people in settings that cannot be adequately monitored. These reasons are all *relevant*. But whether they are *adequate* will depend on whether the procedure of justification can be met.

The paternalistic justification seems insufficient in this case for requiring Mrs. Hollinger to have breakfast. Paternalism may be defined as nonacquiescence in a person's wishes, choices, and actions for that person's own benefit.[22] The alleged benefits in this case may include (a) Mrs. Hollinger's health and (b) her future autonomy of action. It is not clear that breakfast is essential to protect Mrs. Hollinger's health, particularly in view of her eating habits over many years, and requiring her to eat under the circumstances may actually harm her physical and psychological health.

Another benefit that may be invoked in a paternalistic rationale is autonomy itself. One argument is that respecting Mrs. Hollinger's autonomy of choice or decision—to refuse breakfast—would fail to promote her future autonomy of action. A version of this argument appears in the claim that residents would just stay in bed all morning and get lazy if permitted to do so. However, there is no evidence that allowing Mrs. Hollinger to skip breakfast and get ready for the day with less stress would lead her to want to stay in bed all morning or to become lazy. Here again much depends on the interpretation of Mrs. Hollinger's actions. If we interpret her actions narrowly (i.e., as having to do with problems surrounding breakfast), there does not seem to be a sufficient paternalistic justification based on future autonomy of action to override her current autonomous choices by requiring her to be ready for breakfast at the set time.

The staff's stated justification for examining Mrs. Hollinger's personal property and removing some items—shoe polish, a bottle of aspirins, and a decorative candle—was to protect others in the nursing home, not to protect her. (The removal of sherry may have had both justifications.) Personal property falls within the zone of privacy, which can be defended on the basis of the PRA, but neither autonomy nor privacy is absolute. However, as prima facie binding they establish presumptions against interventions that infringe them. Thus, both the examination and the removal of personal property, without Mrs. Hollinger's consent, require justification, along the lines indicated by the process sketched above. In particular, it is important to determine whether there are alternative ways to protect other vulnerable residents without infringing Mrs. Hollinger's autonomy, privacy, and liberty by examining and removing her personal property without her consent. (The case report is not clear about whether the regulations that govern what may be kept in a resident's room are only institutionally developed or also governmentally imposed, but the demands of justification apply to both sources of regulations.)

Among the various infringements of autonomy, there was the staff's threat to put Mrs. Hollinger with the "feeders" in a separate dining room if she continued to be late for breakfast. This coercive threat engendered great stress on her part. Even though she could still make some choices in this context, her range of choices was so limited and the alternatives were so dismal that she was under great stress and had trouble sleeping. Attempts at persuasion and even positive inducement to eat breakfast with the other residents would not have been illegitimate in this case, but it is far from clear that the staff had adequate reasons, despite the appeal to governmental regulations, to justify a coercive threat in this case.

Ethics, politics, and governmental regulation. As Aristotle and others have noted, ethics and politics are inseparably connected. Politics may be internal to the institution of nursing homes or external to it (e.g., through the larger community's interest in what happens in long-term care). It is now necessary to consider the adequacy of the final justification offered by the nursing home staff for the refusal to acquiesce to Mrs. Hollinger's and her son's request that she be exempted from the requirement to take breakfast in view of her dislike for breakfast and the distress the situation is causing her. This justification identifies federal regulations as the culprit in restricting the discretion of nursing homes to provide individualized care that could meet Mrs. Hollinger's needs and wishes.

Although these federal regulations determine much of the context of health care, it is difficult to view them in this case as any more than a rationalization—even a form of "bad faith"—for the nursing home's conduct. As this case is presented, the nursing home's personnel appear to have other reasons for their refusal to accede to Mrs. Hollinger's request. Despite the various reasons proffered—concern for Mrs. Hollinger, concern for other residents, institutional needs, and governmental regulation—it is tempting to view the real reason as an effort to control the independence and individuality of Mrs. Hollinger, who is viewed as a somewhat difficult resident because of several previous "battles."

What about the regulations themselves? First of all, federal and state regulations are needed to ensure the protection of residents of nursing homes, who are often weak and vulnerable and unable to assert their rights. What Robert Dahl notes about government also applies to rules and procedures in nursing homes: "It seems wiser to design a government on the assumption that people will not always be virtuous and at times surely will be tempted to do evil, yet where they will not lack for the incentive and the opportunities to act according to their highest potential."[23]

It is also appropriate that nursing home regulations encompass matters that become the source of concern and distress from Mrs. Hollinger: (1) what residents may keep in their rooms and (2) the frequency of meals for residents. Such regulations may strike an uneasy balance between generality (e.g., safety and adequate nutrition) and specificity (e.g., a detailed list of what may not be permitted in residents' rooms and the number of hours between served meals). Insofar as they tend in the latter direction, as they often do for reasons of clarity and enforceability, they may not permit as much latitude or discretion as would be appropriate to provide individualized care in accord with residents' needs and preferences.

However, in at least one instance, and perhaps others, the nursing home chose a stringent rather that a lenient interpretation of the regulations. The regulation that no more than fourteen hours should elapse between a "substantial evening meal" and the first meal the next day could be interpreted as establishing an option right or a mandatory right on the part of nursing home residents. This distinction, as drawn by Joel Feinberg, affects what the rights-holder can do with his or her rights.[24] An example of a mandatory right is the right to education, which also imposes a duty to obtain an education at least up to a certain level.

At a minimum the federal regulation regarding the time between meals establishes the institution's obligation to *provide* meals at those intervals; but it would be surprising if the regulation also required that residents accept or receive all meals. And yet that is how the nursing home interpreted the regulation. The logic of this interpretation would even require force-feeding of resistant autonomous residents, not only to protect their life and health. The implausibility of such an interpretation is another reason to suspect that the institution is displaying bad faith in its conflict with Mrs. Hollinger.

If the regulation is as strict as the nursing home suggests, then political efforts should be undertaken to create a more satisfactory balance among the principles of respect for autonomy, beneficence, nonmaleficence, and justice.[25] One way to achieve such a balance is to include PRA-based exceptions in the formulation of rules intended to express the last three principles (e.g., the requirement of the provision of food at certain intervals). If the regulation is strictly construed and cannot be altered, however, then selective noncompliance would be justified in order to respect residents' wishes and needs.

If the nursing home follows the plausible interpretation of the regulation as imposing an obligation to provide food at set intervals but not the obligation to force residents to eat, then it will probably need to establish procedures to let residents exercise their option right to refuse to eat at times. For example, the nursing home might require Mrs. Hollinger and her son to sign a statement of their wishes in this matter, rather than awakening her each morning to determine her wishes for that day. Another possibility is to test her regularly (e.g., by weighing her) to make sure that her weight is remaining adequate. At any rate, the moral pressure to find alternatives for protecting Mrs. Hollinger, short of infringing her autonomy, stems from the logic of the prima facie principle of respect for autonomy.

CHAPTER FIVE

□

How Much Should the Cancer Patient
Know and Decide?

with Bettina Schoene-Seifert

INTRODUCTION

How much should the cancer patient know and decide? The question is a matter of great moral complexity, involving conflicts among moral principles and rules on two levels. First, there are substantive "first-level" problems, such as how much information professionals should disclose to patients. Second, there are procedural "second-level" problems, such as who the primary decision maker should be—patient, family, or professional.

This chapter will present a case that raises both types of problems and then sketch a moral framework for analyzing the problems. The guiding principle throughout this discussion is respect for persons, which means that competent patients have the right to know and to decide. This principle and the associated rights are subject to certain internal and external limits. We will argue, however, that family wishes, in general, do not set an acceptable external limit on what the competent patient may know and decide. We will use this framework to illustrate several problems related to disclosing diagnostic and prognostic information, determining treatment options, and selecting the general level of care.

CASE REPORT

Mr. X, a fifty-four-year-old patient with a long history of nodular goiter, presented with recent growth of a thyroid mass and hoarseness. Surgery was performed following a biopsy diagnosis of anaplastic thyroid carcinoma. Only partial removal of the tumor was possible, however, and pulmonary metastases were also suspected. Three weeks after the operation, the patient complained of dyspnea, which proved to be caused by a local recurrence. Further tests revealed lung and bone metastases. Despite irradiation and chemotherapy, as well as palliative treatment, the patient died about three months after the initial diagnosis.

The patient, a hard-driving entrepreneur who dominated both his family and his business, was first told that there was a probability of "malignant cell transformation" in his thyroid gland. He immediately consented to the recommended surgery and was told afterward that the diagnosis had been confirmed and that the tumor had been successfully removed. The likelihood of lung metastases was not mentioned.

During a two-hour conversation, the physician then informed the patient's wife, son, and daughter-in-law of the patient's clinical status and his extremely bleak prognosis. The physician encouraged them to ask everything they wanted to know and answered every question with patience and empathy. The family decided to conceal the diagnosis and prognosis from the patient, and the treating physician accepted their decision. Mr. X was told only that he needed "preventive" treatment. After being informed of possible side effects, he enthusiastically consented to irradiation and chemotherapy. When he later developed dyspnea, it was explained as "postoperative swelling" that would subside with treatment. The true nature of his subsequent back pain was never discussed with him, and he never expressed any concern about its seriousness.

The family inquired daily about the patient's condition, either by phone or in person, and was treated very kindly by the staff. The patient himself complained that his recovery was too slow. During intermittent discharges from the hospital, he continued to work in his business—against medical advice—until he exhausted himself. In the hospital, the patient's complaints and harsh criticism of the nurses generated hostility, causing them to avoid him whenever possible. Mr. X was never offered a chance to talk about his impending death since, until the very end, everyone pretended that he would soon recover.

MORAL FRAMEWORK

A complete moral framework for assessing Mr. X's case would require more space than is available here, but what follows are some of the most important features. Although several grounds have been proposed for the patient's right to know and decide, the firmest involves the principle of respect for persons, including their autonomous choices. This principle generates several rights that are relevant to the case of Mr. X:

- The competent patient's right to decide about his or her own medical care.
- The competent patient's right to information for making an informed decision.
- The competent patient's right to control access to private and confidential information that has emerged in the medical relationship.

It is a form of insult and disrespect to abridge a competent patient's rights when the only reason is to protect what others regard as that patient's own welfare.[1]

The clinician has a moral duty to benefit and not to harm the patient, as well as to respect the patient's autonomous choices. Indeed, the foundation of medical care is patient benefit. However, the provision of medical care for all competent patients, including cancer patients, should be controlled by their own choices. Actions to protect a competent patient—when no one else is at serious risk—by lying or withholding information, by disclosing information to the family or others, or by overriding the patient's choices, are unwarranted insults.

For many, the ideal relationship between clinician and patient involves shared decision making or mutual conversation,[2] but this ideal may be only partially realized for various professional, institutional, and personal reasons. It is important to indicate, therefore, where the final decision rests in cases of conflict. Although the principle of respect for persons applies to both the clinician and the patient, they have different moral vetoes to cast and moral trumps to play. Neither one has absolute sovereignty over the other.

The rights of patients to know and to decide are not absolute and unlimited. First, there are *internal limits*. If a patient cannot make autonomous choices, it is not insulting or disrespectful to override that patient's wishes. And the clinician's duty to benefit and not to harm the patient requires actions to protect the patient from his or her nonautonomous choices. In the case of Mr. X, there was no evidence that he could not make autonomous choices, and the paternalistic nondisclosure, deception, and breach of confidentiality were unwarranted. (We will confine our attention to competent patients, omitting such important questions as how to define and measure competence and incompetence.)

Second, there are also *external limits*. Even a competent patient's decisions about information and treatment are frequently constrained by external factors, some of which are morally legitimate. External limits include public and institutional allocation of resources, research protocols to generate knowledge, public policies to regulate the use of drugs, professional standards regarding the range of acceptable medical treatment, and the clinician's own conscience. Clinicians frequently accept one major external limit—the family's wishes—even though there is no moral or legal warrant for doing so in the case of a competent patient, such as Mr. X.

Sometimes, of course, a family may offer evidence about a patient's inability to make autonomous choices (for example, emotional instability);

such evidence may form part of a paternalistic rationale for not disclosing information to the patient or for overriding his or her choices. There is no moral justification, however, for acceding to the family's request to withhold information from a competent patient such as Mr. X when there is no evidence that the patient declines such information or wants the family to decide. In the case of Mr. X, after the surgery the physician discussed the diagnosis and prognosis with the family, who requested that the information be withheld— probably for paternalistic reasons—and the physician complied. The family, however, should not have received that information in the first place without Mr. X's permission, and the physician had no moral right to respect the family's wishes unless Mr. X had earlier delegated authority to his family. Both actions—breach of confidentiality and deception—were disrespectful and insulting to Mr. X, a competent patient, and his consent to further treatment was invalid because of incomplete and misleading information.

DISCLOSURE OF DIAGNOSIS AND PROGNOSIS

For the purpose of this discussion, there are two major types of information: information about diagnosis and prognosis, and information about diagnostic and therapeutic procedures and alternatives, including their risks and benefits. Disclosure of information so that a patient can consent to or refuse further medical treatment will include both types. The most widely debated cases of disclosing or withholding the first type of information concern the diagnosis of cancer and the prognosis of imminent death where there are no further treatment options.

Over a period of twenty or so years, a dramatic shift occurred in the stated policies of physicians regarding disclosure to cancer patients. In 1961, 88 percent of the physicians surveyed had a policy of not disclosing a diagnosis of cancer to the patient;[3] in 1979, however, 98 percent of those surveyed had a policy of telling the cancer patient.[4] Whatever the actual current practice, it is evident that physicians' stated policies or practical presumptions have certainly changed.

The 1979 study shows an interesting ambiguity about the roles of the patient's wishes and the family's wishes in the withholding of information. For the physicians surveyed, the four factors "most frequently believed to be of special importance" in deciding whether to tell the cancer patient were the patient's expressed wish to be told (52 percent), emotional stability (21 percent), age (11 percent), and intelligence (10 percent). Acting on the patient's wishes is consistent with the principle of respect for persons, while the other factors could justify nondisclosure or partial disclosure to a substantially nonautonomous patient. In the same study, however, the "four most frequent factors considered in the decision to tell the patient" were age (56 percent), relatives' wishes about telling the patient (51 percent), emotional stability (47 percent), and intelligence (44 percent).[5]

Three of the four factors on each list are the same, but the patient's wish is replaced by a relative's wish in the second list. Familial preferences appear to be unjustifiably influential in clinicians' decisions about disclosure to patients of diagnosis and prognosis. According to another study about the disclosure of terminal illness, "the physician is often a family agent, and our observations reveal that 37 percent in the present sample will primarily honor the wishes of the family. Nine percent claim that they will tell the patient what he wants to know regardless of the family's decision. The remainder leave their options more or less open by coordinating with the family what the patient will be told."[6]

In a survey conducted for the President's Commission for the Study of Ethical Problems in Medicine and Biomedical and Behavioral Research, the most common reasons to withhold information from a patient were inability to cope with information (34 percent), inability to understand (38 percent), and the wishes of the patient's family (21 percent).[7] In approximately two-thirds of the cases of nondisclosure or partial disclosure, the decision was rarely or never based on the patient's wishes. Of the physicians surveyed, 80 percent said that they usually discussed the withheld information with another family member.

It is important to note that in the public sample of the President's Commission, 49 percent of the public believed that a physician's decision to withhold information from the patient is justifiable when the patient's family requests nondisclosure or incomplete disclosure, and 8 percent of the public reported having made such requests.[8] In view of what other opinion polls have indicated about the public's desire to receive information (e.g., 82 to 92 percent of adults want to be told),[9] it is possible that the acceptance of familial requests as legitimate by 49 percent of the public reflects an *agent's* rather than a *patient's* perspective—that is, the respondents may have viewed themselves as family members making requests to withhold diagnostic and prognostic information rather than as patients about whom others might make such requests.

One question rarely raised about familial requests to withhold information is: By what right did the family receive the information in the first place? Although some courts have recognized the legal right of a patient's spouse to receive health information about a patient, the legal status of disclosure to family members without the patient's permission is not totally clear.[10] From our perspective, it is not morally acceptable for clinicians to disclose information to family members of a competent patient without the patient's permission. The competent patient has the right to control access to private and confidential information, and the physician should respect this right.

The family provides important care and support for most patients, but the competent patient has the moral right to veto familial involvement. In view of the principle of respect for persons, the best policy is probably to ask the competent patient both at the outset and as the illness progresses about the extent to which he or she wants to receive information, to make decisions, and to share information with others.[11]

SPECIFIC TREATMENT OPTIONS

Having been told of the probability of "malignant cell transformation" in his thyroid gland, Mr. X consented to the recommended surgery, presumably after being informed about the nature of the procedure, its probable benefits and risks, and the alternatives. Such information is now generally considered to be essential for informed consent. Consent that is not informed is not valid; it does not give clinicians a right to proceed with diagnostic or therapeutic procedures. The legal requirement, backed by the moral principle of respect for persons, is for clinicians to disclose at least what a reasonable person would find materially relevant in making decisions about medical treatment; clinicians should also disclose what the particular patient wants to know in making his or her decision.

Even if Mr. X did give valid consent to the surgery, he did not, because of misinformation, give valid consent to irradiation and chemotherapy. He was told that the surgery had removed the tumor, but not that lung metastases had probably developed. The clinicians recommended "preventive" treatment, and Mr. X enthusiastically consented to irradiation and chemotherapy after having been informed about their nature and risks. His consent was not valid, however, because he was misinformed about the diagnosis and prognosis and thus about the limited benefits he could actually expect from the treatments. The family's consent was not sufficient; indeed, the family had no right to make the decision for Mr. X, because he was competent to exercise his own rights to know and to decide and had not delegated those rights to anyone else.

Much of the debate about the disclosure of information regarding specific treatments has focused on their risks. It has been suggested that a risk is material and thus should be disclosed when a reasonable person, in what the physician knows or should know to be the patient's position, would be likely to regard it as significant in deciding whether to accept or to forego a treatment.[12] A clinician's fear that the patient will refuse a treatment when informed about its risks is not a sufficient reason to withhold information about the risks. It is morally acceptable, however, to tell the patient that there are serious but remote risks (for example, one chance in 10,000 of death from anesthesia) and let the patient decide how much additional information he or she wants about those risks. Such a policy respects the patient's wishes regarding the amount and kind of information that should be disclosed.

Disclosure of alternative treatments is frequently controversial in the care of cancer patients. For example, some states have enacted specific laws to require physicians to provide breast cancer patients with complete information on all alternative treatments that are medically acceptable.[13] In this case and others, patients may want to choose among treatment options according to several values, such as both quality and quantity of life. For example, some people with cancer of the larynx may choose irradiation over surgery, trading quantity of life for quality of life since irradiation preserves speech at an increased risk of death from the disease.[14] Beyond tradeoffs between quality and quantity of life, risk-aversive patients with lung cancer may prefer irradiation (which offers a smaller chance of prolonged survival but little risk of early death) over surgery (which offers an increased chance of prolonged survival at the risk of early death).[15]

In general, the physician does not have a legal or moral duty to provide a treatment that he or she believes to be ineffective, harmful, or inferior to other available treatments. The principle of respect for persons implies that the clinician is not a mere tool or instrument of the patient's wishes. But this principle does require the clinician to disclose information about alternatives that, according to professional opinion, are within the range of medical acceptability and to refer or transfer the patient. The duties to disclose alternatives and to refer or to transfer the patient are stronger than the duty to provide a particular treatment.

If an experimental therapy is being tested in a clinical trial, the clinician has a moral duty to disclose its availability if the patient appears to be medically eligible. After all, one legal and moral prerequisite of such trials is that none of the therapies be clearly superior in outcome to the others. There is, however, no moral duty to disclose novel or experimental therapies not part of research protocols if the clinician believes that they are ineffective or harmful.

Legitimate requirements of research protocols set acceptable external limits on patient choices. A patient's request to participate in a clinical trial—involving, for example, a new drug treatment regimen for cancer—may be limited by both eligibility standards and available slots. There may also be some acceptable limits on the disclosure of information as the trial develops—for example, about trends in favor of one treatment—but the patient retains the right to withdraw at any time. In any event, patients have a right not to participate in any research protocol, even though their right to participate is subject to external limits.

Several external limits on patient choices regarding specific treatments have been acknowledged in this section, particularly the limits set by professional opinion on the range of medically acceptable options, the physician's own conscience in not providing (in contrast to disclosing) certain options, and the requirements of research protocols. Another important external limit is set by social and institutional policies in the allocation of resources for and within health care. For example, a patient's request for a heart transplant that might be beneficial in his or her case is subject to standards and procedures of allocation. Some standards and procedures of allocation are unjust, while others are just, but it is not possible here to develop criteria for distinguishing them.

GENERAL LEVELS OF CARE

There are at least four general levels of care of terminally ill patients:
1. Emergency resuscitation.
2. Intensive care and advanced life support.

3. General medical care, including antibiotics, drugs, surgery, chemotherapy, and artificial nutrition and hydration.
4. General nursing care and comfort.[16]

Each level raises important issues about the competent patient's right to know and to decide. In general, within defensible external limits, the competent patient should be able to determine the levels of care, and on each level the particular methods of care should fit the patient's wishes and preferences. Our arguments about specific treatment options apply especially to the third level of care; some of the arguments in this section will also apply to that level as well as to the other levels.

Both the duty to benefit the patient and the duty to respect the patient usually require the provision of general nursing care along with appropriate medications for patient comfort and dignity. Mr. X, for example, received palliative care, but in view of the staff's hostility toward him, that care might not have been adequate. The relief of pain and suffering has not received the attention it merits in the care of patients, particularly those with chronic pain. Even though there is no legal right to pain relief,[17] clinicians have a moral duty, based on both beneficence and respect, to relieve the pain of terminally ill patients in accord with their wishes—even at increased risk of addiction or of earlier death.[18]

The risk of addiction is overestimated and unduly feared in the care of terminally ill patients, and the risk of earlier death may be acceptable to patients with excruciating pain. Clinicians also have a moral duty to indicate the availability of alternative approaches, such as hospice, which assign high priority to the relief of pain and suffering. There are, of course, external limits on patient decision making. For example, in 1924 Congress outlawed the use of heroin for medicinal purposes and in the 1980s defeated legislation to establish a four-year experimental program to allow physicians to prescribe heroin for terminally ill cancer patients. The recent action was justified on the grounds that approved drugs are as effective as heroin and that even this exception would undermine efforts to control heroin.[19]

Intensive care and advanced life support—the second general level of care—were not provided to Mr. X, but since he was never informed about his dismal prognosis he could not have made a valid choice about their use in any case. In general, competent patients should be permitted to choose or to refuse intensive care and advanced life support, such as the respirator, subject to external limits (for example, the equitable allocation of space in the intensive care unit). More questions arise about incompetent patients, including previously competent patients, and we will discuss advance directives for intensive care and advanced life support as well as for other levels and modes of care. First, however, we will discuss cardiopulmonary resuscitation (CPR)—the first level—which patients obviously cannot choose or refuse at the moment of need.

Cardiac arrest, the President's Commission notes, occurs at some point in the dying process of every person.[20] Hence, it would be possible to use CPR in an attempt to prolong, at least briefly, the lives of most patients who die in hospitals. Although death is certain unless CPR is administered promptly, there is debate about whether the success of CPR should be measured by immediate or long-term results. For example, in one hospital three of every ten patients who die receive CPR, but in one study 96 percent of the cancer patients (100 percent of those with metastatic disease) who suffered cardiac arrest and received CPR did not survive to discharge.[21] Although most hospitals have policies regarding CPR, these policies do not always require discussion of "no code" or "DNR" (Do Not Resuscitate) orders with patients or their families. From a legal standpoint, in most states patients may request no code or DNR, but hospitals and professionals are not legally obligated to obtain the patient's or the family's permission for such orders.[22] Since patients cannot make a decision about CPR at the time of a cardiac arrest, the decision must be made in advance and in the absence of meaningful probabilities of cardiac arrest.

According to one physician, the physician should make the decision about code/no code because only he or she has most of the medical knowledge on which major decisions must be based. He also suggests that when the physician has firmly decided that a no-code order is the proper course, it usually works out better for him to explain to family members why resuscitation will not be attempted rather than to ask them whether they want it to be attempted. Family members will rarely disagree if they have placed their trust in the physician. "Such explanations to the patient . . . are thoughtless to the point of being cruel, unless the patient inquires, which is extremely unlikely."[23]

Such a paternalistic view of code/no code decisions is rarely defended so boldly because it patently violates the principle of respect for persons. Nevertheless, it may not be uncommon even among physicians who believe in patient participation in decision making. For example, in a study of 154 patients who were resuscitated at Beth Israel Hospital in Boston in 1981, physicians had an opinion about the patient's wishes in 68 percent of the cases, but only thirty patients (19 percent) had discussed resuscitation before the arrest with their private physicians or a house officer or both. In fifty-one cases (33 percent), however, the families were consulted.[24]

Because of the principle of respect for persons and the derivative rights of patients to know and decide, the morally preferable policy is to raise the issue of CPR at least in general terms, in advance wherever possible, while keeping the decision open to revision as circumstances change.[25] The competent or previously competent patient's wishes, where known, should be determinative, not only about code or no code but also about a limited or partial code. For example, in one case, a competent patient wanted basic CPR but not intubation or mechanical ventilation. It can be argued that a clinician should respect such a request if it is informed and voluntary.[26] The patient should also be informed that a decision about one level or mode of care, such as no code, does not necessarily imply a decision about other levels and modes of care and that he or she can rescind any decision.

This discussion has already raised the issue of advance directives by a competent person to cover periods of incompetence. A person may also want to provide advance directives, however, for other levels and modes of care, not only for CPR. More than one third of the general public have given some instructions, usually oral rather than written, regarding their care if they become incompetent.[27] Most states now recognize advance directives, so-called "living wills," as legally valid; and they protect from legal liability physicians and others who act in accord with them. These states allow a competent person, under certain conditions, to specify standards for decisions about treatment or to designate decision makers to serve as proxies if he or she becomes incompetent. (The latter approach is also possible through some statutes establishing durable powers of attorney.)

A major problem in the specification of standards is their interpretation and application to the actual situation that develops.[28] For example, what does "heroic" or "extraordinary" mean for a particular patient in a particular set of circumstances? Partly because of the difficulty of anticipating the circumstances that actually develop, it is sometimes argued that a physician may be morally more justified in not following a patient's advance directive than in not following a competent patient's current choice in similar circumstances.[29] Designation of a decision maker may be more satisfactory because the proxy can presumably articulate the patient's values and preferences in the actual situation. There may be practical problems, however, such as the availability and competence of designated decision makers.

In most states, although physicians are not legally bound to carry out a patient's advance directive, they are legally bound to make a good faith effort to transfer the patient to the care of a physician who will carry out those wishes.[30] As mentioned, the duties to disclose information and to transfer the patient are stronger than the duty to actually carry out the patient's wishes, partly out of respect for the physician's conscience. The disclosure of information may be indispensable to the patient's informed decision making, but his or her wishes can often (though not always) be carried out by other parties. It is also important to ask whether the physician's obligation to disclose information entails the disclosure of the availability of advance directives. As in the case of CPR, it is morally preferable, even if not morally obligatory, for clinicians to inform competent patients that they can prepare advance directives.

Finally, suppose that Mr. X had been informed of his prognosis and had asked the physician to assist him in committing suicide or even to kill him. In a case described by Lo and Jonsen,[31] Mr. R, sixty-two years old, was hospitalized for colon cancer that had metastasized. When it became clear that he probably would not benefit from either chemotherapy or irradiation, Mr. R refused further laboratory tests, antibiotics, transfusion, and dialysis, but accepted pain medication, nasogastric suction, and intravenous fluids. Four days later, he asked doctors to "speed up" his death, less because of his pain and discomfort than because he could not endure the dying process any longer and did not want to jeopardize his family's limited financial resources. In this case, the house staff gave him substantial doses of morphine and diazepam, with the agreement of the family.

Even though so-called passive euthanasia—allowing a patient to die—is legally as well as morally acceptable under certain conditions, active euthanasia—"mercy killing"—is illegal in every state, even when the patient requests it. The main distinction between active euthanasia and assisted suicide, which is also illegal in most states, is who has the final agency—in assisted suicide, the patient; in active euthanasia, another party. The prohibition of assisted suicide and active euthanasia constitutes another external limit, embodied in professional codes as well as in law. There are plausible (though not necessarily conclusive) moral reasons for such a prohibition, particularly of active euthanasia—for example, the danger of abuse and the importance of maintaining barriers against killing, even from motives of mercy.[32]

CONCLUSION

The answer to the question, "How much should the cancer patient know and decide?" depends, first of all, on whether the patient is competent to understand information and to make decisions. Paternalistic actions to withhold information from the patient, to disclose information to others, or to make decisions for the patient are not warranted unless he or she also suffers from some defect, encumbrance, or limitation in deciding, willing, or acting. When there are no serious internal limits on a patient's capacity to know and decide, the patient's wishes and preferences should be respected, subject to certain external limits.

In general, competent patients have a stronger and broader right to refuse than to choose procedures, because there are more justifiable external limits on choice than on refusal. Apart from a few conceivable situations where the patient's refusal of treatment would substantially harm others, the family's wishes and preferences do not constitute a defensible external limit on the patient's right to know and to decide, as in the case of Mr. X. The principle of respect for persons generates the right to control access to information that emerges in the medical relationship as well as the right to know and to decide. Thus, it is not appropriate, as in the case of Mr. X, to disclose information to the family without the patient's consent or to respect the family's wishes regarding disclosure to the patient or treatment of the patient. It is an insult and indignity to the competent patient to involve the family without his or her consent.

Nevertheless, the question is ambiguous. How much should the cancer patient know and decide? In affirming the competent patient's right to know, to decide about treatment, and to control access to confidential information, we do not deny the patient's right to decline to know or to decide. It is important to distinguish an ideal of patient participation from a right of patient participation. Whatever the appropriate model of the "good" patient, we have construed the right to know and decide as an option right, not a mandatory right. In principle, the competent patient may waive his or her right to know and to decide. Such waivers are usually partial, rather than complete; they transfer part of the right to know or to decide, and they are revocable. Space does not permit us to elaborate the complicated conditions and limits of morally valid waivers. It is sufficient to emphasize that the principle of respect for persons implies a patient's right to know and to decide but not necessarily a duty to know and to decide.

POSTSCRIPT (BY JFC)

Case Narratives and Moral Perspectives

Cases are kinds of narrative—an obvious but sometimes neglected point. Hence, in contrast to some interpreters, cases cannot be separated completely from narrative and from narrative ethics. Some of the uncertainty about the relation between case analysis and the burgeoning interest in narrative ethics stems from the range of topics covered by the phrase "narrative ethics," including such different topics as the narrative of a community (tradition) and the narrative of an individual (a life story with a fitting character and virtues). The neglect of cases as narratives may also stem from some casuists' tendency to structure a case so as virtually to obliterate its narrative qualities. However, the case is, at a minimum, a narrative, at least a mini-narrative, which involves characters, plot, action, etc. Hence, the issues that arise for narratives in general also arise for these mini-narratives we call "cases." And narrators of cases, whether clinicians, ethicists, or others, must face their moral responsibilities in their narration, whether oral or written, because narration is itself a performative act.[33]

In a recent analysis, Tod Chambers imaginatively and incisively examined the relations between various approaches to ethics and various styles of case presentation.[34] Holding that an objective presentation of a case is impossible, he contends that any presentation of a case for moral discussion already structures that case in a way that reflects the author's vision of the moral world. His examples include two cases to which I contributed, however modestly, in writing or rewriting.[35] In focusing on the case featured in the preceding essay, Chambers examines our putative use of dramatic irony in relation to what he supposes to be our approach to ethics. Although his observations and arguments are often quite interesting and important, they are at points overstated and even misleading.

Chambers notes that we tell Mr. X's story twice, with the first paragraph relating the biological drama and the second the drama of the disease's effects on Mr. X and his family. Thus, a story of illness follows a story of disease. The first paragraph "gives readers foreknowledge that creates an ironic twist; as they read the second paragraph they know Mr. X's fate, his inevitable death." The particular type of irony involved is "the dramatic irony of the Greek tragedy, in which the audience knows beforehand the flow of events, the outcome, and watches to see how the story is rendered rather than how it ends." Because they already know from the first paragraph that Mr. X dies, readers "turn to the second paragraph, with its presentation of the impact on the parties involved, fully aware that no miracle occurs in this case and thus focused on how Mr. X moves toward his unavoidable end." "Every line," Chambers stresses, "increases the unique undertone of an ironic telling."

This irony continues in the presentation of the physician's conversation with the family in the "manner of a physician's informing a patient: 'The physician encouraged them to ask everything they wanted to know and answered every question with patience and empathy.'" "Because of readers' foreknowledge of Mr. X's death," Chambers suggests, "they watch as all the patience and empathy that they would expect to be given to Mr. X are given to his family. The writers go on to describe the family's decision not to inform Mr. X of his condition, and the physician's acceptance of this decision." He suggests that dramatic irony continues in our statement: "Mr. X was told only that he needed 'preventive' treatment. After being informed of possible side effects, he enthusiastically consented to irradiation and chemotherapy." According to Chambers, "the use of the phrases 'being informed' and 'enthusiastically consented' makes readers see the situation as a parody of informed consent." In sharp contrast to the family's treatment—their daily inquiries, whether by phone or in person, were treated "very kindly by the staff"—the patient's "complaints and harsh criticism of the nurses generated hostility, causing them to avoid him whenever possible." Finally, Chambers notes, the narrative's conclusion offers the irony of fate, in which the opposite of what one expects happens: "Mr. X was never offered a chance to talk about his impending death since, until the very end, everyone pretended that he would soon recover."

Structuring such a case in dramatic irony, Chambers contends, "creates a special sense of familiarity between the writer and the reader," making "insiders of readers, who can return a knowing wink to the author." Exactly how does this structure allegedly work in the above case? "Schoene-Seifert and Childress's ability to make readers insiders in their world is essential for their ability to argue subsequently for the prima facie principle of respect for persons. The ironic presentation—the establishment of a secret collaboration between reader and writer—makes the presentation of self-evident principles self-evident." He further observes: "The ironic presentation of Mr. X's case places readers as insiders with the writers, and as insiders, they share the writers' view of the world and the self-evident features of that world." And he suggests that some of our statements "are not so much arguments as presentations of previously acknowledged principles to a fellow insider."

Several responses are in order. First, while I am not sure I can reconstruct precisely how our case description evolved, I am confident that we did not write and rewrite the case, which originated with Schoene-Seifert, in a deliberately ironic way. However, since authors' intentions do not always control interpretation, Chambers may well be right in much of his analysis of the significance and effects of the ironic structure he discerns in the case presentation.

Second, there is some tension in Chambers' claims about the relation between our case presentation and our ethical vision. On the one hand, he seems to hold that "mak[ing] readers insiders in [our] world is *essential for [our] ability to argue subsequently for the prima facie principle of respect for persons*" (emphasis added). On the other hand, he suggests that our case presentation through dramatic irony makes "*the presentation of self-evident principles self-evident*" (emphasis added). However, these two points are in tension, because the appeal to self-evidence hardly counts as an argument for the principles—it's more a call for recognition.

Third, in fact, in view of the limited space we had—and we even exceeded what was allotted—we did not so much *argue* for the principle of respect for persons, and its derivative rules or rights, as simply *assume* them for purposes of this essay. Without in any way speaking for my coauthor, some of the arguments that I find cogent appear elsewhere, but those arguments are not presupposed in this essay—in contrast to what Chambers supposes.[36] Acceptance of the principle of respect for persons by the two authors of this paper could be heuristic and hypothetical, merely for purposes of this particular discussion, or it could be the result of a convergence or overlap of our respective theories. Nor do we suppose that the principle of respect for persons and various derivative rights are self-evident. Our invocation of W. D. Ross does not imply a commitment to intuitionism and its construal of self-evidence. In this essay, our use of Ross is limited to his category of prima facie weight and stringency. In some of my later work, for example, in the fourth edition of *Principles of Biomedical Ethics,* an appeal to Ross's "common morality" does play a significant role—but not in this essay. And my appeals to common morality in later works involve argument, not simply self-evidence from an intuitionist perspective. In short, it is possible to assume principles for purposes of a particular (and limited) ethical analysis without mounting an argument for them—indeed, even without presupposing an argument for them—and without maintaining that they are self-evident and need no argument.

Finally, Chambers suggests that by making our readers insiders, and the characters in the case outsiders who cannot see "either the irony of the situation or the principles that make their actions ironic," we expect readers to be outraged, as we are, by this case as a form of insult and disrespect. Most of the cases I present as a (co)writer or (co)rewriter—or report as narrated by others—involve conflicts of moral principles, rules, etc., which create genuine perplexities and perhaps even dilemmas. However, it is also quite appropriate to present a case in which there is no dilemma, or at least no felt dilemma, from the standpoint of the case authors in light of their assumed moral norms. Our effort to justify our ethical assessment did involve arguments about the implications of the principle of respect for persons. However, in this particular analysis we assumed the principle of respect for persons but then traced its implications and argued that it could not be overridden in these circumstances—or relevantly similar circumstances—by competing principles; hence, we concluded that the actions by the physicians and family were not ethically justified.

What does Chambers propose as an alternative to our (and others') use of cases? He proposes "virtuous" case narration. He is right, I believe, that it is impossible to find recourse in a thoroughly objective presentation, because all communication is value-laden—even the apparently objective clinical case history "subtly communicates a particular way of seeing the ill person" through its language, which often uses the passive voice, dehumanizes the patient through its metaphors, and the like. I accept his claim that "all forms of narration" inevitably express "a particular way of viewing reality," and that narrative is thus particularly risky and potentially deceptive—both self-deceptive and other-deceptive—because it presents one particular vision of the world, including its moral dimensions, "as natural and self-evident, as though there were no other way of seeing."

In "a virtuous style of case presentation in bioethics," as Chambers sees it, authors would acknowledge the "situatedness" of their cases, particularly by confessing their role as "authors" in the sense of assuming responsibility for the case. By so doing, "authors" would "destroy the illusion that an objective presentation of a case exists." Unfortunately, it is just not clear what else a "virtuous style of case presentation" involves.

Chambers does criticize what he sees as our (and others') "detachment" of the case presentation from the subsequent analysis and evaluation—an arrangement that implies an objective case presentation. While I concur that it is important to avoid "detachment" as he describes it, I would also argue that it does not characterize Schoene-Seifert's and my case presentation and essay, for the very language we use indicates quite clearly the moral deficiencies of some of the actors and actions in the case, thereby invoking the assumed principle of respect for persons (and its derivative rights). And the opening two paragraphs of the essay identify, again quite clearly, the substantive moral principles that will be featured in the whole essay and, hence, at least implicitly, in the "case report" that immediately follows. So, in effect, we confessed all that was actually operative in our case presentation.

However, Chambers reads much more into the implications of our use of dramatic irony than is warranted, particularly about our "approach." And, somewhat ironically, he neglects other contextual factors that frequently shape how a case is presented in relation to authors' ethical frameworks. These include such mundane matters as severe space limitations, assigned topics, and even assigned titles, all of which were important in our preparation of this particular essay, including the "case report." Such explanations are less interesting than the authors' overall ethical framework, but they may sometimes be more accurate.

For authors to take seriously the rhetorical features of narrative and to acknowledge that their own style of case presentation is influenced by their moral perspective (though not always as decisively as Chambers supposes) does not go very far in cultivating a virtuous style. More can and should be done from a moral standpoint in attending to language, metaphors, analogies, and the like in case narration and, perhaps also, in considering different narrations in order to ensure balance, given the impossibility of neutral presentations. Discussion of a case from various moral standpoints entails testing its various descriptions or narrations as well as assessing agents and their actions from those different standpoints. If the cases are intended to present real agents, events, etc., the narratives are subject to additional constraints.

Other issues arise in the presentation of cases such as the one Schoene-Seifert and I used. Several details were changed in order to protect the privacy and confidentiality of the parties involved in the real case. Such changes are often morally imperative; and counter to the views of some proponents of thick descriptions, it is not clear that, or how much, the omission or alteration of these details changes—or should change—the moral presentation and evaluation of the case. At any rate, one fundamental question is how thick case descriptions must be.[37] And that may depend in part on their purposes. Similar questions of moral authorship arise for hypothetical cases—whether realistic or unrealistic—as for real cases, but there are additional constraints in the narration of a real case.

Cultural Differences in Knowing and Deciding

Several recent studies have challenged what they take to be the dominant Western bioethical stance on respect for patient autonomy and its implications for the kind of question addressed in this essay: "How Much Should the Cancer Patient Know and Decide?" (a title assigned by our editors). Here I will only indicate some important results of these studies, which enrich the principle of respect for autonomy rather than provide a warrant to abandon it, as some of the researchers appear to suppose. Some of these challenges to what is too often called "patient autonomy" confuse rights, duties, and ideals of autonomy, and once this confusion is cleared up the principle of respect for [patient] autonomy remains strong, even though it never stands alone and is far from absolute.

Several UCLA researchers sought to study the differences in the attitudes of elderly subjects (sixty-five years or older) from different ethnic backgrounds toward (a) disclosure of the diagnosis and prognosis of a terminal illness, and (b) decision making at the end of life.[38] According to the researchers' summary of their main findings, based on 800 subjects (200 from each ethnic group), "Korean Americans (47%) and Mexican Americans (65%) were significantly less likely than European Americans (87%) and African Americans (88%) to believe that a patient should be told the diagnosis of metastatic cancer. Korean Americans (35%) and Mexican Americans (48%) were less likely than African Americans (63%) and European Americans (69%) to believe that a patient should be told of a terminal prognosis and less likely to believe that the patient should make decisions about the use of life-supporting technology (28% and 41% vs. 60% and 65%). Instead, Korean Americans and Mexican Americans tended to believe that the family should make decisions about the use of life support." Ethnicity was the primary factor correlated with attitudes toward truth telling and patient decision making, with years of education and personal experience with illness and with withholding or withdrawing treatment playing secondary roles.

Stressing that "belief in the *ideal* of patient autonomy is far from universal" (my emphasis), the researchers contrast that ideal with a "family-centered model," which emphasizes the individual's web of relationships and tends to place a higher value on "the harmonious functioning of the family than on the autonomy of its individual members." The researchers conclude that, in light of their finding that some groups are more likely to hold a family-centered model of medical decision making than the patient autonomy model, "physicians should ask their patients if they wish to receive information and make decisions or if they prefer that their families handle such matters." What is required is not abandonment of the commitment to individual autonomy or to its legal expression (the doctrine of informed consent) but rather "broadening our view of autonomy so that respect for persons includes respect for the cultural values they bring with them to the decision-making process."

In a study of Navajo values and ways of thinking regarding "negative information," the kind of information that may emerge in the disclosure of risk and "bad news" and in the practice of advance care planning, two researchers sought to learn how health care providers "should approach the discussion of negative information with Navajo patients" in order to provide "more culturally appropriate medical care."[39] Frequent conflicts emerge, according to these researchers, between the Western bioethical principle of autonomy and the traditional Navajo conception that "thought and language have the power to shape reality and to control events." According to that traditional conception, telling a Navajo patient who has been recently diagnosed with diabetes the potential complications of that disease may actually produce the complications, because "language does not merely describe reality, language shapes reality." Hence, traditional Navajo patients may actually "regard the discussion of negative information as potentially harmful," in contrast to "positive ritual language" that may actually maintain or restore health. One middle-aged Navajo nurse reported that the risks of bypass surgery had been explained to her father in such a way that he refused to undergo the procedure: "The surgeon told him that he may not wake up, that this is the risk of every surgery. For the surgeon it was very routine, but the way that my Dad received it, it was almost like a death sentence, and he never consented to the surgery." Because of the evident Navajo discomfort with negative information, the researchers found "ethically troublesome" those policies which, in compliance with the Patient Self-Determination Act, attempt to "expose all hospitalized Navajo patients to the idea, if not the practice, of advance care planning."

These two important studies—and numerous other studies over the last several years—greatly enrich our understanding of diverse cultural beliefs and values that affect what particular communities and individuals hold about "how much cancer patients (and other patients) should know and decide." However, many of these studies also reflect a major *mis*understanding of what many bioethicists and many laws and policies—though by no means all—propose regarding patient autonomy. And they thus often mistakenly view their studies as opposing, rather than enriching, the principle of respect for autonomy. Part of what is needed is a better, fuller interpretation of the principle of respect for autonomy, particularly over against some excessively individualistic, rationalistic, legalistic, and formalistic conceptions that have often appeared in the bioethical literature and in laws, policies, and practices in health care.

As a first matter, the principle involved is better understood as a *principle of respect for [patient] autonomy* than as a principle of autonomy. Its requirements entail respecting what patients autonomously choose regarding information and their own decision making. Properly understood, it establishes ethical (and often legal) rights to know, that is, to receive or to decline information, and to decide, that is, to choose and to refuse treatments. When, however, the principle is stated as a *principle of [patient] autonomy*, it also suggests an *ideal* or even a *duty for patients*. Even in studies that reach what I regard as ethically defensible conclusions, the language is potentially misleading. For instance, one chart presents "measures of patient autonomy" rather than measures of respect for patient autonomy.[40] The chart's formulation seems to presuppose a right and a wrong way of exercising patient autonomy.

Joel Feinberg draws an important distinction between an option or discretionary right and a mandatory right.[41] The right to education is a good example of a mandatory right: Not only does the citizen have a right to an education, but he or she also has a duty to receive education up to a certain level. By contrast, "I have a discretionary right in respect to X when I have an open option to X or not to X correlated with the duties of others not to interfere with my choice." The position taken will have an impact on what the right-holder can do with his or her right. However one resolves the dispute about rights in general, there may be debates about the characterization of particular rights, such as the right to life, and the right to know and decide. If the right to know and decide is connected to human interests and benefits rather than to human choices, it may be construed paternalistically.

In effect, the principle of respect for [patient] autonomy, which was sharply formulated to avoid many forms of paternalism, has sometimes been subtly reformulated to justify a particular kind of paternalism, one based on the exercise of autonomy in particular ways. This new, autonomy-based paternalism is all the more insidious because, in the blinding light of the language of autonomy, it may not be recognized for what it is. Some bioethicists and some laws and policies present an *ideal* or *duty* of patient autonomy, understood in terms of the ideal of or duty to accept all relevant information (where relevance is established by impersonal, general standards) or to make one's own decisions, rather than merely drifting along or implicitly delegating decision making authority to others. What was originally—and rightly—conceived as an option right has now become a mandatory right. Furthermore, autonomy is conceived in excessively rationalistic, formalistic, legalistic, and individualistic ways.

What some critics, including some researchers into different cultural values, fail to see is that they do not challenge the Western bioethical principle of respect for autonomy within a pluralistic context, but rather confirm it, at least when it is properly understood. Patients should have the *right* to know (and not to know) and to decide (to accept or to decline). The practical ethical problem (and, increasingly, the practical legal problem) is to figure out ways to inform patients of their rights without forcing them into particular molds. For instance, there is a potentially serious tension between the two studies noted above: One recommends inquiring in advance about patients' preferences regarding information and decision making, while the other suggests, but somewhat ambiguously, that even being informed of a right to decide if problems develop is itself harmful. The tricky practical ethical question is whether it is possible to inform patients of their rights, under the principle of respect for patient autonomy, including their rights to know and to decide, without compromising particular systems of belief and value. At the very least, it may be possible to inquire in the most general terms about their own desires to receive information and to make decisions relative to their own desires for disclosure to other parties or decision making by those parties.

CHAPTER SIX

□

Mandatory HIV Screening and Testing

Faced with epidemics societies frequently curtail individual rights and liberties, often in ways later considered to be excessive and unnecessary. Even liberal societies have protected the public health by such measures as compulsory screening and testing, quarantine and isolation, and contact tracing. For a liberal society—that is, a society that recognizes and protects individual rights and liberties—the difficult question is when the public health justifies overriding these rights and liberties.[1] For several years—roughly from the late 1950s to the early 1980s—the United States considered itself immune to major public health crises. Even though debates lingered about policies to control sexually transmitted diseases, these diseases were not viewed as major threats to individual or communal survival. During the same period, laws, public policies, and social practices, including health care, increasingly protected various individual rights and liberties. Then came AIDS with its major threat to the public health and, upon the development in mid–1985 of tests for antibodies to the human immunodeficiency virus (HIV), calls to use these diagnostic tools to identify seropositive individuals through consensual and/or mandatory screening and testing.[2] Numerous bills were introduced in state legislatures to require testing in various contexts, such as application for marriage licenses, and the federal government instituted mandatory screening of selected groups, including immigrants, inmates in federal prisons, and military personnel.

Advocates of mandatory screening and testing, who often invoke the metaphor of war against AIDS, frequently fail to indicate what actions would, or should, follow the identification of HIV antibody positive individuals. Yet it is not possible to isolate mandatory (or even consensual) screening and testing from other possible interventions because, quite in contrast to a vaccine, screening and testing by themselves have no direct impact on an epidemic. What is done with the information, and with the people who test positive, will determine its effect on the AIDS epidemic. Hence, a first question for all proposed HIV antibody tests is why the information is wanted and what will be done with that information (as well as with the people who test positive). Debates have raged about whether sexual contacts should be traced, whether current sexual partners should be warned, whether certain employers should be notified, whether seropositive individuals should be quarantined, and the like. Such questions provide the backdrop for much of the debate about mandatory screening and testing.

In this essay I will assume the social and political principles and rules embedded in a liberal society, then try to determine when such a society, under the public health threat from AIDS, may justifiably resort to mandatory screening and testing for HIV antibodies, thereby overriding some of its important principles and rules. After sketching the major presumptive or prima facie principles and rules that may constrain public health efforts in a liberal society, I will identify some conditions that need to be met to justify infringing those principles and rules in the pursuit of public health. Then I will develop a typology of screening/testing policies, stressing the distinction between consensual and mandatory screening/testing, and examine the issues raised by different policies. Throughout this discussion I will note the central role of analogical reasoning, and I will conclude with an analysis of the impact of the metaphor "war against AIDS" on our society's continuing debates about mandatory screening and testing.

A SOCIAL-ETHICAL FRAMEWORK FOR ASSESSING POLICIES OF SCREENING AND TESTING

It is not necessary to belabor why control of HIV infection and AIDS is a public health concern rather than merely a private, individual matter. AIDS is an infectious disease; people may be infected—and infectious—for many years before they become aware of their condition; there is no known cure for AIDS, which has an extremely high death rate (apparently 100 percent, given enough time); the suffering of AIDS patients and others is tremendous; the cost of caring for AIDS patients is very high; and so on. Protecting the public health, including the health of individuals, is a legitimate moral concern—even a moral imperative—based on fundamental moral principles affirmed by a liberal society, including beneficence, nonmaleficence, and justice, as well as respect for persons and their autonomy.[3] "Public health is what we, as a society, do collectively to assure the conditions in which people can be healthy. This requires that continuing and emerging threats to the health of the public be successfully countered."[4] AIDS is a paradigmatic instance of the public health threats that must be effectively countered by the society.

If the goal of controlling AIDS is a strong moral imperative for the society as well as for individuals, we have to determine which measures may and should be adopted to achieve this goal. "The AIDS virus has no civil rights"— that rhetoric often suggests that the moral imperative to control AIDS cancels all other moral imperatives. Of course, few who use this rhetoric attempt to justify compulsory universal screening accompanied by mass quarantine or even mass slaughter. Even so, their position rarely takes seriously enough other significant moral principles and rules, also embedded in a liberal society's laws, policies, and practices. I want to examine some of those principles and rules and sketch what they imply for public policies to control this infectious disease. This argument hinges on the best available scientific and medical information, including evidence that the spread of AIDS usually involves consensual, intimate contact, in the form of sexual activity or sharing intravenous drug needles and syringes, and that casual contact is not a mode of transmission. A different set of facts could justify different screening and testing policies. (Below I will indicate where my judgments about particular policies have been revised since the publication of the original version of this essay because of changes in scientific and medical evidence; the main example appears in maternal-fetal screening and testing because of the evidence that AZT substantially reduces the risk of maternal-fetal HIV transmission.) Indeed, different policies may be justified for HIV infection and tuberculosis because of differences in modes and ease of transmission, in risks to more casual contacts, and in treatments.[5]

The principle of respect for persons is a primary principle in the ethical framework of a liberal society.[6] It implies that we should not treat people merely as means to ends. From this principle (and others) we can derive more specific rules that also direct and limit policies. Three such rules are especially important for this analysis: rules of liberty, including freedom of association; rules of privacy, including bodily integrity and decisional space; and rules of confidentiality. The principle of respect for persons and its derivative rules can be stated as individual rights or as societal obligations and duties—for example, the individual's right to privacy or the society's obligation not to violate an individual's privacy. Even if some of these rules are independent rather than derived, they are closely related to personal autonomy—for instance, the individual can autonomously waive the rights expressed in the rules or, perhaps more accurately, can exercise those rights by yielding liberty or privacy or by granting access to previously confidential information. Furthermore, in a liberal society, overriding the principle of respect for autonomy and these rules can be justified more easily to protect others than to protect the agent himself or herself. Thus, there is a strong suspicion of paternalism, which infringes these principles and rules to protect the agent rather than others.[7] (Of course, the goals of intervention are often mixed.) Rather than offering a theoretical foundation for the principle of respect for autonomy and various derivative rules, I will assume that they can be discerned in our constitution, laws, policies, and practices.—in short, in the social ethics or political morality operative in a liberal society.

CONDITIONS FOR OVERRIDING PRIMA FACIE PRINCIPLES AND RULES

Which, if any, policies of mandatory screening and testing for HIV antibodies can be ethically justified in a liberal society in view of its moral obligation to control AIDS and its other moral principles and rules? In order to answer this question we have to determine both what these principles and rules mean and how much weight they have relative to other principles and rules. (For brevity I will sometimes use "principles" or "rules" to cover both principles and rules.) Some apparent violations of rules may not really be violations; upon closer inspection they may turn out to be consistent with the rules properly interpreted. And none of these rules is absolute—each one can be justifiably overridden in order to protect the public health under some conditions. However, in a liberal society these principles and rules are more than mere maxims or rules of thumb that yield to any and every utilitarian objection. If these principles and rules are not absolute or mere maxims, is there an alternative conception of their weight or stringency? They may be construed as prima facie binding, that is, they have heavy moral weight or strong binding power in and of themselves. Because they are morally weighty and binding, other things being equal, it is necessary to justify any departures from them, and the process of justification involves meeting a significant burden of proof.[8]

At least five conditions must be met to justify infringements of these rules.[9] These conditions, which will be called *justificatory conditions*, are effectiveness, proportionality, necessity, least infringement, and explanation and justification to the parties protected by the rules. They express the logic of prima facie duties or rights, that is, how principles and rules with such weight or strength are to be approached, as well as the substance of liberal principles and rules. A presupposition of these justificatory conditions is that the goal— both expressed and latent—is to protect public health, that is, the health of the community and the individuals within it, rather than to express moralistic judgments about individuals and their conduct or to exclude seropositive individuals from the community by subjecting them to discrimination or denying them access to needed services.[10]

First, it is necessary to show that infringing these rules will probably realize the goal of protecting public health. This first condition is one of effectiveness.

A policy that cannot realize its public-health goal and also infringes these moral rules simply has no justification—it is arbitrary and capricious. An ineffective policy to control HIV infection that does not infringe these (or other) moral rules may be wasteful, unwise, and even stupid, but if it infringes moral rules, there is, in addition, a decisive moral argument against it.

Second, it is necessary to show that the probable benefits of a policy infringing these moral rules outweigh both the rules infringed and any negative consequences. This condition, which may be called proportionality, is complex, for it involves considering not only the weight of the infringed moral rule but also other harms, costs, and burdens that may flow from the infringement. For example, it is necessary to consider not only the weight or strength of the rule of privacy but also other probable negative effects, such as discrimination, that may befall the one whose privacy is infringed. (As I will argue later, in order to justify policies that *impose* community, it may also be essential to *express* community through measures to avoid or reduce such negative consequences. Indeed, if community is expressed through such measures, it may be unnecessary to impose community.)

Third, it is not sufficient to meet the first two conditions of effectiveness and proportionality and thus to show that infringement of these rules will produce better consequences for more people. As prima facie binding, these rules direct us to seek alternative ways to realize the end of public health before infringing the rules. For example, if it is possible to protect the public health without infringing liberty and privacy (and other moral rules), then the society should do so. This condition is one of necessity, last resort, or lack of a feasible alternative. Priority belongs to policies that do not infringe the rules of liberty, privacy, and confidentiality. Thus, policies that seek to *educate* people about acting in certain ways have moral priority over policies that *force* people to act in certain ways. However, under some circumstances, where coercive policies would be effective and proportionate, they can be justified if they are also necessary means to protect the public health. Many of the proposed policies of mandatory screening fail this third test.

Fourth, even when a liberal society is justified in infringing its own moral rules to protect the public health, it is obligated to seek policies that least infringe its rules. Only the degree or extent of infringement that is necessary to realize the important end can be justified. For example, when liberty is at stake, the liberal society should seek the least restrictive alternative; when privacy is at stake, it should seek the least intrusive and invasive alternative; and when confidentiality is at stake, it should disclose only the amount and kind of information needed for effective action. This fourth condition may be called that of "least infringement."

Finally, even when a public policy that infringes one or more moral rules satisfies all four prior conditions, the underlying principle of respect for persons may generate additional requirements. This principle may, for instance, require informing those whose liberty, privacy, and confidential relations are infringed. In many cases, such as coercive screening for HIV antibody, where blood has not already been drawn for other purposes, the infringements will be evident to the parties affected. However, this may not be true in all cases, especially if secrecy or deception is involved. In some contexts, as Sissela Bok notes, secret or deceptive actions may be more disrespectful and insulting to the parties affected than coercive actions.[11] Hence, the society may have a duty to disclose, and even to justify, the actions to the person and perhaps even sometimes to undertake compensatory measures. Even if it is essential to infringe a person's rights to protect the public health, that person should not be reduced to a mere means to the goal of public health. This crucial point can also be stated through an image drawn from Robert Nozick: Overridden moral principles and rules leave "moral traces."[12] They do not evaporate or simply disappear when they are justifiably overridden—they are outweighed, not canceled. And they may require subsequent corrective or compensatory actions.

As is true of much moral, political, and legal reasoning, this process of reasoning about the justification of mandatory HIV antibody tests is to a great extent analogical. The principle of universalizability or formal justice requires treating similar cases in a similar way. Hence it is necessary to consider the relevant similarities and differences among candidate targets for screening, not only for HIV antibodies but also for other conditions, such as genetic ones. Reasoning in a liberal society certainly considers precedent cases—for example, the relevant similarities and differences between HIV antibody screening and other mandatory screening policies that have been held to be justifiable—but also the precedents mandatory HIV antibody screening would create. "HIV and AIDS must not be treated in isolation," Margaret Somerville argues; "comparison with analogous situations is mandatory."[13]

In addition to formal justice, material criteria of justice, also embedded in a liberal society, are relevant to assessments of public policies of screening and testing that override the prima facie rules of liberty, privacy, and confidentiality. Standards of justice, in both broad and narrow senses, serve as the backdrop for debates about such policies and provide specific challenges to some actual and proposed policies, particularly because of discrimination, procedural deficiencies, the maldistribution of benefits and burdens, and the like. One of the major negative social effects of breaches of confidentiality about HIV infection is discrimination in various settings, such as housing.

MAJOR TYPES OF SCREENING/TESTING POLICIES

In order to apply the principle of respect for personal autonomy and its derivative rules—all conceived as prima facie binding—along with the conditions for justifying their infringement, it is necessary to identify the main types of proposed policies of mandatory screening and testing for HIV antibodies. The following chart indicates some of the most important options: Policies of screening/testing may be consensual or compulsory, and universal or selective.

SCREENING/TESTING POLICIES

	Degree of Voluntariness	
	Consensual	Compulsory
Universal	1	2

Extent of Screening

Selective	3	4

The term "screening" usually refers to testing groups, while the term "testing" usually refers to testing individuals. However, even in mass screening, the individual is tested as part of the targeted group. Thus, I will continue to use both terms as appropriate without drawing a sharp distinction between them. A more important distinction concerns the *scope* of the screening or testing: Is it universal or selective?

Universal screening. There is simply no adequate justification for *universal screening,* whether voluntary or compulsory. Even serious proposals for widespread screening usually fall short of universal screening. For example, Rhame and Maki have recommended "HIV testing vigorously to all U.S. adults under the age of sixty regardless of their reported risk history."[14] The arguments against universal screening are compelling: Universal screening is not necessary to protect the public health; HIV infection is not widespread outside groups engaging in high risk activities; screening in groups or areas with low seroprevalence produces a high rate of false positives; universal screening would be very costly and would not be cost effective—the funds could be spent more effectively on education; and its potential negative effects, including discrimination, would outweigh any potential benefits. It is not even justifiable to encourage everyone to be tested.

If universal consensual screening is not justifiable, there is, a fortiori, no justification for universal compulsory screening since it would infringe the principle of respect for autonomy and the rules of liberty and privacy (and probably confidentiality) without compensating benefit. The main rationale for compulsory universal screening is that seropositive individuals could then reduce risks to others. While there is evidence of some individual behavior change following disclosure of seropositivity, especially with counseling,[15] there is no evidence that mandatory testing would produce equivalent voluntary change in taking risks and in subjecting others to risk.[16]

Identification of and disclosure to the individuals who are seropositive do not necessarily translate into benefits for others, without additional interventions. And appropriate education and counseling, apart from mandatory testing, could be effective ways to realize the same end, especially because universal precautions are recommended in risky situations whether a person is seropositive or seronegative.[17] In light of the current scientific and medical evidence, compulsory universal screening thus fails to meet the five conditions identified earlier for justified breaches of relevant moral principles and rules. Universal precautions, pursued through education, are morally preferable to mandatory universal screening, and the latter is not demonstrably more effective and would, in addition, produce some serious negative consequences.

Selective screening. Consensual or voluntary selective screening may appear to pose no moral problems. However, it does raise some important questions—for example, who should be encouraged to be tested, who should bear the costs, what sort of pretest and posttest counseling should be provided, and what conditions make the decision to have the test a rational one. Individuals making rational choices will consent to testing only when they perceive a favorable risk-benefit ratio. And "more than most medical tests, HIV screening has major benefits and harms that must be weighed."[18] In all risk-benefit analyses, the comparison is between the probability of harm of some magnitude and the probability of benefit of some magnitude. Perhaps because "risk" is a probabilistic term, while "benefit" is not, risk-benefit analyses sometimes neglect to determine both the probability and the magnitude of both benefits and harms of proposed policies, which any adequate analysis must include.

If we assume the accuracy of the test results, the possible *benefits* of testing to *seronegative* individuals include reassurance, the possibility of making future plans, and the motivation to make behavioral changes to prevent infection, while the possible *benefits* to *seropositive* individuals include closer medical follow-up, including prognoses at various stages, earlier use of AZT (and any other treatments), prophylaxis or other care for associated diseases, protection of loved ones, and clearer plans for the future. There appear to be no *harms* to *seronegative* individuals, if we assume the accuracy of the test results, while there are major *risks* to *seropositive* individuals.[19] These risks may be identified as both psychological and social (with considerable overlap and interaction). The psychological risks include anxiety and depression—followed by a higher rate of suicide than for the population at large[20]—while the social risks include stigma, discrimination, and breaches of confidentiality.

Clearly the society can have a major impact on the risk-benefit analyses of potential candidates for voluntary testing. On the one hand, the medical benefits of early diagnosis are possible only if seropositive individuals have adequate access to health care. Hence, even on the benefit side, the problem is not merely medical, because of the social problem of access to health care. On the other hand, the risks can be greatly reduced by societal decisions to allocate funds and to establish strong rules to protect individual rights and liberties. The society should provide resources for pretest and posttest counseling both to help the individual and also to reduce risks to others, and it should protect seropositive individuals from breaches of privacy and confidentiality and from discrimination in such areas as housing, employment, and insurance. Without such societal support and protection, the risks of the test may outweigh its benefits, including any medical benefits, for rational individuals.

If voluntary or consensual testing of selected groups can be effective, in some contexts, then there is no justification for compulsory screening of those groups. The liberal justificatory framework imposes this necessary condition:

Moral principles and rules are not to be infringed if the same important ends—
in this case, the public health—can be realized without their infringement.
Furthermore, the society may even have to bear additional costs to protect its
important principles and rules.

The requirement, expressed by the terms "mandatory" and "compulsory,"
may be imposed on the one to be tested or on the tester or on both. For
example, in the context of screening blood donated for transfusions, the term
"mandatory" refers in the first instance to the obligation imposed on organizations collecting blood. And such "mandatory" screening may be imposed or
only permitted by the state. The one area of mandatory selective screening that
is morally settled and uncontroversial is screening all donated (or sold) blood,
organs, sperm, and ova, in part because recipients cannot take other measures
to protect themselves. In this area universal precautions cannot work without
screening and testing.

Many policies of screening are actually mixes of voluntary actions and
compulsory actions, as is evident in programs of screening donated blood.
While individuals can choose to donate blood or not, their blood will be tested
if they do donate. Another good example is the policy of screening recruits into
the military. The U.S. armed forces now consist of volunteers, who are tested
for HIV antibodies upon entry (and then again at least once a year). It could
be argued that volunteers for the armed forces consent to HIV testing, because
they voluntarily enter an institution where screening is compulsory, and that
the mandatory screening thus does not violate any of the volunteers' rights.
This argument has merit, but it does not adequately address the question
whether the society can justify its policy by the conditions identified earlier.
The major express justification is that each member of the armed forces is a
potential donor of blood for transfusions on the battlefield; hence the rationale
is comparable to the one that governs screening donated blood. Even if this
rationale is solid, along with the rationale of protecting HIV-infected people
from live virus vaccines used in the armed forces, there is also reason to suspect
that the military's homophobia was a factor in the decision. Furthermore, there
is concern about the precedent this mandatory screening creates, especially
when, as sometimes happens, proponents also appeal to the cost of future
health care for HIV-infected former members of the armed forces.[21]

Voluntariness may be compromised in various ways. We tend to think of
coercion as the major compromise of voluntariness—for example, forcing
someone to undergo a test for HIV antibodies. However, conditional requirements—if you want X, then you have to do Y—may also be morally problematic. For example, requiring HIV antibody testing as a condition for some
strongly desired benefit, such as a marriage license, may constitute an undue
incentive, even if it is not, strictly speaking, coercive because the person can
choose to decline the benefit.[22] One important question then is whether it is
fair to impose the condition, even if the person is free to take it or leave it. This
question expresses the important general point that selection of targets for
mandatory screening—or even for voluntary screening—is in part a matter of

justice in the distribution of benefits and burdens[23] and that not all important moral issues are reducible to a determination of the voluntariness of choices. The justificatory conditions identified earlier serve as important selection criteria.

Nevertheless, because of the centrality of the principle of respect for personal autonomy in a liberal framework, it is not surprising that consent plays such an important role in the analysis and assessment of screening policies. When an individual gives valid consent to testing, there is no violation of his or her personal autonomy and liberty of action. Nor does consensual testing violate the rule of privacy, for the individual voluntarily surrenders some of his or her privacy. And, finally, if a person grants others access to the information about his or her HIV antibody status, the rule of confidentiality is not violated. In a liberal framework, valid consent creates rights that did not previously exist. (I will use the terms "voluntary" testing and "consensual" testing interchangeably to refer to testing with the individual's voluntary, informed consent—it is irrelevant, for our purposes, whether he or she requested the test or only accepted a recommended test.)

What counts as consent? This question becomes important because people have been held to consent in various ways, not simply by express oral or written statements. And some varieties of consent have even been invoked to override a person's express wishes and choices. I shall discuss this topic in the context of screening hospital patients for HIV antibodies. One difficult question is whether the institutional rules of consent should be structured to authorize HIV antibody testing without express, specific consent.

VARIETIES OF CONSENT IN THE CONTEXT OF SCREENING HOSPITAL PATIENTS

The term "routine" often covers several types of screening and testing. However, this term is seriously misleading without further qualification, for its ethical significance is unclear until we can determine whether the screening will be done routinely, without notice or the possibility of refusal, or whether it will be offered routinely with the possibility of refusal.[24]

When a hospital patient consents to HIV antibody testing after being informed about the risks and benefits of the test—including whether the intended benefits are primarily for the patient or for care givers—there is no breach of moral principles. But are there forms of consent in addition to express and specific oral or written consent that can create rights on the part of hospitals to test patients who have not provided specific, express consent?

Consider, first, *tacit consent,* a favorite tool of political theorists in the contract tradition. Tacit consent is expressed silently or passively by omissions or by failure to indicate or signify dissent.[25] If a newly admitted hospital patient is silent when told that the HIV antibody test will be performed, along with other tests, unless he objects, his silence may constitute valid (though tacit) consent, as long as there is understanding and voluntariness—the same conditions that are important for express oral or written consent. According to some interpreters, even if a patient is not notified of or given any information about the test, his or her failure to dissent from the test may be *presumed* to be consent on the basis of a presumed general understanding of hospital testing policy. However, its moral validity as consent is suspect, despite the patient's right to dissent or opt out.

Whereas tacit consent or presumed consent is expressed through failures to dissent, *implied* or implicit consent is, in part, inferred from actions. Consent to a specific action may be implied by general consent to professional authority or by consent to a set of actions. Does a person's voluntary admission to the hospital imply consent to the HIV antibody test without express consent? Again much will depend on the patient's understanding. However, rather than relying on a general consent or consent to several tests, hospitals should be obligated to seek specific, express consent for HIV antibody tests because of the psychological and social risks of the tests and their results.

There are two major reasons for explicit, specific disclosure and explicit, specific consent. On the one hand, the diagnostic or therapeutic procedure may be invasive—for example, drawing blood. However, if a patient has consented to blood drawing for various tests to determine his or her medical condition, there is no additional invasion of his or her body to test that blood for HIV antibodies. Then the second major reason for specific disclosure and consent enters—the risk to the patient. Even where the blood has already been drawn, the test has the psychological and social risks noted above, and explicit, specific consent should be sought.[26] (This argument does not exclude the possibility and justifiability of unlinked testing of blood samples in the hospital in order to determine seroprevalence as part of epidemiological studies, since such studies do not involve psychological and social risks.)

A Virginia law (Virginia Code 32.1–45.21) invokes "deemed consent" in permitting a health care provider to test a patient's blood, without specific consent, following the provider's exposure to the patient's body fluids under circumstances where HIV infection might be spread. In such a case, "the patient whose body fluids were involved in the exposures shall be deemed to have consented to testing for infection with human immunodeficiency virus (and) to have consented to the release of such test results to the person who was exposed." The law assigns health care providers the responsibility, in other than emergency circumstances, to inform patients of this provision of deemed consent prior to the provision of health care services. Presumably, the patient's acceptance of health care following disclosure of this statutory provision counts as, or is deemed to be, consent. Even if such consent is valid, the fairness of imposing this condition remains an important question.

Because consent is so important an implication of the principle of respect for personal autonomy, we often resort to fictions such as deemed consent or another variety of consent even when they do not appear to constitute valid consent. We tend to extend the meaning of rules of consent and of the principle of respect for personal autonomy in order to avoid conflicts with our other moral principles. Such fictions may, however, obscure important moral conflicts. It may be more defensible, for instance, for the society to indicate that it believes it can justifiably *override* a patient's autonomy, liberty, privacy, and confidentiality in order to obtain and provide information about an individual's HIV antibody status to a health care provider who has been exposed to the risk of infection in the provision of care. A fiction such as deemed consent may appear defensible in part because it incorporates the patient into a larger moral community of shared concern, but this does not negate the cost of the fiction.

There is evidence that hospitals sometimes test patients for HIV antibodies without their consent (and without providing information about actions to reduce risks).[27] In general, hospitals are not justified in testing patients for HIV antibodies without specific consent, either to benefit the patients themselves or to enable caregivers to take extra precautions. Information that a patient is seronegative may create false security on the part of the professionals in view of the long period that may elapse between exposure to HIV and the development of antibodies. Thus, the best protection, though not an inexpensive or perfect one, is offered by universal precautions. According to one study, there is no evidence "that preoperative testing for HIV infection would reduce the frequency of accidental exposures to blood."[28]

Where exposure has already occurred, and the blood is not available for testing, it is very difficult to justify forcible extraction of the patient's blood to test for HIV antibodies to reduce the caregiver's anxieties. Where the blood is already available, it is easier to justify testing after exposure, even against the patient's wishes. However, the analogy with the Hepatitis B virus argues against mandatory testing, whether before or after exposure.[29] One ethically acceptable possibility would be to obtain patients' advance express consent—not merely deemed consent—to testing if accidental exposure occurs. (I will discuss testing health care professionals below.)

PREMARITAL SCREENING

State-mandated premarital screening for HIV antibodies also mixes voluntary choices with compulsory screening: While individuals choose to apply for marriage licenses, the test is required as a condition of application. Debates about such a policy often invoke historical analogies. For example, Gary Bauer, former assistant to President Reagan for Policy Development, uses mandatory premarital screening as an example of the "routine testing" that is similar to measures taken in the past to deal with threatening epidemics. And he argues that mandatory premarital screening for syphilis contributed to a "sharp reduction in the infant mortality rate from syphilis."[30] By contrast, Larry O. Gostin contends that "statutes for syphilis screening were largely regarded as a failed experiment" and notes they have been repealed by most states.[31]

The analogy with screening for syphilis is interesting in part because, in contrast to HIV infection, syphilis is treatable and infected individuals can be rendered noninfectious. Many argue that it is even more imperative to require premarital HIV tests as a condition for marriage licenses because there is no cure for AIDS, a fatal disease. By contrast, opponents contend that the absence of effective treatment is a good reason not to require the tests. Because antibiotics can cure syphilis, it is justifiable to withhold a marriage license until there is proof of a cure, but "it would be contrary to public policy to bar marriage to seropositive individuals," who cannot now be cured.[32]

Legislatures (Illinois and Louisiana) that passed statutes mandating premarital screening subsequently rescinded those statutes, largely on grounds that they were not cost effective. A report on the first six months of experience in Illinois indicated that only eight of 70,846 applicants for marriage licenses were found to be seropositive, while the cost of the testing program for that period was estimated at $2.5 million, or $312,000 for each seropositive individual identified.[33] Furthermore, half of those identified as seropositive admitted having engaged in risky behavior and could probably have been identified more efficiently through voluntary programs aimed at populations with a higher seroprevalence rate. Illinois also experienced a 22.5 percent decrease (a total of approximately 10,300) in the number of marriage licenses, while neighboring states granted licenses to a significantly larger number of Illinois residents than usual. Since the applicants had to cover the costs of the tests, the state of Illinois did not have to determine the most cost-effective ways to spend public money and make trade-offs.[34] However, Illinois did lose the revenue from marriage licenses, estimated at $77,250 for six months. The authors conclude what policy analysts had predicted prior to the experiment: "The Illinois experience with premarital testing provides a strong argument against widespread or publicly supported HIV antibody screening of low prevalence populations."[35]

While many of the proposed policies of compulsory selective screening fail the test of necessity because there are viable alternatives, the policy of mandatory premarital screening also fails to meet the conditions of effectiveness and proportionality. There is no evidence that the screening program prevented any additional illnesses or that it was a "rational or effective public health policy."[36] It mistakenly assumed that sexual relations only begin after the marriage, and it may even have been counterproductive in driving some couples away from the institution of marriage. Furthermore, the costs may have kept some low-income people from applying for marriage licenses. Finally, the important public health objective of protecting spouses (and future offspring) can be pursued in other ways that will not compromise respect for personal autonomy and yet will probably be more effective and cost effective— for instance, providing information about HIV risks and providing of voluntary testing with counseling, perhaps free of charge, to all applicants for marriage licenses. The context of applying for a marriage license appears to be an appropriate one "for promoting individual HIV risk assessment with educational materials."[37]

SCREENING PREGNANT WOMEN AND NEWBORNS

One major goal of premarital screening is to prevent infected offspring of the marriage. In the United States approximately four million women give birth each year. Of those, an estimated 7,000 are infected with HIV, and perhaps 2,000 of their children will be HIV infected. Studies investigating the mechanisms of maternal-fetal HIV transmission identify two major factors: the level of virus in the pregnant woman's blood, and birth events that expose infants to maternal blood.[38]

Because mandatory premarital screening programs are failures, because voluntary premarital screening is limited, and because HIV infections may occur after marriage, many have proposed screening pregnant women and newborns at least in selected settings. Since all fifty states and the District of Columbia mandate neonatal screening for several diseases, the question arises: Why not treat HIV "just like any other disease"? Even if HIV infection were strictly analogous to these other medical conditions, such as phenylketonuria (PKU), Kathleen Nolan notes that it would be difficult to determine what it would mean to treat HIV infection "just like any other disease," because states have a wide variety of laws—for example, there is no condition for which all states mandate screening, and many states allow parental refusal.[39] Furthermore, HIV infection differs from most other diseases because of its social risks. For example, so-called "boarder babies" cannot leave the hospital because no one will take these HIV-infected babies. Hence, HIV infection, Nolan argues, is best viewed as a "separate case."

Still within the context of analogical reasoning, Nolan identifies three criteria for justifying neonatal *genetic* screening: (1) the seriousness of the genetic condition; (2) availability of presymptomatic interventions that effectively prevent serious injury; and (3) an acceptable benefit-cost ratio. While HIV infection clearly satisfies the first criterion, it did not clearly satisfy the other two in neonatal cases, at least until the mid-1990s, except where there were specific reasons for testing—and parents usually consent under those circumstances. (Furthermore, neonatal and infant screening, for over a year, will reflect maternal antibodies without specifically indicating neonatal or infant infection.) In summary, Nolan argued, "calls for mandatory neonatal screening emerge primarily from beneficent clinical attitudes towards newborns, and they are rejected primarily on the grounds that not enough benefit accrues at present to justify overriding parental autonomy and family values."[40] The situation changed significantly with convincing evidence about ways to reduce the risk of maternal-fetal HIV transmission, but there is still considerable debate about whether it changed enough to warrant mandatory neonatal screening or prenatal testing.

Up to 1994, the following argument against mandatory prenatal testing, here presented in summary form, was quite persuasive: Mandatory prenatal testing to protect offspring appears relatively ineffective because nothing can be done to prevent the infection of the fetus; it is morally controversial because abortion is the only way to prevent another HIV-infected infant and yet only 25–40 percent of the offspring of HIV-infected pregnant women are also infected; and it is potentially counterproductive because it drives pregnant women away from settings where HIV screening is compulsory. The most defensible policy—because the most respectful of pregnant women's autonomy and also the most productive of desirable consequences—was to offer pregnant women, in high seroprevalence areas or with risk factors, prenatal testing for HIV with adequate information so that they could make their own decisions, with appropriate pretest and posttest counseling and support services.[41]

Then in 1994 a clinical trial established a way to reduce the risk of maternal-fetal HIV transmission. In a major multicenter, randomized, double-blind, placebo-controlled trial, HIV-infected women with CD4 counts greater than 200, who were symptom-free and who had not received zidovudine (AZT) earlier, received AZT (or a placebo) during pregnancy and during labor, and their infants received AZT for the first six weeks of life. This trial was terminated early (at the first interim analysis) on ethical grounds because of the great disparity between HIV transmission in the placebo and AZT groups: Thirteen babies (8.3 percent) in the AZT group and forty (25.5 percent) in the placebo group were HIV infected, as determined by at least one viral culture. Hence, AZT reduced the risk of maternal-fetal transmission by 67.5 percent.[42]

These scientific data presented challenges to the dominant approaches to both neonatal and prenatal screening and testing. Much of the subsequent controversy centered on anonymous newborn screening. For several years the Centers for Disease Control and Prevention (CDC) had supported state health departments around the country—forty-five states in all participated—in the epidemiological surveillance of HIV infection by testing, anonymously, blood samples from patients in hospitals, clinics, and emergency rooms. This program did not require specific informed consent because the blood samples were taken for other purposes (and thus involved no additional bodily invasions) and because they were stripped of identifiers (and thus did not subject patients to additional social risks). Furthermore, efforts to secure specific consent would have precluded universal testing in the relevant institutions, and mandatory testing could not be ethically justified because of the conditions identified earlier. Many concurred that such blinded surveillance satisfied both ethical and legal standards.

Furthermore, this blinded surveillance program produced important data about the incidence of HIV infection, including the rate of maternal-fetal HIV transmission, which could then be used to shape more effective policies and practices. Neonates' blood samples are available because of various state-mandated neonatal tests. However, because these samples are tested anonymously, the women whose babies are HIV infected (and who are themselves infected) cannot be notified, in contrast to the approach taken to most other diseases. (Of course, these women may already know—or may subsequently learn—that they are HIV positive from other sources, including voluntary testing.) One rationale for CDC's approach is that testing infants is a surrogate test for their mothers because all babies born to HIV-infected women test positive at birth because of the presence of their mother's antibodies, even though many of the babies are not actually HIV infected.

Critics charged that this information could and should be used to help infants and their mothers and thus needed to be unblinded. As a consequence, a bill ("The Newborn Infant Notification Act") was introduced into Congress to prohibit anonymous tests of blood samples drawn from newborns and to authorize screening only when it would be possible to notify mothers of HIV-positive babies. In response, the CDC announced that it would suspend its tests.

The congressional bill generated wide support as a way to protect mothers and children, and CDC's recent practice, according to many critics, appeared to be similar to the notorious Tuskegee syphilis experiment, in which African-American men were left untreated, even after the availability of penicillin, a safe and effective treatment, in order to determine the natural history of untreated syphilis. Although the Nazi analogy is more widely used to oppose various research projects, the Tuskegee syphilis experiment analogy is also powerful where information is generated that either cannot be or is not shared with patients or other subjects.

Pediatrician Arthur Amman, for instance, insists that the "maintenance of anonymous test results at a time when treatment and prevention are readily available will be recorded in history as analogous to the Tuskegee 'experiment.'"[43] However, Ronald Bayer disputes the analogy: "In the case of Tuskegee, individuals who were known to be afflicted with syphilis were willfully deprived of that knowledge," and public health officials conspired to "prevent those impoverished African American men from knowing about their situation or about the availability of therapy." Such a situation differs greatly from blinded surveillance of HIV infection. "To compare blinded surveillance for HIV infection to Tuskegee," Bayer continues, "is to threaten the interest of the mothers and babies who could benefit from carefully targeted efforts to identify those in need of care. Clever political slogans about the right of mothers to know about their babies' lethal infections should not serve to justify an attack on studies that have proved so invaluable in the past and that remain crucial today."[44] We should, Bayer argues, vigorously encourage voluntary HIV testing and yet also retain the anonymous surveillance studies to guide effective public health efforts. Furthermore, since AZT can reduce the risk of transmission of HIV from pregnant women to their offspring, it is important to use resources to encourage testing prior to birth. And women and babies who are identified as seropositive through voluntary testing need to have access to various clinical and social services.

Even with the new data about the value of AZT to reduce maternal-fetal HIV transmission, the most defensible approach to protect offspring remains voluntary prenatal (and neonatal) testing, with all the necessary support, for all pregnant women. In contrast to its earlier recommendation of voluntary testing for pregnant women at high risk of infection, such as intravenous drug users and prostitutes, the CDC in mid-1995 issued guidelines to set a new standard of prenatal care: All pregnant women should be offered a test, following counseling, to determine whether they are HIV infected. And women who do not receive prenatal care should be offered HIV testing for themselves or their infants after delivery.

Such efforts to reduce maternal-fetal HIV transmission are not unproblematic, particularly in relation to other possible priorities, such as preventing HIV infection in women and detecting HIV infection in women early and facilitating pregnancy prevention. One major concern involves the unknown long-term effects on children, especially since 60 to 75 percent of the children of HIV-infected mothers would not be infected even without AZT. In addition, from a public health standpoint, concerns emerge about producing AZT-resistant strains of HIV through its increased use. And, finally, the cost of AZT is beyond the reach of many women who are HIV infected.[45]

Such concerns, along with persistent social risks, further support a pregnant woman's voluntary decision to undergo or decline testing and to accept or refuse AZT therapy. In addition, her decision to accept AZT therapy will obviously depend, in many cases, on the availability of insurance or other financial resources to cover its costs. Voluntary prenatal testing followed by a recommendation of AZT therapy will be ineffective without the necessary funds. Here again it is crucial for public policies to *express* community. Even efforts to *impose* community through mandatory prenatal and neonatal testing will be ineffective without the expression of community through policies to fund AZT and other needed care.

Much of the debate about mandatory and voluntary testing hinges on claims that some women, perhaps especially black and Hispanic women, who account for 75 percent of all AIDS cases among women in the United States, will avoid care if they know that they will be subjected to mandatory testing for HIV infection. Whatever the outcome of this debate, which often follows ideological lines, it is difficult to justify mandatory testing without asking about the availability of resources for AZT and other care, and it is difficult to justify mandatory testing when there is evidence that pregnant women will accept prenatal testing when it is offered in certain contexts and certain ways. Variations in the rate of acceptance appear to be correlated with such factors as perceived risk of infection, the risk of social discrimination, and whether testing is universally and routinely offered to all or only to a targeted high-risk population. According to one study of poor pregnant women in Baltimore, 91 percent consented to be tested when they could view it as part of routine treatment, which they could accept or decline.[46]

In conclusion, it is also important—as this analysis suggests and as Nolan reminds us—to consider whether we are following defensible precedents (historical analogies) for genetic diseases and what precedents our policies regarding HIV testing may set for other genetic diseases that in the future can be detected in utero. Indeed, she notes, our different responses to cystic fibrosis and HIV infection suggest that morally extraneous factors such as race and geography play significant parts in our societal and professional judgments, particularly in the movement from nondirective to directive counseling (for example, at least prior to 1994, counselors appeared to be more directive in urging HIV-infected women to abort).

PRISONERS AND OTHERS IN STATE CUSTODY

Mandatory screening might appear to be justified in several institutional settings, mainly custodial settings, such as prisons and institutions for the mentally infirm or mentally ill. I will concentrate on prisoners, with brief attention to arrested prostitutes and patients in public psychiatric facilities.

Many prisoners have been subjected to mandatory screening because of the risk of rape and the frequency of consensual sexual intercourse in prisons where there is also high seroprevalence. Evidence indicates that HIV transmission is not uncommon in prisons. According to the authors of one study of hundreds of prisoners incarcerated in the Florida state prison system since the late 1970s, "the discovery of HIV-positive individuals in a long-term, continuously incarcerated population is strong, though presumptive, evidence for intraprison transmission of HIV infection."[47] Even if mandatory screening of prisoners, followed by quarantine or isolation, could reduce the spread of HIV infection, it is not clearly cost-effective and it imposes risks of injustice on seropositive prisoners, "ranging from unequal facilities for inmates to violations of a number of hard-won rights of prisoners."[48] Furthermore, it is not the least intrusive or restrictive alternative. Indeed, on grounds of respect for persons, as well as other moral principles, it is crucial for the society to reduce nonconsensual sexual intercourse in prisons.[49] It is also important to educate prisoners—and perhaps even to provide condoms—to reduce the risk of transmission of HIV through consensual sexual activities. Even providing voluntary testing may be useful. According to one study in Oregon, "two-thirds of all inmates, including those at highest risk for HIV, sought HIV counseling and testing when given the opportunity."[50] Nevertheless, as Ruth Macklin rightly argues, "it would not be unethical to subject to a blood test any prisoner who has sexually assaulted another inmate, and if he is found seropositive, to isolate him permanently from the general prison population."[51]

Similarly, it would not be unethical, in terms of the principles of a liberal society, to confine a recalcitrant prostitute who puts others at serious risk, for example, by continuing to practice unsafe sex. A more difficult question is whether it is justifiable to require prostitutes to undergo HIV testing. Although prostitution is widely viewed as a major route for the spread of HIV infection among heterosexuals, Martha Field argues that the overall HIV infection rate among prostitutes does not justify targeting them for mandatory testing, that prostitutes tend to be well informed about safe sex practices, and that over 99 percent of all prostitutes are never arrested. Thus, mandatory testing of prostitutes does not appear to satisfy the conditions identified earlier. However, if states pass laws to require prostitutes to be tested when in state custody, Field views two conditions as essential. First, testing should be required only of those who are convicted, because of the legal presumption of innocence. Second, patrons should be tested if prostitutes are tested—"although patrons are seldom arrested or convicted, both parties are guilty of criminal behavior that is capable of transmission."[52]

Patients in state psychiatric institutions may also be at risk of transmitting HIV infection through sexual intercourse. As in the case of prisoners, the state should attempt to prevent rape in psychiatric institutions—and not only because of the risk of HIV infection. Since many psychiatric patients are unable to benefit from education about HIV infection and safe sex, there may be a stronger argument than in some other contexts for identifying and isolating HIV-infected patients. Even so, it is necessary to make sure that there are no other effective alternatives and to choose the least restrictive alternative.

IMMIGRATION AND INTERNATIONAL TRAVEL

"Seeking to secure our national borders against an 'invasion' of AIDS," Carol Wolchok writes, "the United States now requires HIV antibody testing of 500,000 immigrants, nonimmigrants, and refugees annually. In addition, in what may be the most massive use of HIV testing in this country, more than 2.5 million aliens living in the United States must be tested in order to qualify for legal residence."[53] Historically, societies have responded to epidemics by closing their borders, but the current U.S. policy is ironic because the United States is probably a net exporter of HIV infection. Nevertheless, HIV infection was added to the list of designated diseases (now eight, including infectious leprosy, active tuberculosis, gonorrhea, and infectious syphilis) for which aliens can be excluded from the United States. In view of the precedent set by listing the other exclusionary diseases, it did not seem to many to be unreasonable to add HIV infection. Now temporary visitors to the United States have to indicate their HIV antibody status when they apply for a visa, and the immigration officer may require a test. Applicants for permanent residence must undergo a serologic test for HIV infection (and syphilis) and if positive cannot receive permanent residence. Because the costs of the program fall on the applicants, the society does not have to face the question of trade-offs in the use of public funds.

Analogical reasoning is also significant here, not only in considering historical precedents but also in considering the implications of the mandatory screening of aliens for other groups in the society. As Margaret Somerville notes, viewing prospective immigrants who are HIV infected as "a danger to public health and safety would necessarily set a precedent that all HIV-infected people [already in the country] could be similarly characterized."[54] In addition, it sets precedents for genetic screening as more and more genetic tests become available.

"A just and efficacious travel and immigration policy," according to Gostin and his colleagues, "would not exclude people because of their serologic status unless they posed a danger to the community through casual transmission."[55] Indeed, the list of excludable conditions should be revised because only active tuberculosis poses a threat of casual transmission. Immigrants are thus not a major threat to the public health, and screening them will have only a modest impact on the course of the epidemic in the United States, which already has over one million HIV-infected persons. In addition, the screening program discourages travelers from being tested and "drives further underground undocumented aliens who live in the United States and reduces their incentive to seek counseling and preventative care."[56] Finally, current U.S. policy does not contribute to efforts to control HIV infection in the international community.[57]

Since the current screening program does not contribute significantly to public health objectives, the major argument for excluding seropositive immigrants appears to be the cost to the society of providing health care. It may not be intrinsically unjust for a society to exclude immigrants on grounds of the costs of providing health care, but several commentators cogently argue that "it is inequitable . . . to use cost as a reason to exclude people infected with HIV, for there are no similar exclusionary policies for those with other costly chronic diseases, such as heart disease or cancer."[58] This argument rests on the requirement of justice to treat similar cases in a similar way. And, as is the case for each proposed screening target, screening immigrants raises larger philosophical and ethical questions about the boundaries of the moral communities.

OTHER AREAS OF SELECTIVE SCREENING

I have concentrated on a few areas to indicate how the liberal framework of principles, rules, and justificatory conditions applies to proposed policies of mandatory HIV antibody screening and testing. The arguments extend, mutatis mutandis, to other proposed policies. What we know about the transmission of HIV renders mandatory screening in the workplace, for example, utterly inappropriate, unless there is exposure to bodily fluids under circumstances that would transmit the virus.

Concern has been expressed about the possibility that health care professionals, particularly dentists and surgeons, might transmit the virus to their patients during invasive procedures. In July, 1990, the Centers for Disease Control (CDC) reported a case of "possible transmission" of HIV to a patient during the removal of two of her teeth by a dentist who had AIDS. Although the "possibility of another source of infection cannot be entirely excluded," the CDC noted the absence of any other reported risk factors and a close relation between the viral DNA sequences from the patient and the dentist.[59] Subsequent investigations concluded that this dentist infected at least six of his patients, including Kimberly Bergalis, who made a powerful appeal for mandatory HIV testing of health care professionals in poignant testimony before Congress—what George Annas called "the most dangerous two minutes in the history of the AIDS epidemic." Controversy has persisted about these cases, in part because of the absence of a cogent explanation of the mode of transmission, in part because the CDC's investigation did not pursue other acknowledged risk factors in these patients, and in part because of the absence of evidence of transmission in many other interactions between HIV-infected dentists and surgeons and their patients in the context of health care.

During the same week as the first report of dental-patient transmission, researchers reported that there was no evidence that a Tennessee surgeon with HIV infection had infected any of his surgical patients, many of whom consented to HIV tests.[60] Other studies have reached similar conclusions. According to the researchers involved in a study of the patients of an HIV-infected general dentist in Florida, "the risk for transmission of HIV from a general dentist to his patients is minimal in a setting in which universal precautions are strictly observed. Programs to ensure compliance with universal precautions would appear preferable to programs for widespread testing of adults."[61]

Nevertheless, in view of the mode of transmission of HIV "it would be unexpected if HIV transmission" did not sometimes occur in some surgical procedures.[62] It is important, however, to keep the risk of such transmission in perspective. The probability is very low, perhaps between one in 100,000 to one in one million operations.[63] While the risk is "exceedingly remote" of infection for any single patient, "the cumulative risk over a surgical career is real."[64] The main risk appears to be in seriously invasive procedures, particularly vaginal hysterectomies or pelvic surgery, where there is "blind" (i.e., not directly visualized) use of sharp surgical tools.[65]

If there are grounds for restricting the activities of any surgeon—or dentist—who is known to be HIV infected, then there are probable grounds for mandatory testing to determine which ones are HIV infected. This area continues to require careful attention, as well as appropriate monitoring, and the conclusions—at least for some forms of surgery—are likely to be applicable to both health care professionals and patients, since each party may put the other at risk of HIV infection.

Sometimes the goal of screening or testing is to reduce economic costs rather than to prevent transmission of HIV infection. Both goals appear in some screening policies—for example, both immigration and military screening. The goal of saving funds is often primary in workplace screening, and it is clearly the only concern in insurance screening. It is not unreasonable or even unjust for insurance companies to screen and for states to allow them to screen applicants for life and health insurance for HIV antibodies, just as they screen for other conditions. However, it is a failure of justice as well as of compassion and care, for the society, through the federal and state governments, not to provide funds so that HIV-infected individuals and other sick people can obtain needed health care. In short, the fundamental problem is not the actions of insurance companies but the larger societal failure to respond effectively and justly to the health care needs of its citizens, including those with HIV infection.

CONCLUSION: THE WAR AGAINST AIDS

Not only do societies often react—and, at least in retrospect, overreact—through coercive measures to epidemics of communicable diseases, they frequently do so in the name of *war* against the diseases. The metaphor of "the war against AIDS" has been very prominent in justifications of mandatory screening and testing. "Under the guise of a war against AIDS," one commentator notes, "American politics have recently become enamored of an argument over testing citizens" for HIV antibody.[66] Further reflecting the prominence of military metaphors, Gostin observes that it often appears that "the first line of defense in combating AIDS is to identify carriers of the virus by systematic screening."[67] And, in a sensitive discussion of major vocabularies of concern about AIDS, Monroe Price notes that "the crisis of epidemic is a natural substitute for the crisis of war," and that "the question is whether the AIDS epidemic will become such a serious threat that, in the public's mind, it takes on the stature of war."[68]

The metaphor of war is natural in our sociocultural context when a serious threat to a large number of human lives requires the mobilization of societal resources, especially when that threat comes from biological organisms, such as viruses, which attack the human body. For example, the military metaphor is one way to galvanize the society and to marshal its resources for an effective counterattack, and AIDS activists may even exploit it for this purpose.[69] However, the metaphor has other entailments that need to be questioned and perhaps even opposed, especially in our sociocultural context.[70]

From the beginning of the war against AIDS, identification of the enemy has been a major goal. Once the virus was identified as the primary enemy, it became possible to develop technologies to identify human beings who carry or harbor the virus. This led to what Ronald Bayer calls the "politics of identification."[71] How are antibody-positive individuals viewed? As carriers of HIV, are they enemies to be fought? Should the society try to identify them? And how should it act on the information that a particular individual carries or harbors the virus? The line between the virus and the carrier becomes very tenuous, and the carrier tends to become an enemy just as the virus he or she carries. However much Surgeon General C. Everett Koop could argue that this war is against the virus, not against people, the distinction is too subtle for many in the community. Furthermore, it is not surprising that this metaphor of war often coexists with metaphors of AIDS as punishment or as otherness, because many deny that several actions that lead to exposure to HIV are "innocent" and, furthermore, view the associated lifestyles as threats to dominant social values.[72]

The military metaphor tends to justify coercive measures, such as quarantine and isolation of internal threats. In World War II the United States sent Japanese-Americans to internment camps, without due process, and with the approval of the U.S. Supreme Court in *Korematsu v. United States* in 1944. These coercive policies were later discredited, and Congress even approved reparation to those who were interned. However, the coercive policy of identification and internment "demonstrates how, in times of war, like times of public health crisis, the actions of government become clothed with an unusual inviolability."[73] And it becomes even worse when the public health crisis is itself construed as warfare, because of the perceived disjunction between "peacetime procedures" and "wartime needs."[74]

The metaphor of war against AIDS would not be so dangerous if our society had a better appreciation of the moral constraints on resort to and conduct of warfare, represented in particular in the just-war or limited-war tradition.[75] The justification and limitation of war, including the means employed, follow the pattern of the prima facie principles and conditions for justified infringements identified earlier. In general, like a dinosaur, the United States has been slow to engage in war but then hard to control once it starts to act. After AIDS appeared, the early societal response was limited in part because the disease was viewed as a threat mainly to those on the margins of society, especially gays, but then when AIDS was viewed as threat to the larger society, the response was conceived in terms of war. In general, the United States tends to engage in total war, with unlimited objectives and unlimited means, as expressed in our willingness to destroy cities in Vietnam in order to save them. In the war against AIDS it is important to recognize both limited objectives and limited means. I do not believe, in contrast to Susan Sontag,[76] that we must retire the metaphor of war, but we should explore its logic carefully, limit its application by just-war criteria, and supplement and correct it with other metaphors.

Caring has often been viewed as an alternative metaphor.[77] If the military metaphor tends to conflate the virus and the carriers of the virus, the caring metaphor tends to focus concern on the individuals who carry the virus. Even a liberal society need not be a society of mere strangers—it can at least be a society of "friendly strangers." This friendliness can be expressed in care, compassion, and empathy. Insofar as the metaphor of war against AIDS tends to divide the community by casting HIV-infected individuals in the role of enemies, it thereby undermines some of the conditions that sustain voluntaristic policies. After all, trust is indispensable for voluntaristic policies—otherwise the social risks are significant—and it presupposes communal commitments to provide funds for health care and to enforce rules against discrimination, breaches of confidentiality, and the like.

These communal efforts are critically important, because the groups most affected by AIDS—gay men and intravenous drug users—exist on the "margins" of the community.[78] They thus tend to view coercive policies to identify antibody-positive individuals as analogous to the Nazi efforts to identify Jews and others for nefarious purposes.[79] When the "politics of identification" reflects the socio-cultural metaphor of war, it is easy to understand the fears of HIV-infected individuals, especially those in marginalized sub-communities. The war metaphor tends to exclude HIV-infected individuals, as enemies, from the larger community, while the metaphor of care tends to include them in the community. Compulsory measures, such as mandatory screening, appear to *impose* community, but, unless the society also *expresses* community—for example, through the allocation of funds for health care and protection of individual rights and liberties—it largely excludes coerced individuals from the community.

If the society expresses solidarity with HIV-infected individuals, it is less likely to need coercive policies in place of voluntaristic ones, which can be made more effective through the reduction of social risks and the provision of essential means. And if the society fails to extend solidarity to all, coercive policies are likely to be ineffective and even counterproductive, because they too presuppose voluntary cooperation at many points, often for individuals to enter situations where testing is encouraged or required and usually to implement measures to protect others (e.g., disclosure of sexual contacts).

Mandatory HIV screening, in most settings, would set a precedent of overriding rights in a crisis—in a war against disease—even when it produces no benefits, when the burdens outweigh the benefits, or when alternative ways exist to protect the public health. We need to respond out of metaphors other than—or at least in addition to—war, with careful attention to the moral commitments of a liberal society and the justificatory conditions for overriding, under some circumstances, our prima facie principles and rules. How we respond to HIV's threat to public health will shape and express our "identity and community in a democracy under siege."[80]

Part Three

■

TERMINATION OF LIFE-SUSTAINING TREATMENT

CHAPTER SEVEN

□

"Who Is a Doctor to Decide Whether a Person Lives or Dies?"

REFLECTIONS ON DAX'S CASE

with Courtney S. Campbell

The fact that Dax Cowart is alive today, having finished law school and living a productive life, may be viewed as a triumph and vindication of medical paternalism over patient autonomy. After all, Cowart, who suffered severe burns as a result of an accidental explosion of natural gas in 1973, was given medical treatment for his own benefit despite his numerous refusals and repeated requests to die. Dax's case has become so prominent in biomedical ethics because it poses such an important and dramatic conflict between moral duties to benefit patients and to respect their wishes. The original videotape, *Please Let Me Die*,[1] which presents Dr. Robert White's psychiatric interview with Dax Cowart, effectively challenges viewers to consider how they would balance the principles of respect for persons and patient benefit when a patient refuses life-prolonging treatment even though he or she is not terminally ill, that is, is not irreversibly and imminently dying and life could be prolonged indefinitely with reasonably good quality. (A brief summary of this videotape appears in the appendix to this chapter.)

In our use of this videotape in courses, the most common response is that Dax should have been allowed to exercise his moral and legal right to refuse life-sustaining treatment. Viewers are convinced by what they see and hear—Dax Cowart's powerful and lucid voice emerging from his badly burned, scarred, and pain-racked body to assert his right to make his own choices and to ask, "Who is a doctor to decide whether a person lives or dies?" Of course, they always want to know the outcome—what happened to the patient? And they are often stunned by the report that Dax Cowart subsequently accepted treatment and is doing very well, even though he is totally blind and permanently disfigured. This report forces them to reconsider their original judgment that he should have been allowed to die. We find that this case illuminates several issues in the debate about medical paternalism and patient autonomy. It does not serve merely as an illustration of a preset balance of patient benefit and patient autonomy; its very complexities force qualifications, or at the very least, clarifications of theories of paternalism.[2]

THE NATURE OF MEDICAL PATERNALISM

It can be assumed that the end of medicine and health care is patient benefit. Even if the traditional first duty of physicians and other health care professionals is *primum non nocere*—first of all, or at least, do no harm—there would be no point to medicine and health care if professionals did not aim at patient benefit, including cure, prolongation of life, and relief of pain and suffering. However, recognition that the end of medicine is patient benefit does not take us very far. Patient benefit obviously includes many different specific ends, and those ends usually cannot be obtained without means that have risks, burdens, and costs. Hence, even if patient benefit cannot be defined apart from medical competence, it also cannot be defined apart from a framework of values that can identify some outcomes as benefits and others as harms or burdens and assign weights to different positive and negative outcomes.

But if judgments about patient benefit necessarily presuppose values, then the question immediately emerges: Whose values should be determinative? In Dax's case, we might ask whether his values or the values of his mother, the physicians, and other health care professionals involved in his care, or those of others should determine actions. It is an oversimplification to pose the conflict as one between beneficence and autonomy, because it is also in part a conflict about whose interpretation of beneficence will triumph. As Dax Cowart has noted, physicians often try "to benefit the patient on their own terms rather than the patient's. My case was an example of where the two are not the same."[3] Nevertheless, because the claim of medical authority to override the patient's wishes rests on the professional's commitment and competence to benefit the patient, and the claim of the patient rests on the right of self-determination, we will state the conflict in terms of beneficence and respect for persons, with respect for autonomy as a subset of the principle of respect for persons. Paternalism may be defined as nonacquiescence to a person's wishes, choices, or actions for that person's own benefit. In general it assigns primacy to beneficence over respect for persons, including their autonomy.

According to the principle of respect for persons, autonomous patients should have the right to make their own choices to accept or to refuse medical treatments, including lifesaving medical treatments, as long as their choices do not impose serious harms or burdens on others. To deny or override this right is disrespectful because it insults the autonomous patient by imposing on him or her someone else's conception of patient benefit and of the good life and death. Nietzsche once suggested that all assistance is an insult, but it is more defensible to hold that all assistance against an autonomous person's informed and voluntary refusal is an insult. A basic objection to appeals to the principle of respect for persons, rooted in the Kantian and Millian traditions, is their failure to attend to time and to community. We want to show that the principle of respect for persons can and should be interpreted to allow for time and community, but in ways that will not justify treating an autonomous patient against his or her informed and voluntary choices where those choices do not impose serious risks or burdens on others.

Our formulation of what the principle of respect for persons implies for autonomous patient choices is subject to both internal and external limits, some of which are expressed in the following chart:

		Adverse Effects	
		On Self	On Others
	Autonomous	1	2
Patient's Choices			
	Nonautonomous	3	4

Even the autonomous patient's choices are constrained by several *external* limits, some of which can be morally justified. Such limits include societal and institutional policies of allocating resources, regulations on the use of drugs, professional standards of care, and the professional's conscience. Many of these limits and others restrict a patient's right to choose treatments, rather than his or her right to refuse treatments. But even the negative right to refuse treatments may be justifiably limited when those actions seriously harm others. Thus, when serious adverse effects of a patient's choices fall on others (2 and 4 in the above chart), the patient's choices may sometimes be justifiably restricted. However, one external limit that clinicians often recognize—the family's wishes—is morally unwarranted in the care of autonomous patients, apart from the few imaginable cases where the patient's refusal of lifesaving treatment would subject the family to serious harms or burdens.

In addition, there are *internal* limits. If a patient cannot make autonomous choices (see 3 on the chart), it is not insulting or disrespectful to override his or her choices. And the clinician's duty to benefit and not to harm the patient requires actions to protect the patient from his or her nonautonomous choices.

These points can be explicated by reference to Dax's case. Most actions against a patient's wishes involve efforts to protect both the patient and others; hence they are rarely purely paternalistic. Impure paternalistic actions aim at the patient's welfare but also at the welfare of others. In Dax's case, the clinicians identified his welfare as the primary reason for their refusal to acquiesce in his expressed wish to die, but they also invoked harms and burdens to others. For example, Dr. White expressed concern about the "unfair burden" that would be imposed on Cowart's mother if treatment were discontinued. Rex Houston worried about the negative effect of Dax's death on the lawsuit against the company that maintained the pipeline that had leaked the gas; Dax's death could reduce the amount of the settlement for the family. And some clinicians, such as Dr. Duane Larson, felt that Dax's refusal placed them in a situation where their own moral values would be compromised. The more paternalism is "impure" and the foreseen harms or burdens to others are real, rather than mere rationalizations, the more defensible the intervention is.

Paternalism may involve active or passive nonacquiescence in the patient's choices. In *active* nonacquiescence, the paternalist refuses to accept a patient's request for nonintervention or noninterference, while in *passive* nonacquiescence, a paternalist refuses to carry out the wishes or choices of a patient or to assist the patient in his or her action. It is easier, *ceteris paribus,* to justify passive paternalism than active paternalism, in part because passive paternalism reflects the professional's autonomy, which means that he or she is not a mere instrument or tool of the patient's wishes, and in part because passive paternalism leaves the patient other options.

In Dax's case, the professionals were actively paternalistic when they refused to honor his request for nonintervention and discontinuation of treatment. Yet some of their arguments were more complex. For example, Dax's request was not honored, because to some practitioners it implied a "willingness to participate in his suicide" (White) or because "to not treat this patient is in a sense to kill him" (Larson). Of course, their objections to participating in Dax's suicide or in killing him may have rested on their interpretation of societal rules against assisting suicide and mercy killing, as well as on their judgments about his welfare. However, questions could be raised about their interpretation of these rules.

Finally, paternalism may be *hard* or *soft,* depending on the role the patient's own values play in the decision-making process. This distinction focuses on the *source* of the values that are invoked to justify paternalism. In Dax's case, his clinicians, his mother, and other relevant parties, such as the family attorney, often displayed hard paternalism in their appeal to values (such as religious salvation) and rankings of values (such as the priority of life over quality of life) that were alien to Dax himself. Hard paternalism is usually more difficult to justify because it discounts the patient's own values in favor of someone else's values, often presented as "objective" values. Interestingly, when Dax's *own* values—particularly the values of freedom and self-determination—were invoked as the reason for not acquiescing in his choice, he decided to accept further surgery (e.g., Larson's challenge to Dax in *Dax's Case*). Soft paternalism, which takes the patient's own values seriously, even against his or her expressed wishes, is more defensible, *ceteris paribus,* than hard paternalism, which imposes alien values.

Because of the principle of respect for persons, there is a strong moral presumption against paternalistic actions by physicians and other health care professions. However, such paternalistic actions can be justified under some circumstances. The pattern of justification that seems acceptable combines beneficence and respect for persons; it may be characterized as a principle of *limited beneficence*. It builds on professional beneficence toward the patient but limits and constrains that beneficence by respect for the patient's autonomous wishes, choices, and actions. According to one suggestive image, beneficence provides the engine—the motivation and direction—of medical care, while the patient's wishes, choices, and actions determine the tracks along which it runs.

In order to justify paternalistic interventions without violating the principle of respect for persons, it is necessary, first of all, to rebut the presumption of an adult patient's competence to make his or her own choices. Second, it is necessary to show that the patient would suffer serious harm without actions on his or her behalf. The third condition for justified paternalism is proportionality: The intervention cannot be morally justified unless it would probably be effective and its positive results would probably outweigh its negative results. Fourth, even when paternalistic interventions are justified, the least restrictive, least humiliating, and least insulting means should be employed. In general, the more *impure* the paternalistic action is (that is, the more it prevents harm to parties other than the patient), the *softer* it is (that is, the more it appeals to the patient's own values), and the more its nonacquiescence is *passive* rather than active, the more easily it can be justified. (The above conditions for justification apply mainly to active paternalism.)

According to this analysis, paternalism takes many different forms, not all of which are equally objectionable. In particular, limited active paternalism, which applies beneficence within the limits of respect for persons, can be defended when the above conditions are met. However, most commentators agree that the first necessary conditions of that model were not met in Dax's case, at least at the time of the interview with Dr. White (ten months after the accident), for Dax was an autonomous decision maker. If this interpretation is accurate, then active paternalistic interventions would not have been limited by the principle of respect for persons and would have been unjustified. However, a closer analysis shows the complexity of this case as well as the difficulty of interpreting the limits set by the principle of respect for persons, particularly in determining whether people are making autonomous choices and what their choices really are. Our analysis of these complexities will consider both time and community, which, according to critics, are neglected by proponents of the principles of respect for persons and limited paternalism.

THE AMBIGUITY OF AUTONOMY

The principle of respect for persons entails respect for their autonomous wishes, choices, and actions. However, as noted earlier, this principle is subject to both internal and external limits. Questions about internal limits emerged in several ways in Dax's case.

As Bruce Miller has suggested, it is important to distinguish two senses of autonomy: effective deliberation and freedom of action.[4] When Cowart began to refuse treatment, the attending physicians raised questions about his autonomy. Focusing on "effective deliberation," Dr. Charles Baxter maintained that in the "shock phase," burn survivors are "incompetent" to make decisions about their treatment. Dax's mother concurred in this assessment, concluding that "his condition was so bad that he could not make judgments about what to do with the rest of his life, and whether or not to have treatment."

By contrast, Sharon Imbus and Bruce Zawacki, members of the burn team for the Los Angeles County–USC Medical Center, hold that "during the first few hours of hospitalization . . . even the most severely burned patient is usually alert and mentally competent."[5] Thus, their burn team takes "an *aggressive* approach to decision making to preserve patient autonomy," giving patients sufficient information about their condition and prospects and asking them whether they wish to choose between full therapy and ordinary care. They contend that such patients both exercise more "self-determination" and receive more "empathy" and that the mortality rates have not increased in the burn unit. They do not extend this aggressive approach to autonomy to all burn patients, but only to those for whom survival is unprecedented.

Critics of this approach contend that even in the early period the severely burned patient suffers from such physical and emotional shock that he or she cannot participate in decision making and, furthermore, that early maximal treatment is warranted because of uncertainties about which patients can be salvaged, particularly as treatments improve.[6] Thus, critics contend that, at least in the early period, the first two conditions for justified paternalism are met in the case of severely burned patients. Such patients are not competent to decide because of physical and emotional shock, and, because physicians cannot be certain about which patients can be salvaged, they should provide maximal treatment to all, for withholding treatment at that point guarantees death.

Dax Cowart refused treatment from the time of the accident. He first asked a farmer who approached to get him a gun so that he could end his life; he then asked the rescue squad not to take him to the hospital and the physicians in the emergency room not to treat him. Interventions against his wishes at those points were justified in order to gain time to be able to determine his competence and to make an accurate diagnosis and prognosis, which could not be made immediately. Only when these conditions were met could his choices have been autonomous in the sense of effective deliberation with adequate information resulting in voluntary choices. Dax also indicates that he had nightmares and hallucinations during one period—believing, for example, that the staff was using him as a guinea pig for their experiments— but that by the time of the interview with Dr. White he could look back with what he describes as "a clear mind" and admit that it didn't happen. Choices made under those earlier conditions would not have been autonomous.

However, at some point during the next ten months, he became autonomous in the sense of effective deliberation. Most commentators and viewers of *Please Let Me Die* agree that by the time of this interview Dax Cowart was autonomous in this sense. Indeed, the efforts of the physicians to have him declared incompetent, so that they could continue corrective surgery as well as other treatments, were rebuffed by the psychiatric reports. Still, one question is *when* during the ten months did he become competent and have adequate information to make his own decisions about treatment/nontreatment.

The second sense of autonomy—freedom of action—raises some interesting questions. Are persons who display effective deliberation to be considered autonomous when, as in Dax's case, they are physically unable to act freely on their choices? If we hold that the principle of respect for persons engenders obligations to acquiesce in the wishes, choices, and actions of patients, it is reasonable to ask what this principle requires when a patient's deliberative choices cannot be realized in actions.

In considering whether Cowart was autonomous, most commentators have found that he passed the threshold for autonomy because of his capacity for effective deliberation, along with the authenticity of his choices, that is, their consistency with and reflection of values that he had affirmed throughout his life. Yet it is important not to neglect the problems raised by the fact that he was incapable of autonomy as freedom of action because he was physically unable to act on his wishes and choices. (We will return to authenticity later.) For example, Dax indicated that he did not "intend to die from the infection" that would result from the discontinuation of the daily tankings. However, his unwillingness to die this way meant that to die required the assistance of others. Dr. White insists that, "in essence, he was asking others to participate in his suicide." Hence, the refusal of physicians and other health care professionals to act on his wishes may have stemmed in part from their refusal to be mere instruments of a patient's chosen ends, particularly in violation of professional codes and criminal laws that prohibit assisted suicide. If he had planned to rely on his family or friends for assistance, they would have faced similar questions.

It is also important to distinguish, as Robert Nozick suggests, autonomy as a "side-constraint" from autonomy as an "end-state."[7] As a side-constraint, autonomy limits what we may do to and for others even to benefit them. Autonomy in this sense determines the channels along which benefits flow to patients from medical actions; it limits paternalistic interventions. However, because Dax Cowart was not autonomous in the sense of *free action,* in contrast to effective deliberation, it might be argued that continuation of treatment was essential to restore his autonomy as free action. In such an argument autonomy is understood not as a side-constraint but as an end-state to be realized through the continuation of therapeutic procedures. This understanding of autonomy as the end-state of free action appears in Dr. Larson's recollection of what he told Cowart: "If you want to die, at least let me fix those hands, at least you can do something with them. Then if you want to commit suicide, that's for you to decide."

It is tempting to construe this pursuit of the end-state of autonomy as nonpaternalistic, but viewing autonomy as a benefit to be sought simply expands the traditional understanding of medical benefits, and allowing it to override autonomy as a side-constraint still fits the paternalistic model.

There are good reasons to be suspicious of this effort to recast the problem in order to avoid an apparent conflict with the principle of respect for persons. Nevertheless, our analysis is intended to identify the ambiguous nature of autonomy—and appeals to autonomy—in decisions about treatment and nontreatment, in Dax's case as well as other such cases. How the professional responds may depend on which feature of autonomy—effective deliberation or free action—is emphasized, and on whether autonomy is conceived as a side-constraint or as an end-state.

THE TEMPORAL DIMENSION: DISCERNING PATIENT PREFERENCES OVER TIME

Dax Cowart's initial requests to die were "literally ignored" because of a judgment that the physical and emotional "shock" of the accident and burns had rendered him incompetent to engage in effective deliberation. Often in emergency situations or life-threatening refusals, temporary interventions and treatments are necessary and morally justified in order to gain time to discern with greater accuracy the patient's wishes and choices. Even John Stuart Mill's classic essay *On Liberty* justified a temporary intervention to make sure that a person about to cross a dangerous bridge was acting autonomously—that is, was competent, understood the risks, and chose voluntarily to cross the bridge.[8] The temporal dimension is very important in an analysis of Dax's case, and such an analysis may suggest further complexities because of the different parties' views of how *time* could play a role in confirming, ratifying, or altering Dax's wishes and choices and their own responses.

Too often crisis-oriented medicine and ethical analysis arbitrarily cut off the present from the past and the future, even though the self exists in time and develops over time.[9] Thus, one important index of autonomy is what has been termed *authenticity*, particularly the consistency of a person's wishes, choices, and actions with the values he or she has represented or expressed over time. The intuitive idea of authenticity is "acting in character," and we wonder whether choices or actions are autonomous if they are "out of character" with what we know of the person over time. If they are "in character," we are less likely to suspect that they do not represent genuine autonomy.

This idea of authenticity has been a principal and pervasive feature in Cowart's interpretation and defense of his refusal of life-sustaining treatments. His desire to die was formed by values that had shaped his life in the past, especially his conception of the good life as the free, independent, and active life, which, for him, included sports, rodeo, flying, etc.: "He always wanted to do things for himself and in his own way." Thus, Dax's refusal of life-sustaining treatment was an "authentic" choice when measured against the values he had consistently affirmed, both prior to and after his accident.

By contrast, the burn therapist, Dr. Larson, appealed to the temporal dimension to suggest that Dax's request was inauthentic and out of character: "If you're the *kind of person* I've been led to believe you were *before* you were burned, then don't ask us to let you die, because that means we're killing you." Although this move from "letting die" to "killing" is highly problematic in the circumstances, the main point is to appeal to Dax's values over time in order to justify continuing treatment. Significantly, it was this self-described "attack" by Larson that resulted in Cowart's decision to drop his objections to further therapeutic surgery on his hands.

Dax's refusal up to that time had been based on a further temporal dimension of the case—his vision of the future. Although Dax had repeatedly requested to be allowed to die, Dr. Robert Meier reports that his serious refusal of treatment began after an "off-handed remark" by a surgical resident that the therapy, and consequent pain, could involve a "number of years." By contrast, for Larson, the most important aspect of time in Dax's case was the time already "invested." Cowart had already invested twelve months in a most excruciating ordeal, and to give up then would be like a marathon runner quitting with one mile to go. It is unclear whether Larson's opinion would have changed significantly if less time had been invested.

Often paternalistic actions are initiated or continued, even if presently resisted, on the basis of an appeal to the patient's future consent or subsequent ratification, similar to what parents express when they say to their children— "You'll thank me for this later." Dax attributes this belief to the physicians and nurses who tried to keep him alive: "I also feel that they probably were under the impression that in the end I would be glad that they did and that I would be grateful in the long run that they did force me to undergo this treatment."[10] Ada Cowart, Dax's mother, believes that *time* has indeed ratified her decision to continue treatment against her son's wishes: "Looking back over the last ten years, I think I made the right decision. . . . Now that he's married, enjoying life and his business, I know it's right now."

Dax himself has refused to offer retrospective ratification of the decisions on his behalf but against his wishes. And this refusal makes his case even more interesting than it might have been otherwise. This refusal to say "thank you for saving my life" troubles many physicians and other health care professionals, even those not involved in his care. For example, in a Medical Center Hour at the University of Virginia in March 1984,[11] Dax presented his case, contending that he had been treated wrongly even though he was now glad to be alive and had achieved a quality of life that he had not anticipated at the time of his accident and during the course of his treatments. Several questions from the audience focused on what was perceived to be his inconsistency in being glad to be alive while being ungrateful to professionals for their efforts to keep him alive against his wishes. But, in his view, the decision to continue treatment against his autonomous wishes was morally unjustifiable paternalism then and now, because "the ends do not justify the means." It was disrespectful to him as a person.

This discrepancy in perspectives indicates that it is insufficient to appeal to future ratification as a basis for continuing treatment against a competent patient's wishes in the present. In this case, the person who was most affected by the decision has never ratified it. And it is quite conceivable that things could have been very different—for example, Dax did attempt to commit suicide at least twice because of various difficulties. Furthermore, even though he is pleased with the outcome, that is, where he is today, Dax contends that even if he could have foreseen that enjoyable outcome he would still have refused treatment because of the pain and suffering it involved. And he would make the same decision under the same circumstances again, but, he concedes, he "might not make the decision [to refuse treatment] as readily."[12] (Incidentally, he justified his name change from Donald to Dax because Dax was easier for him to hear, but it may also serve another function: It marks a sharp break in his identification of himself over time and a separation of the present and the future from the painful past.)

An appeal to future ratification is insufficient to justify paternalistic interventions, and it is also unnecessary to justify such actions. Such appeals at the time of decision making simply reflect the agent's hope; they are unnecessary for justification. To make justification hinge on *actual* ratification is inappropriate because many patients never gain or regain the capacity to ratify others' decisions and actions on their behalf. And *predicted* ratification is redundant and restrictive. It is redundant because the prediction is based on the other criteria identified earlier for justified paternalism; it is restrictive because some paternalistic interventions may be justified even when it cannot be predicted with confidence that the patient will ever ratify them. For example, we believe that temporary paternalistic interventions were justified in order to gain time to determine and restore Dax's competence and to determine his prognosis as well as to discern his true wishes in the face of adequate information. Even though Dax now appears to ratify some of those decisions but not others, their justification does not depend on his actual or his predicted ratification.

We have already suggested that paternalistic interventions are sometimes important to "gain time." Dr. White suggested that Dax "wait and see." And Mrs. Cowart believed it was important to "gain time" for other reasons as well. She wanted her son's treatment continued so that he could "realize his responsibility to God and to realize what he should be doing." Furthermore, she expressed concern that her son might subsequently change his mind and accept or even request treatment but that his change of mind might come too late and "there would be nothing anyone could do to help him" if his early refusal was honored. In considering refusals of lifesaving treatments, it is important to allow for the possibility of changes over time in the wishes and choices of patients; hence, irreversible actions, including omissions, must be taken with the utmost caution. But mere possibilities should not be allowed to outweigh the reality of present autonomous dissent in determining whether treatment should be continued.

Past consent now repudiated by refusal may raise questions about the authenticity of the patient's choice, particularly if past consent made sense in terms of the patient's long-term values. The idea of authenticity captures our sense that selves develop over time with persistent and enduring patterns and are not mere collections of choices and actions. And yet it would be a mistake to view authenticity as a criterion of autonomy. At most, considerations of authenticity should alert us to ask further questions. If the choice or action, such as refusal of life-sustaining treatment, is inconsistent with what we know of a person and his or her character, then we should seek justifications or explanations.

It is not only difficult to respect persons as persons because they change over time but also because they give conflicting signals at the same time. Often their wishes, choices, and actions are ambiguous. Then it is necessary to determine which wishes, choices, and actions to respect. It is difficult to determine at this point how clear-cut Dax's directives were at different times during the treatment. For example, in 1984 Dax indicated that his lawyer was probably right when he indicated that Dax had "wanted to live long enough to get the settlement from the oil company that we had sued in an action of negligence" regarding the maintenance of the pipeline that had exploded.[13] The suit was worth more while Dax was alive, and he wanted his family to be taken care of. However, once the suit was settled this concern ceased to be relevant, and most reports of the case acknowledge Dax's firm and persistent rejection of treatment.

Nevertheless, it is sometimes possible to point to a person's signals other than his or her explicit statements. The process of communication is very complex and requires attention to various verbal and nonverbal signs, as well as time for interaction. Attention to those other signs may tempt health care professionals and others who want to save the patient to believe that they have discerned the patient's "real" will to live, rather than to die. It is also tempting to read some other message into the patient's expressed wish to die. Although claims about the patient's real wishes or messages are not always mistaken, they need to be taken with caution when they are at odds with the patient's expressed wish to die. Because Dax finally submitted to treatment when he had apparently gained his victory, in part as a way to extricate himself from the system, it might be supposed that he had not really wanted to die and that his request to die was actually an attempt to expiate his feelings of guilt over his father's death, or a protest against fate, against the loss of control over his destiny, against his dependence and his being "at the mercy" of others, against inadequate attention to his pain and suffering, or against what he perceived as depersonalized and even incompetent care. At the very least, he wanted increased *control* over his life. Rights enable patients to become agents and, within limits, to direct their care. And, Dax emphasizes, he wanted the right to control, that is, to make his own decisions, including the decision to die.[14]

We believe that it was imperative to acknowledge Dax's right to decide for himself whether to accept or refuse treatment after enough time had elapsed to determine his prognosis, his competence, and his settled wishes. At some point paternalistic interventions were unjustified. Such interventions can be justified against the wishes of nonautonomous patients on grounds of both beneficence and respect for persons. But when the patient is sufficiently autonomous, respect for persons requires that the patient be allowed to make his or her own determinations of benefits. Nevertheless, familial or professional beneficence does not evaporate under pressure from patient autonomy. Respect for persons unleavened by care can appear as indifference to the life and death of others.

RESPECT FOR PERSONS IN THEIR COMMUNITIES

Appeals to the principle of respect for persons are often viewed with suspicion not only because they appear to remove people from time but also because they appear to remove people from their communities. The original presentation of Dax's case in the videotape *Please Let Me Die* has been criticized because of the absence of the family, as well as other significant parties; *Dax's Case* adds these other characters. Certainly it is important to consider the social context of any patient's request to be allowed to die.

People are not as distinct and as separate as some interpretations of the principle of respect for persons appear to suggest. This point not only suggests that "no man is an island" because actions have so many effects on others; it also implies that an individual's wishes often reflect the social context in ways that are sometimes overlooked.

Robert Burt argues that if we ignore "the emotional context" of Dax's invocation of the right to die, "we would find ourselves purporting to obey the wishes of a caricature," for his dependence led him to inquire about the wishes of others, not only to express his own wishes. His plea to be allowed to die, Burt notes, was perhaps "more a question to others about their wishes toward him" because he could not die, at least as he wanted to, without their active collaboration.[15] Hence it is morally important for all parties to express, and for the patient to perceive, care. Such care by professionals, family members, and friends may require altering features of the patient's environment, for example, by reducing pain and gesturing concern. Without this social context of care, acquiescence in the patient's wishes may be (rightly) perceived as indifference rather than respect.

Expression of community through gestures of care and concern is an important but often neglected aspect of ethical policies, practices, and actions. For example, what do a relative's or a professional's actions symbolize about the patient's place in and significance for the community? Our chart above included some attention to the community, but mainly through the impact of a person's actions on others, and we argued that it is easier to justify overriding a person's wishes, choices, or actions when they have adverse effects on others than when those adverse effects fall only on the patient. But protecting the community, including individuals within the community, is not the same as expressing community. Dax recognizes that the people who wanted to keep him alive "had an honest concern for [his] well-being."[16] In short, they included him in their community. But such an expression of community, whether by the family or by health care professionals, may emphasize care or beneficence at the expense of respect. After unwarranted neglect because of fascination with individual autonomy, community is now rightly receiving a great deal of attention in biomedical ethics. But the turn toward community, as important and essential as it is, may become tyrannical if it is not limited by respect for persons.

It is easy to affirm community, but such affirmations should only be made in the spirit of realism. First, realism recognizes that community is often an ideal rather than a reality. Even if we are not strangers, we may not constitute a genuine community of shared values or we may not have enough knowledge of each other to know whether we constitute such a community. In the absence of community, rights are important to protect individual self-determination. Second, realism is needed not only because community is often only an ideal and because human beings are imperfect, but also because "moral passions," as Lionel Trilling notes, "are even more willful and imperious and impatient than the self-seeking passions." Dangers lurk in "our most generous wishes." Trilling continues, "Some paradox of our nature leads us, when once we have made our fellow men the objects of our enlightened interest, to go on to make them the objects of our pity, then of our wisdom, ultimately of our coercion."[17] Thus, it is important to insist that a true community, involving equals, cannot exist without respect for persons and that such respect requires procedures to protect rights as well as needs. Dax's case poses for us the challenge of trying to affirm both care and respect, both community and individual rights, and to indicate how they can be combined and which has priority when their conflict cannot be avoided.

POSTSCRIPT: LISTENING TO DON'S/DAX'S STORY (BY JFC, 1996)

The case of Donald, or Dax, Cowart has become a focal point for continuing debates about the respective roles of narratives and principles in medical ethics. Some recent discussants juxtapose this essay and two writings by William May on this case.[18] Ethicists Sumner B. Twiss and Paul Lauritzen[19] contrast May's "narrative" approach with our "principlist" approach, which Twiss views as liberal individualistic in substance (in contrast to an ethic of care or of the common good) and Lauritzen construes as rational demonstrative (in contrast to persuasive). Our respective works, writes Lauritzen, "could not be more different" and, according to May's "implied critique," our position is, at best, thin and incomplete. May's own approach, continues Lauritzen, depends on his ability to "tell Dax's story in a certain way" and thus to "shift the interpretive frame altogether." While such contrasts are often useful and sometimes even illuminating, they can be overdrawn. Even though May and Campbell and I start at different points—narratives and norms, respectively—there is considerable convergence.

Using a dramatic narrative approach to illuminate the burned person's shattered world, May argues that the Dax case "challenges the conventional analysis of great moral issues in medical ethics," with its customary focus on conflicts between two rival values of life and quality of life and between two rival principles of paternalism and autonomy for selecting the right decision maker. In his rich and powerful discussion, May contends that conventional frameworks fail in such a world-shattering case, mainly because the catastrophe is so devastating that the terms life/quality of life cannot adequately express it. The dramatic, narrative language of life/death/rebirth, May suggests, better expresses what is involved. However, he later returns to debates about paternalism and autonomy, observing that the image of rebirth, of the construction of a new identity, does not eliminate—and may even intensify—the problem of paternalism.

By contrast, Campbell and I start with debates about the meaning and weights of the principles of beneficence and respect for autonomy in conflicts about paternalistic interventions; but we then attempt to show how difficult it is to respect persons in their concreteness, because they are temporal creatures (living in the past, present, and future) and because they are also social creatures. We cannot respect other persons' autonomy without listening to their narratives in order to determine what their wishes, choices and actions really are. (Even though we did not use the language of narrative, it is consistent with and further explicates what we argued.)

So May begins with narratives but moves to norms; we begin with norms but move to narratives. And, substantively, we come to the same conclusion about what ought to have been done. Campbell and I concluded that "it was imperative to acknowledge Dax's right to decide for himself whether to accept or refuse treatment after enough time had elapsed to determine his prognosis, his competence, and his settled wishes. At some point paternalistic interventions were unjustified." And May concluded: "Whatever the team can do for the victim, it cannot bring him to new life without his consent. It cannot provide him with such parenting. The domestic analogy is wrong" (p. 35).

Furthermore, just as May recognizes the dangers of both paternalism and antipaternalism—paternalists are tempted by the sins of the overbearing, and antipaternalists by the sins of the underbearing—so we stressed both principles of respect for persons and of beneficence in order to avoid similar temptations. As I have argued elsewhere, "both paternalists and their critics are susceptible to various temptations. Paternalists are tempted by pride and self-righteousness . . . the arrogance of benevolence. . . . Critics of paternalism are often tempted by sloth and even indifference. They stand back, frequently denying any responsibility for what happens to others, sometimes remaining unmoved by the needs of others." Both risks can be reduced "by properly defining and maintaining a tension between the principles of beneficence and respect," both of which should be expressed in health care of all kinds.[20]

The fact that May's rich, narrative discussion also concludes that Cowart should have been allowed to make his own decision about treatment is, for some commentators, very important—a different conclusion would have called into question his narrative analysis. In short, principles (theory) can check narratives (experience), just as narratives can check principles.[21] I affirm such a dialectical relation between norms and cases as mininarratives. (See chapter 2 above.) And some commentators, such as Lauritzen, miss the way Campbell and I argue that the theory of paternalism may have to be qualified, or at the very least clarified, in light of such cases as this one.

It is also important to consider how May reaches his conclusion. He identifies three versions of the pro-choice or pro-autonomy approach: (1) pro-happiness based on mere preference, (2) pro-liberty as a necessary precondition of moral community, and (3) pro-choice as a resolute will to make good on a primary decision. He rejects the first because it is uncritically relativistic and the second because it recognizes only minimal duties, mainly noninterference. However, he accepts the third as a necessary condition for the constitution or reconstitution of moral identity, particularly following a catastrophic loss, and he uses it in a consequentialist justification for allowing Cowart to choose/refuse treatment: "Perhaps . . . leaving the door open to the possibility of his refusing treatment may make it easier for him to accept treatment and to persevere in the consequences of that acceptance and make good" on it. The health care team, as already noted, cannot bring Cowart to a new life "without his consent." Thus, the severely burned patient "resembles not the balky toddler in the bosom of the family, but rather the agonist in Greek tragedy or the stricken religious figure cut off from the safeties of family and city." If Cowart chose to live, he could not simply "take up his old life. He must become a new man. Don Cowart becomes Dax. No parentalist can force him down that road." What is required is an "interior transformation . . . the reordering of one's identity from the ground up." While the community can assist this "heroic movement" in various ways, "without consent to transformation the patient cannot move from saying 'please let me die' to 'I am glad to be alive'" (pp. 34–35). The paternalist metaphor is misleading when it is extended to patients who have suffered "major loss" far outside the boundaries of experience of the professionals themselves. Paternalistic interventions are wrong in such cases largely because they are ineffective for the end that is sought.

By contrast to May's consequentialist argument, Campbell and I accept a nonabsolutist deontological position that justifies Cowart's (and others') prima facie right to make decisions in both catastrophic and noncatastrophic cases because of the importance of respecting persons and, I might add, their own stories. To respond otherwise would insult Cowart and treat him with disrespect. Why our position is more adequate as a baseline than May's, even though it too needs a narrative perspective, emerges more clearly in the light of some problems that May's approach encounters.

May appeals to some broad interpretive categories to "locate" and illuminate Dax's experience, his ordeal, as a severely burned patient. Especially important are the categories of life/death/rebirth and Christian conversion as well as the hero in Greek tragedy. Such broad narrative categories can be illuminating for certain purposes, but confusing and even dangerous for others. In particular, they are confusing and even dangerous when used to illuminate others' particular narratives in order to determine how caregivers ought to respond to their refusal of (or request for) treatment. Archetypal narratives may obscure individual narratives and thus seriously distort what a particular patient is saying—for instance, May's use of religious, often Christian, narrative patterns may be quite out of context for a severely burned patient who is atheistic. Any interpretive categories must be sufficiently broad and neutral to encompass a wide range of conceptions of a good life and death.[22]

Physicians and other health care professionals must attend to the patient's own narrative, as required by the principle of respect for persons, rather than construing that narrative as a mere instance of a grand pattern of life/death/rebirth or even of a set of substantive beliefs, such as a religious community's core convictions. This point holds even if the individual explicitly identifies in some way with the larger narrative. The implications are significant: We have to respect, for instance, a particular Jehovah's Witness's position on blood transfusions, even if it is wholly rejected by his/her larger religious community—it is his or her own position that matters. Or, to take an example from chapter 5 above, a physician or other health care professional should respect a particular Navajo's own views about information and decision making, rather than simply inferring them from the traditional Navajo world view, which this particular Navajo may largely reject.

To be sure, interpretive categories drawn from representative narratives can be very helpful in approaching particular burn victims and survivors. In *Journeys through Hell: Stories of Burn Survivors' Reconstruction of Self and Identity*, Dennis J. Stouffer attempts to understand the lives of burn victims and survivors "as articulated within their biographical accounts or stories."[23] He treats burn survivors' "accounts and personal stories . . . not as ancillary data but . . . instead as the preeminent vehicle through which their lives can be understood." Using an open-ended question interview, he finds among many burn survivors an image of "being cast into hell," of punishment and destruction, along with another "less overwhelming, less dramatic image" of "purification, transformation, and rebirth." The insights we gain from such analyses are largely consistent with May's own analysis, and they flag important personal, professional, and social issues that must be addressed in adequate— and thus in caring and respectful— responses to such patients. While the narrative perspective can help us see what went wrong, from the standpoint of both care and respect, over the first ten months of treatment (i.e., up to the time of *Please Let Me Die*), it may be unjustifiably paternalistic, or imperialistic, to interpret a particular patient's story through a larger narrative pattern that blurs his or her own identity and individuality.

That this is a serious risk in May's arguments is evident in his comments about Dax as a "new man," with a new identity. May bases his claim in part on the narrative patterns of life/death/rebirth and in part on the fact that Donald changed his name to Dax. While conceding that Cowart had a manifest, functional reason for changing his name—he couldn't write his full name again (and he has also stated that he could hear Dax more clearly than Don)—May finds latent, symbolic meaning in this act: Cowart "knows that Donald Cowart has, if not biologically, existentially died; he must assume a new identity" (p. 16). It is not clear, however, that Dax constituted a new identity for Donald— he still insists that he was wrongly treated. At most, Twiss notes, Cowart "appears only to have deepened his original sense of willful independence and determination to live his life on the terms that he chooses. To interpret the forced changes in his life-plan as a reconstruction or even revision of identity may be to impose on Cowart a moral vision that he does not share and that misinterprets his existential response to his ordeal."[24] I have similar reservations about Lonnie Kliever's important and somewhat similar effort to reenvision "Dax's case" as a "'chronic illness' story" rather than a "'right-to-die' story."[25]

What finally is the upshot of the shift in perspective from norm to narrative (May and Kliever), from quality of life to life/death/rebirth (May), or from a "'right-to-die' story" to a "'chronic illness' story" (Kliever)? What should Cowart's caregivers have done when deciding whether to allow him to choose to die at the point of the original videotape, that is, ten months after the accident? Should they have acquiesced to his stated desire to die or should they have forcibly treated him against his stated preference? Focusing on narratives may simply push the procedural question of decisional authority to a different level—away from the patient's particular decision to the patient's story— without fundamentally altering the question or even the answer. "On the face of things," Kliever writes, "the Cowart story is a classic case of autonomy versus paternalism. Certainly that clash is the perceived message of the film 'Dax's Case' *and represents Dax's own construal of his experience as a patient.*"[26] Even though Dax's own story is not incorrigible, and respect does not require others to accept it at face value, the shift to narrative simply relocates the fundamental question so that now it becomes "whose story is it anyway?"[27] Problems of respect for persons and paternalism arise on the level of narrative, just as they do on the level of concrete decision-making.

This case raises some important issues about the ethics in the narrative encounter—not only about narration, such as whose story, from what perspective, etc., but also about listening, a term I will use to encompass all forms of receiving narration. Without being able to develop the important points here,[28] I will simply note that both narration and listening are *performative*. Analyses of narrative often draw, explicitly or implicitly, on J. L. Austin's analysis of performative speech and suggest that "narration" is the "performative function of story telling."[29] J. Hillis Miller writes that "storytelling, the putting together of data to make a coherent tale, is performative. *Oedipus the King* is a story about the awful danger of storytelling. Storytelling in this case makes something happen with a vengeance. It leads the storyteller to condemn, blind, and exile himself, and it leads his mother-wife, Jocasta, to kill herself."[30]

As Miller's example suggests, listening is performative, just as telling is. At the very least, listening is active, not merely passive, because it involves interpretation. It is very important in biomedical ethics to ask exactly what constitutes ethical listening when the physician or other health care professional listens to the narrative of "the wounded storyteller."[31] To listen ethically means to listen with respect but, as noted, respectful listening does not entail that the listener merely accept at face value what he or she hears. And Campbell and I suggested the difficulty of interpreting what is heard, especially when the temporal and communal dimensions of the self and his/her stories are recognized. Furthermore, when the patient is ambivalent, it may be very difficult to determine exactly where to put the interpretive weight.

Listening is indispensable, whether we start with narratives or start with principles, such as beneficence or care and respect for persons and their autonomous choices. Within either approach, the listener cannot avoid interpretive frameworks but, nevertheless, must be as careful as possible to hear what this particular person is saying, what his/her story is, without acting under a larger story that putatively illuminates the patient's particular story.

Different media for presenting the same narrative may evoke very different responses from listeners. I have now used the famous videotape interview between Dr. Robert White and Donald Cowart numerous times in teaching at various levels and in different formats and media. The medium of narrative presentation often makes a tremendous difference in audience response.

Consider the following possible ways of presenting the material in the original videotape, *Please Let Me Die*, whose very title already provides for the case a strong interpretive perspective that may or may not be warranted: (1) an oral or written summary of the case; (2) a full written transcript of the conversation between Dr. Robert White and Donald Cowart; (3) an audiotape of the conversation but without their visual images; and (4) the videotape. Moving from the first to the fourth medium increases the likelihood, at each stage, that the audience will agree to let Cowart die as he wishes. However, in the movement from the first to the fourth medium, audience responses also tend to become more anguished because they experience Cowart more and more personally, especially when they actually hear his voice or see him on videotape or both.

An interesting phenomenon then occurs when a fifth medium is employed to present the narrative. In using an interactive videotape, students have to take responsibility for their particular judgments. They become participants, even agents, as well as spectators or observers. At this point, according to my impressions, it is harder for them to let him die.

I have rarely used the second videotape, *Dax's Case*, in teaching, mainly because of its length. It interweaves Dax's story, from his own standpoint, with the stories of various other involved parties, such as his mother and his lawyer as well as several caregivers. These stories raise the question of who narrates the story to which caregivers have to respond. However, as Kliever notes (*Dax's Case*, p. 73), even though he and others involved in making the second film tried to avoid tilting the "case" one way or the other,

> the sheer power of the visual images of Dax Cowart—before his accident, during his treatment, and after his rehabilitation—carries more weight than the viewer can dispassionately handle. The very *sight* of Dax Cowart stops argument dead in its tracks over whether he should or should not have been allowed to die. The viewer's mixed feelings of pity and revulsion at the sight of Dax's massive suffering and disfigurement are too overwhelming to hear any other voice in the film than his insistent demands that he wanted to die. The voices of family, friends, nurses, and especially doctors are drowned out by the cries in the treatment room, thereby silencing those very persons whose intimate ties to the patient would raise more sharply the question of whether he ought to kill himself and even whether he really wanted to die. To put it bluntly, 'Dax's Case' *as viewed* has a protagonist but no serious antagonist. This one-sidedness is even more overwhelming in the videotape 'Please Let Me Die,' where Dax is the only voice and where his ravaged body in treatment fills the screen from start to finish.

Kliever goes on to note, as Campbell and I also observed, what Dax Cowart's own physical presence does to the discussion, particularly when the audience tries to put together his claim that he was treated wrongly by the physicians and others who saved his life and yet is glad to be alive now. While the updated story, which is not yet over, may make some difference in our emotional responses, both care and respect for persons require that we focus clearly and sharply, and yet sensitively, on Cowart's own story and recognize that his moral right to make his own decision should have been recognized. The complexities introduced by our analysis, or by various narrative analyses, should not obscure this essential moral point.

APPENDIX: SUMMARY OF COWART CASE AS PRESENTED IN *PLEASE LET ME DIE*

Donald Cowart, a single, twenty-seven-year-old athlete and former air force pilot, was severely burned in July 1973 when his car triggered an explosion of natural gas from a leaking main nearby. His father, who was with him at the time of the explosion, died on the way to the hospital. Mr. Cowart was admitted with about two-thirds total-body burns—mostly third degree burns. As he later recalled, he didn't want to be taken to the hospital at all; and when he got there, "I told them again that I was burned bad enough that I didn't want them to try to do anything for me and only keep me out of pain. But they did go ahead and treat me and although they did not think I was going to make it at the time, they pulled me through." For the next nine months after his admission, Mr. Cowart persisted in his desire to discontinue his treatment—which included skin grafts, several surgeries, and painful daily "tubbings" and dressing changes to prevent otherwise inevitable infection. He recognized that stopping treatment would mean certain death.

In April of 1974 Mr. Cowart was interviewed by a hospital psychiatrist because he had adamantly refused further corrective surgery on his hands and insisted that he be allowed to go home to die. In that interview Mr. Cowart stated that he only wanted "a brief visit" at home and did not intend to die from infection but would "use some other means"—a difficult prospect, since he had been blinded by the explosion, his fingers had been grafted together, and his arms and legs were not fully functional. He described the excruciating pain of his daily treatments, his total loss of independence in doing anything for himself, his apparently slim chance of ever regaining much independence, and his nightmarish dreams that he was being used as a human guinea pig. He had always been athletic, he said, and even rehabilitation would mean that he could only enjoy life by "changing completely the things I am interested in." And in his own view, that end wasn't "worth the pain involved to be able to get to the point where I could try it out." Therefore, he was demanding to see his attorney in order to request a court order for his release from the hospital.[32]

CHAPTER EIGHT

□

Must Patients Always Be Given
Food and Water?

with Joanne Lynn

Many people die from the lack of food or water. For some, this lack is the result of poverty or famine, but for others it is the result of disease or deliberate decision. In the past, malnutrition and dehydration must have accompanied nearly every death that followed an illness of more than a few days. Most dying patients do not eat much on their own, and nothing could be done for them until the first flexible tubing for instilling food or other liquid into the stomach was developed about a hundred years ago. Even then, the procedure was so scarce, so costly in physician and nursing time, and so poorly tolerated that it was used only for patients who clearly could benefit. With the advent of more reliable and efficient procedures in the past few decades, these conditions can be corrected or ameliorated in nearly every patient who would otherwise be malnourished or dehydrated. In fact, intravenous lines and nasogastric tubes have become common images of hospital care.

Providing adequate nutrition and fluids is a high priority for most patients, both because they suffer directly from inadequacies and because these deficiencies hinder their ability to overcome other diseases. But are there some patients who need not receive these treatments? This question has become a prominent public policy issue in a number of cases. In May 1981, in Danville, Illinois, the parents and the physician of newborn conjoined twins with shared abdominal organs decided not to feed these children. Feeding and other treatments were given after court intervention, though a grand jury refused to indict the parents.[1] Later that year, two physicians in Los Angeles discontinued intravenous nutrition to a patient who had severe brain damage after an episode involving loss of oxygen following routine surgery. Murder charges were brought, but the hearing judge dismissed the charges at a preliminary hearing. On appeal, the charges were reinstated and the physicians remanded to trial.[2]

In April 1982, a Bloomington, Indiana, infant who had tracheoesophageal fistula and Down syndrome was not treated or fed, and he died after two courts ruled that the decision was proper but before all appeals could be heard.[3] When the federal government then moved to ensure that such infants would be fed in the future,[4] the Surgeon General, Dr. C. Everett Koop, initially stated that there is never adequate reason to deny nutrition and fluids to a newborn infant.

While these cases were before the public, the nephew of Claire Conroy, an elderly incompetent woman with several medical problems, petitioned a New Jersey court for authority to discontinue her nasogastric tube feedings. Although the intermediate appeals court has reversed the ruling,[5] the trial court held that he had this authority since the evidence indicated that the patient would not have wanted such treatment and that its value to her was doubtful.

In all these dramatic cases and in many more that go unnoticed, the decision is made to deliberately withhold food or fluid known to be necessary for the life of the patient. Such decisions are unsettling. There is now widespread consensus that sometimes a patient is best served by not undertaking or continuing certain treatments that would sustain life, especially if these entail substantial suffering.[6] But food and water are so central to an array of human emotions that it is almost impossible to consider them with the same emotional detachment that one might feel toward a respirator or a dialysis machine.

Nevertheless, the question remains: Should it ever be permissible to withhold or withdraw food and nutrition? The answer in any real case should acknowledge the psychological continuity between feeding and loving and between nutritional satisfaction and emotional satisfaction. Yet this acknowledgment does not resolve the core question.

Some have held that it is intrinsically wrong not to feed another. The philosopher G. E. M. Anscombe contends: "For wilful starvation there can be no excuse. The same can't be said quite without qualifications about failing to operate or to adopt some courses of treatment."[7] But the moral issues are more complex than Anscombe's comment suggests. Does correcting nutritional deficiencies always improve patients' well-being? What should be our reflective moral response to withholding or withdrawing nutrition? What moral principles are relevant to our reflection? What medical facts about ways of providing nutrition are relevant? And what policies should be adopted by the society, hospitals, and medical and other health care professionals?

In our efforts to find answers to these questions, we will concentrate upon the care of patients who are incompetent to make choices for themselves. Patients who are competent to determine the course of their therapy may refuse any and all interventions proposed by others, as long as their refusals do not seriously harm or impose unfair burdens upon others.[8] A competent patient's decision regarding whether or not to accept the provision of food and water by medical means such as tube feeding or intravenous alimentation is unlikely to raise questions of harm or burden to others.

What then should guide those who must decide about nutrition for a patient who cannot decide? As a start, consider the standard by which other medical decisions are made: One should decide as the incompetent person would have if he or she were competent, when that is possible to determine, and advance that person's interests in a more generalized sense when individual preferences cannot be known.

THE MEDICAL PROCEDURES

There is no reason to apply a different standard to feeding and hydration. Surely, when one inserts a feeding tube, or creates a gastrostomy opening, or inserts a needle into a vein, one intends to benefit the patient. Ideally, one should provide what the patient believes to be of benefit, but at least the effect should be beneficial in the opinions of surrogates and care givers.

Thus, the question becomes: Is it ever in the patient's interest to become malnourished or dehydrated rather than receive treatment? Posing the question so starkly points to our need to know what is entailed in treating these conditions and what benefits the treatments offer.

The medical interventions that provide food and fluids are of two basic types. First, liquids can be delivered by a tube that is inserted into a functioning gastrointestinal tract, most commonly through the nose and esophagus into the stomach or through a surgical incision in the abdominal wall and directly into the stomach. The liquids used can be specially prepared solutions of nutrients or a blenderized version of an ordinary diet. The nasogastric tube is cheap; it may lead to pneumonia and often annoys the patient and family, sometimes even requiring that the patient be restrained to prevent its removal.

Creating a gastrostomy is usually a simple surgical procedure, and once the wound is healed, care is very simple. Since it is out of sight, it is aesthetically more acceptable and restraints are needed less often. Also, the gastrostomy creates no additional risk of pneumonia. However, while elimination of a nasogastric tube only requires removing the tube, a gastrostomy is fairly permanent and can be closed only by surgery.

The second type of medical intervention is intravenous feeding and hydration, which also has two major forms. The ordinary hospital or peripheral IV, in which fluid is delivered directly to the bloodstream through a small needle, is useful only for temporary efforts to improve hydration and electrolyte concentrations. One cannot provide a balanced diet through the veins in the limbs: To do that requires a central line, or a special catheter placed into one of the major veins in the chest. The latter procedure is much more risky and vulnerable to infections and technical errors, and it is much more costly than any of the other procedures. Both forms of intravenous nutrition and hydration commonly require restraining the patient, cause minor infections and other ill effects, and are costly, especially since they ordinarily require the patient to be in a hospital.

None of these procedures, then, is ideal; each entails some distress, some medical limitations, and some costs. When may a procedure be foregone that might improve nutrition and hydration for a given patient? Only when the procedure and the resulting improvement in nutrition and hydration do not offer the patient a net benefit over what he or she would otherwise have faced.

Are there such circumstances? We believe that there are; but they are few and limited to the following three kinds of situations: (1) the procedures that would be required are so unlikely to achieve improved nutritional and fluid levels that they could be correctly considered futile; (2) the improvement in nutritional and fluid balance, though achievable, could be of no benefit to the patient; (3) the burdens of receiving the treatment may outweigh the benefit.

WHEN FOOD AND WATER MAY BE WITHHELD

Futile treatment. Sometimes even providing "food and water" to a patient becomes a monumental task. Consider a patient with a severe clotting deficiency and a nearly total body burn. Gaining access to the central veins is likely to cause hemorrhage or infection, nasogastric tube placement may be quite painful, and there may be no skin to which to suture the stomach for a gastrostomy tube. Or consider a patient with severe congestive heart failure who develops cancer of the stomach with a fistula that delivers food from the stomach to the colon without passing through the intestine and being absorbed. Feeding the patient may be possible, but little is absorbed. Intravenous feeding cannot be tolerated because the fluid would be too much for the weakened heart. Or consider the infant with infarction of all but a short segment of bowel. Again, the infant can be fed, but little if anything is absorbed. Intravenous methods can be used, but only for a short time (weeks or months) until their complications, including thrombosis, hemorrhage, infections, and malnutrition, cause death.

In these circumstances, the patient is going to die soon, no matter what is done. The ineffective efforts to provide nutrition and hydration may well directly cause suffering that offers no counterbalancing benefit for the patient. Although the procedures might be tried, especially if the competent patient wanted them or the incompetent patient's surrogate had reason to believe that this incompetent patient would have wanted them, they cannot be considered obligatory. To hold that a patient must be subjected to this predictably futile sort of intervention just because protein balance is negative or the blood serum is concentrated is to lose sight of the moral warrant for medical care and to reduce the patient to an array of measurable variables.

No possibility of benefit. Some patients can be reliably diagnosed to have permanently lost consciousness. This unusual group of patients includes those with anencephaly, persistent vegetative state, and some preterminal comas. In these cases, it is very difficult to discern how any medical intervention can benefit or harm the patient. These patients cannot and never will be able to experience any of the events occurring in the world or in their bodies. When the diagnosis is exceedingly clear, we sustain their lives vigorously mainly for their loved ones and the community at large.

While these considerations probably indicate that continued artificial feeding is best in most cases, there may be some cases in which the family and the caregivers are convinced that artificial feeding is offensive and unreasonable. In such cases, there seems to be no adequate reason to claim that withholding food and water violates any obligations that these parties or the general society have with regard to permanently unconscious patients. Thus, if the parents of an anencephalic infant or of a patient, like Karen Quinlan, in a persistent vegetative state feel strongly that no medical procedures should be applied to provide nutrition and hydration and the caregivers are willing to comply, there should be no barrier in law or public policy to thwart the plan.[9]

Disproportionate burden. The most difficult cases are those in which normal nutritional status or fluid balance could be restored, but only with a severe burden for the patient. In these cases, the treatment is futile in a broader sense—the patient who is competent can decide the relative merits of the treatment being provided, knowing the probable consequences, and weighing the merits of life under various sets of constrained circumstances. But a surrogate decision maker of a patient who is incompetent to decide will have a difficult task. When the situation is irremediably ambiguous, erring on the side of continued life and improved nutrition and hydration seems the less grievous error. But are there situations that would warrant a determination that this patient, whose nutrition and hydration could surely be improved, is not thereby well served?

Though they are few, we believe there are such cases. The treatments entailed are not benign. Their effects are far short of ideal. Furthermore, many of the patients most likely to have inadequate food and fluid intake are also likely to suffer the most serious side effects of these therapies.

Patients who are allowed to die without artificial hydration and nutrition may well die more comfortably than patients who receive conventional amounts of intravenous hydration.[10] Terminal pulmonary edema, nausea, and mental confusion are more likely when patients have been treated to maintain fluid and nutrition until close to the time of death.

Thus, those patients whose "need" for artificial nutrition and hydration arises only near the time of death may be harmed by its provision. It is not at all clear that they receive any benefit in having a slightly prolonged life, and it does seem reasonable to allow a surrogate to decide that, for this patient at this time, slight prolongation of life is not warranted if it involves measures that will probably increase the patient's suffering as he or she dies.

Even patients who might live much longer might not be well served by artificial means to provide fluid and food. Such patients might include those with fairly severe dementia for whom the restraints required could be a constant source of fear, discomfort, and struggle. For such a patient, sedation to tolerate the feeding mechanisms might preclude any of the pleasant experiences that might otherwise have been available. Thus, a decision not to intervene, except perhaps briefly to ascertain that there are no treatable causes, might allow such a patient to live out a shorter life with fair freedom of movement and freedom from fear, while a decision to maintain artificial nutrition and hydration might consign the patient to end his or her life in unremitting anguish. If this were the case a surrogate decision maker would seem to be well justified in refusing the treatment.

INAPPROPRIATE MORAL CONSTRAINTS

Four considerations are frequently proposed as moral constraints on foregoing medical feeding and hydration. We find that none of these dictate that artificial nutrition and hydration must always be provided.

The obligation to provide "ordinary" care. Debates about appropriate medical treatment are often couched in terms of "ordinary" and "extraordinary" means of treatment. Historically, this distinction emerged in the Roman Catholic tradition to differentiate optional treatment from treatment that was obligatory for medical professionals to offer and for patients to accept.[11] These terms also appear in many secular contexts, such as court decisions and medical codes. The debates about ordinary and extraordinary means of treatment have been interminable and often unfruitful, in part because of a lack of clarity about what the terms mean. Do they represent the premises of an argument or the conclusion, and what features of a situation are relevant to the categorization as "ordinary" or "extraordinary"?[12]

Several criteria have been implicit in debates about ordinary and extraordinary means of treatment; some of them may be relevant to determining whether and which treatments are obligatory and which are optional. Treatments have been distinguished according to their simplicity (simple/complex), their naturalness (natural/artificial), their customariness (usual/unusual), their invasiveness (noninvasive/invasive), their chance of success (reasonable chance/futile), their balance of benefits and burdens (proportionate/disproportionate), and their expense (inexpensive/costly). Each set of paired terms or phrases in the parentheses suggests a continuum: As the treatment moves from the first of the paired terms to the second, it is said to become less obligatory and more optional.

However, when these various criteria, widely used in discussions about medical treatment, are carefully examined, most of them are not morally relevant in distinguishing optional from obligatory medical treatments. For example, if a rare, complex, artificial, and invasive treatment offers a patient a reasonable chance of nearly painless cure, then one would have to offer a substantial justification not to provide that treatment to an incompetent patient.

What matters, then, in determining whether to provide a treatment to an incompetent patient is not a prior determination that this treatment is "ordinary" per se, but a determination that this treatment is likely to provide this patient benefits that are sufficient to make it worthwhile to endure the burdens that accompany the treatment. To this end, some of the considerations listed above are relevant: Whether a treatment is likely to succeed is an obvious example. But such considerations taken in isolation are not conclusive. Rather, the surrogate decision maker is obliged to assess the desirability to this patient of each of the options presented, including nontreatment. For most people at most times, this assessment would lead to a clear obligation to provide food and fluids.

But sometimes, as we have indicated, providing food and fluids through medical interventions may fail to benefit and may even harm some patients. Then the treatment cannot be said to be obligatory, no matter how usual and simple its provision may be. If "ordinary" and "extraordinary" are used to convey the conclusion about the obligation to treat, providing nutrition and fluids would have become, in these cases, "extraordinary." Since this phrasing is misleading, it is probably better to use "proportionate" and "disproportionate," as the Vatican now suggests,[13] or "obligatory" and "optional."

Obviously, providing nutrition and hydration may sometimes be necessary to keep patients comfortable while they are dying, even though it may temporarily prolong their dying. In such cases, food and fluids constitute warranted palliative care. But in other cases, such as a patient in a deep and irreversible coma, nutrition and hydration do not appear to be needed or helpful, except perhaps to comfort the staff and family.[14] And sometimes the interventions needed for nutrition and hydration are so burdensome that they are harmful and best not utilized.

The obligation to continue treatments once started. Once having started a mode of treatment, many caregivers find it very difficult to discontinue it. While this strongly felt difference between the ease of withholding a treatment and the difficulty of withdrawing it provides a psychological explanation of certain actions, it does not justify them. It sometimes even leads to a thoroughly irrational decision process. For example, in caring for a dying, comatose patient, many physicians apparently find it harder to stop a functioning peripheral IV than not to restart one that has infiltrated (that is, has broken through the blood vessel and is leaking fluid into surrounding tissue), especially if the only way to reestablish an IV would be to insert a central line into the heart or to do a cutdown (make an incision to gain access to the deep large blood vessel).[15]

What factors might make withdrawing medical treatment morally worse than withholding it? Withdrawing a treatment seems to be an action, which, when it is likely to end in death, initially seems more serious than an omission that ends in death. However, this view is fraught with errors. Withdrawing is not always an act: Failing to put the next infusion into a tube could be correctly described as an omission, for example. Even when withdrawing is an act, it may well be morally correct and even morally obligatory. Discontinuing intravenous lines in a patient now permanently unconscious in accord with that patient's well-informed advance directive would certainly be such a case. Furthermore, the caregiver's obligation to serve the patient's interests through both acts and omissions rules out the exculpation that accompanies omissions in the usual course of social life. An omission that is not warranted by the patient's interests is culpable.

Sometimes initiating a treatment creates expectations in the minds of caregivers, patients, and family that the treatment will be continued indefinitely or until the patient is cured. Such expectations may provide a reason to continue the treatment as a way to keep a promise. However, as with all promises, caregivers could be very careful when initiating a treatment to explain the indications for its discontinuation, and they could modify preconceptions with continuing reevaluation and education during treatment. Though all patients are entitled to expect the continuation of care in the patient's best interests, they are not and should not be entitled to the continuation of a particular mode of care.

Accepting the distinction between withholding and withdrawing medical treatment as morally significant also has a very unfortunate implication: Caregivers may become unduly reluctant to begin some treatments precisely because they fear that they will be locked into continuing treatments that are no longer of value to the patient. For example, the physician who had been unwilling to stop the respirator while the infant, Andrew Stinson, died over several months is reportedly "less eager to attach babies to respirators now."[16] But if it were easier to ignore malnutrition and dehydration and to withhold treatments for these problems than to discontinue the same treatments when they have become especially burdensome and insufficiently beneficial for this patient, then the incentives would be perverse. Once a treatment has been tried, it is often much clearer whether it is of value to this patient, and the decision to stop it can be made more reliably.

The same considerations should apply to starting as to stopping a treatment, and whatever assessment warrants withholding should also warrant withdrawing.

The obligation to avoid being the unambiguous cause of death. Many physicians will agree with all that we have said and still refuse to allow a choice to forego food and fluid because such a course seems to be a "death sentence." In this view death seems to be more certain from malnutrition and dehydration than from foregoing other forms of medical therapy. This implies that it is acceptable to act in ways that are likely to cause death, as in not operating on a gangrenous leg, only if there remains a chance that the patient will survive. This is a comforting formulation for caregivers, to be sure, since they can thereby avoid feeling the full weight of the responsibility for the time and manner of the patient's death. However, it is not a persuasive moral argument.

First, in appropriate cases discontinuing certain medical treatments is generally accepted despite the fact that death is as certain as with nonfeeding. Dialysis in a patient without kidney function or transfusions in a patient with severe aplastic anemia are obvious examples. The dying that awaits such patients often is not greatly different from dying of dehydration and malnutrition.

Second, the certainty of a generally undesirable outcome such as death is always relevant to a decision, but it does not foreclose the possibility that this course is better than others available to this patient.[17] Ambiguity and uncertainty are so common in medical decision making that caregivers are tempted to use them in distancing themselves from direct responsibility. However, caregivers are in fact responsible for the time and manner of death for many patients. Their distaste for this fact should not constrain otherwise morally justified decisions.

The obligation to provide symbolically significant treatment. One of the most common arguments for always providing nutrition and hydration is that it symbolizes, expresses, or conveys the essence of care and compassion. Some actions not only aim at goals, they also express values. Such expressive actions should not simply be viewed as means to ends; they should also be viewed in light of what they communicate. From this perspective food and water are not only goods that preserve life and provide comfort; they are also symbols of care and compassion. To withhold or withdraw them—to "starve" a patient—can never express or convey care.

Why is providing food and water a central symbol of care and compassion? Feeding is the first response of the community to the needs of newborns and remains a central mode of nurture and comfort. Eating is associated with social interchange and community, and providing food for someone else is a way to create and maintain bonds of sharing and expressing concern. Furthermore, even the relatively low levels of hunger and thirst that most people have experienced are decidedly uncomfortable, and the common image of severe malnutrition or dehydration is one of unremitting agony. Thus, people are rightly eager to provide food and water. Such provision is essential to minimally tolerable existence and a powerful symbol of our concern for each other.

However, *medical* nutrition and hydration, we have argued, may not always provide net benefits to patients. Medical procedures to provide nutrition and hydration are more similar to other medical procedures than to typical human ways of providing nutrition and hydration, for example, a sip of water. It should be possible to evaluate their benefits and burdens, as we evaluate any other medical procedure. Of course, if family, friends, and caregivers feel that such procedures affirm important values even when they do not benefit the patient, their feelings should not be ignored. We do not contend that there is an obligation to withhold or to withdraw such procedures (unless consideration of the patient's advance directives or current best interest unambiguously dictates that conclusion); we only contend that nutrition and hydration may be foregone in some cases.

The symbolic connection between care and nutrition or hydration adds useful caution to decision making. If decision makers worry over withholding or withdrawing medical nutrition and hydration, they may inquire more seriously into the circumstances that putatively justify their decisions. This is generally salutary for health care decision making. The critical inquiry may well yield the sad but justified conclusion that the patient will be served best by not using medical procedures to provide food and fluids.

A LIMITED CONCLUSION

Our conclusion—that patients or their surrogates, in close collaboration with their physicians and other caregivers and with careful assessment of the relevant information, can correctly decide to forego the provision of medical treatments intended to correct malnutrition and dehydration in some circumstances—is quite limited. Concentrating on incompetent patients, we have argued that in most cases such patients will be best served by providing nutrition and fluids. Thus, there should be a presumption in favor of providing nutrition and fluids as part of the broader presumption to provide means that prolong life. But this presumption may be rebutted in particular cases.

We do not have enough information to be able to determine with clarity and conviction whether withholding or withdrawing nutrition and hydration was justified in the cases that have occasioned public concern, though it seems likely that the Danville and Bloomington babies should have been fed and that Claire Conroy should not. It is never sufficient to rule out "starvation" categorically. The question is whether the obligation to act in the patient's best interests was discharged by withholding or withdrawing particular medical treatments. All we have claimed is that nutrition and hydration by medical means need not always be provided. Sometimes they may not be in accord with the patient's wishes or interests. Medical nutrition and hydration do not appear to be distinguishable in any morally relevant way from other life-sustaining medical treatments that may on occasion be withheld or withdrawn.[18]

CHAPTER NINE

□

When Is It Morally Justifiable to Discontinue
Medical Nutrition and Hydration?

When, if ever, is it morally justifiable to withhold or to withdraw medical nutrition and hydration from patients? I want to raise this question in the context of two cases.

In the first case, Mrs. X, a seventy-nine-year-old widow, had been a resident of a nursing home for several years. In the past she had experienced repeated transient ischemic attacks. Because of progressive organic brain syndrome, she had lost most of her mental abilities and had become disoriented. She also had episodes of thrombophlebitis as well as congestive heart failure. Her daughter and grandchildren visited her frequently and obviously loved her deeply. One day she was found unconscious on the bathroom floor of the nursing home. She was hospitalized, and the diagnosis was a "massive stroke." She made no recovery, remaining obtunded and nonverbal, but she continued to manifest a withdrawal reaction to painful stimuli and some purposeful behaviors. Mrs. X refused to allow a nasogastric tube to be placed in her stomach; at each attempt she thrashed about violently and pushed the tube away. After the tube was finally placed, Mrs. X pulled out of restraints and managed to remove it. After several days, her IV sites were exhausted. The question for the staff was whether to do further "extraordinary" or "heroic" measures to maintain fluid and nutritional intake for this elderly patient who had made no recovery from a massive stroke and who was largely unaware and unresponsive. After much mental anguish and discussion with the nurses on the floor and with the patient's family, the physicians in charge decided not to provide further IVs, cutdowns, or a feeding tube, and to allow Mrs. X to die. Her oral intake was minimal, and she died quietly the following week.[1]

In the second case, Dr. Hilfiker reports being awakened at 3:00 A.M. by the nurse at the nursing home where Mrs. T, an eighty-three-year-old woman, has been confined since her stroke three years earlier.[2] Mrs. T is now bedridden and aphasic, weighs only sixty-nine pounds, and has decubitus ulcers on her back and hip; but Dr. Hilfiker recalls her condition before her stroke, "her dislike and distrust of doctors and hospitals, her staunch pride and independence despite her severe scoliosis, her wry grin every time I suggested hospitalization for some problem." While talking to the nurse over the telephone and deciding what he should do, he also recalls admitting her to the hospital after her stroke, "one side completely paralyzed, globally aphasic, incontinent, and reduced to helplessness." And he recalls "aggressively" treating the pneumonia that developed, providing intravenous antibiotics despite her apparent desire to die. He rationalized that she was depressed and would get over it. Now the nurse reports that she has a fever (103.5° F), hasn't been eating much the last few days, and has a little cough. Dr. Hilfiker decides that he should not wait until morning, and he goes to the nursing home to examine Mrs. T, who looks out at him from behind her blank face. His tentative diagnosis is pneumonia, and he asks the nurse to call the technician out of bed for a chest x-ray and also orders a urine culture. As Dr. Hilfiker and the nurse discuss the case, they note that Mrs. T's only friend in the nursing home, who probably knows better than anyone else what Mrs. T would want, has said that she "hoped there wouldn't be any heroics if Mrs. T . . . got sick again." But the only relative (a distant niece who lives far away) had called to request that "everything possible" be done for her aunt. Dr. Hilfiker reflects:

> There in the middle of the night I consider "doing everything possible" for Mrs. T: transfer to the hospital, intravenous lines for hydration and antibiotics, thorough laboratory and x-ray evaluation, twice-daily rounds to be sure that she is recovering, more toxic antibiotics, and even transfer to our regional hospital for evaluation and care by a specialist. None of it is unreasonable, and another night I might choose just such a course. But tonight my human sympathies lie with Mrs. T and what I perceive as her desire to die. . . . In any event I decide against the heroics. But I can't just do nothing, either. My training and background are too strong. I do not allow myself to be consistent and just go home. Compromising (and ultimately making a decision that makes no medical or ethical sense at all), I write orders instructing the staff to administer liquid penicillin, to encourage fluid

intake, and to make an appointment with my office so that I can reexamine Mrs. T in 36 hours.[3]

A FRAMEWORK FOR ANALYSIS

In the first case, the staff decided to let Mrs. X die even though they could have prolonged her life for some time through medical nutrition and hydration. In the second case, Dr. Hilfiker decided to encourage fluid intake and to administer liquid penicillin but not to provide an intravenous line for hydration and antibiotics. He felt that his "compromise" made no "medical or ethical sense at all." Nevertheless, it represents one way to draw lines in withholding and withdrawing treatment. As both cases suggest, one major issue in drawing lines is whether all medical treatments, depending on the circumstances, can be construed as "heroic" or "extraordinary" rather than "ordinary." In particular, are medical nutrition and hydration by peripheral or central intravenous lines, nasogastric tubes, or gastrostomies morally similar to other medical and surgical procedures that are sometimes construed as "heroic" or "extraordinary"? Is it appropriate to view nutrition and hydration as "medical treatments" when they involve medical procedures or require surgery?

In accord with many commentators over the last several years, I believe that the language of ordinary and extraordinary or heroic means of treatment should be replaced because it is misleading.[4] It is misleading because it obscures the principles that should govern decisions about withholding or withdrawing medical treatments. If we replace "ordinary" with "obligatory," and "extraordinary" or "heroic" with "optional," we can then examine the principles by which to determine when a particular medical treatment is obligatory and when it is optional. Apart from constraints set by scarcity and principles of justice in the allocation of resources (to which I will return later), the fundamental moral principles for decisions about withholding or withdrawing medical treatments that prolong life are (1) beneficence, often stated in terms of patient-benefit or the patient's best interests, and (2) respect for persons, often stated in terms of respect for autonomy.[5] Because of these principles, health care professionals should always seek the best interests of their patients, subject to the constraints and limits set by the competent (or previously competent) patient's wishes, choices, and actions. In general, a competent patient should be free to refuse any medical treatment as long as that refusal does not violate the rights of others. Only limited or constrained paternalism—beneficence constrained and limited by respect for persons and their autonomy—is morally defensible.[6] More difficult and problematic are the sorts of cases presented at the outset, where the patient is incompetent (or not clearly competent) or where his/her wishes and preferences are unclear. In decisions about an incompetent patient who has not previously expressed his or her wishes, the relevant moral consideration, based on beneficence, is the patient's benefit or best interests.

This moral consideration establishes a presumption in favor of all medical treatments that prolong life. Such treatments may be presumed to be in the patient's interests (and even in accord with his/her wishes if they could be known). However, this presumption is rebuttable. It can be rebutted, as I have noted, when a competent patient refuses treatment or a previously competent patient has refused treatment—for example, through an advance directive. But it can also be rebutted when the appropriate decision makers for an incompetent patient determine (a) that the treatments are futile or useless, (b) that they provide no benefits to the patient, or (c) that their benefits are outweighed by their burdens to the patient. Thus, the same moral principles that establish a presumption in favor of treatment also indicate when the presumption may be rebutted. We may believe that this presumption should not be rebutted too quickly, easily, or lightly; we may even believe that decision makers should experience anguish, as in our first two cases. But whatever burden of proof and whatever anguish is appropriate in such decisions, any medical treatment may be withheld or withdrawn as morally optional in some circumstances.

One response to this search for principles for decisions about life-sustaining medical treatment is to look at medical practice. This response is quite appropriate because we can expect to find the principles of beneficence and respect for persons at work in medical practice; to a greater or lesser extent, these principles are already embodied in medical practice. They are part of the ethos that ethics as critical reflection examines. However, this response tends to fall into the same trap that has ensnared the language of ordinary and extraordinary or heroic means of treatment. That language is misleading because it directs attention toward customary medical practice and away from underlying principles. Thus, treatments are frequently viewed as "ordinary" if it is "usual" or "customary" for physicians to use them for certain diseases, such as pneumonia, or certain problems, such as malnutrition. Treatments are considered "extraordinary" or even "heroic" if it is "unusual" or "uncustomary" for physicians to use them for certain diseases or problems. The terms even become attached to technologies themselves, as they now become considered as "ordinary" or "extraordinary" forms of treatment. The patient tends to disappear from view, even though it is the patient's disease or problem that the physician treats with the technology in question. The criteria for distinguishing ordinary from extraordinary means of treatment tend to be located in the customary practice of medicine, which usually but not always reflects the patient's interests and preferences—the principles of beneficence and respect for persons.

Criteria other than usual and unusual, or customary and uncustomary, medical practice have also been proposed for determining which treatments are ordinary (obligatory) and extraordinary or heroic (optional). For example, some practitioners and commentators have considered whether the treatment is simple or complex, natural or artificial, noninvasive or invasive, inexpensive or costly. If a treatment is simple, natural, noninvasive, or inexpensive, it is more likely to be viewed as "ordinary" or "obligatory" than if it is complex, artificial, invasive, or costly. But if such criteria are relevant it is only because they express other moral principles, particularly beneficence and respect for persons. For example, if a complex treatment is available and in accord with the patient's wishes and interests, it is difficult to see why morally it should be handled differently than a simple treatment also in accord with the patient's wishes and interests. To take another example, several oddities emerge when the criteria of natural and artificial are invoked. According to one study conducted after the natural death act was implemented in California, physicians in that state generally viewed respirators, dialysis, and resuscitators as artificial, but split evenly on intravenous feeding, while two-thirds viewed insulin, antibiotics, and chemotherapy as natural.[7] In general, they viewed mechanical systems as more artificial than drugs and other treatments. Nevertheless, physicians' construals of some treatments as artificial and others as natural are not morally relevant to judgments about whether and when those treatments may be withheld or withdrawn, because they do not connect in any significant way with the moral considerations that should govern such decisions. Other criteria, such as the degree of invasiveness (noninvasive-invasive) and cost (inexpensive-costly), may be morally relevant in view of the patient's overall condition, interests, and preferences. But their relevance stems from their established connection with beneficence and respect for persons.

MAJOR ARGUMENTS AGAINST FORGOING MEDICAL NUTRITION AND HYDRATION

If this analysis is correct, then no medical treatment as such is always morally obligatory; whether it is morally obligatory depends on the patient's condition, interests, and wishes. Two main arguments have been presented to exempt medical nutrition and hydration from this general claim. In different ways, these arguments try to establish that medical nutrition and hydration are not relevantly similar to other medical treatments and that the formal principle of universalizability (treat similar cases in a similar way) does not require or even permit us to make the same judgments about medical nutrition and hydration that we make about other medical procedures.

The first argument holds that medical nutrition and hydration are always required because they are always necessary for comfort (patient benefit) and dignity (patient respect as well as benefit). Such an argument probably appears in the claim of the nurse whose actions led to the indictment against Nejdl and Barber, the two California physicians who discontinued intravenous nutrition to a patient who had severe brain damage after the loss of oxygen following routine surgery: "Food is an ordinary means. And everyone has a right to ordinary treatment."[8] This argument clearly undergirds the rule proposed by the Department of Health and Human Services (July 5, 1983) for treatment of handicapped newborns: "the basic provision of nourishment, fluids, and routine nursing care is a fundamental matter of human dignity, not an option for medical judgment."[9] That rule includes medical as well as non-medical means of providing nutrition and hydration.

Whether this first argument appeals to beneficence or to respect for persons, to comfort or to dignity, it is difficult to defend an absolute requirement to provide nutrition and hydration by any possible means. For example, a central IV involves some risks.[10] There is also some evidence that patients who are allowed to die without artificial hydration may even die more comfortably than patients who receive intravenous hydration.[11] Whenever medical nutrition and hydration are required for the incompetent patient's comfort and dignity, they should be provided; the competent patient may, of course, decline them. But we should not suppose that the comfort and the dignity afforded by the relief of parched lips always necessitate the provision of medically adequate nutrition and hydration, which may not offer comfort and dignity in some circumstances.

A second argument that medical nutrition and hydration are always obligatory focuses on their symbolic significance: what they symbolize about caregivers and patients and their relationship. This argument from symbolic values stresses the similarities between *nonmedical* and *medical* acts of providing food and water, rather than the similarities between medical nutrition and hydration and other medical treatments. According to this line of argument, the medical provision of nutrition and hydration symbolizes, expresses, and conveys care and compassion. By contrast, "starvation" can never express care and compassion. The provision of food and water is a central symbol of care in both secular and religious contexts. We respond with care to the newborn by providing food and water, and we continue to extend care in this way. Daniel Callahan has noted that feeding the hungry is the most fundamental of all human relationships and "the perfect symbol of the fact that human life is inescapably social and communal."[12] Our interdependence combines with our actual experiences of thirst and hunger to make this symbol even more powerful—thirst and hunger are uncomfortable and we view severe malnutrition and dehydration as extreme agony. It is also important to note the religious rituals that involve sharing food and drink and the religious admonitions to feed the hungry and to give drink to the thirsty.

This argument from the symbolic value of medical nutrition and hydration hinges on assumptions about human interests and preferences for life and for the food and water essential to sustain life, about the net benefit of nutrition and hydration and particular means of providing them, and about the discomfort of thirst and hunger and of dehydration and malnutrition. Those assumptions are solid enough to establish a presumption in favor of the provision of medical nutrition and hydration; but as I have argued, they do not hold in all cases. Indeed, sometimes medical nutrition and hydration may not be required by the patient's interests and preferences and may even violate those interests and preferences. In such cases, caregivers may be nourishing a symbol that has little or nothing to do with the interests and preferences of actual patients. Still there might be arguments for nourishing such a symbol.

One such argument appears in Daniel Callahan's brief statement, "On Feeding the Dying."[13] He admits ("cannot deny") "the moral licitness of the discontinuation of feeding under some circumstances (for example, those specified by Lynn and Childress . . .)."[14] Even if withholding or withdrawing medical nutrition or hydration is not always morally wrong, it is always repulsive and repugnant. And we should continue to experience revulsion, repugnance, and repulsion "at the stopping of feeding even under legitimate circumstances" and "even in those cases where it [stopping the feeding] might be for the patient's own good." Callahan does not use the moral language of guilt, which might be occasioned by the violation of moral duties or rights in a particular case. His major themes are the symbol of feeding and its attendant emotions or sentiments, particularly the ones evoked by not feeding. His main thesis is that it is a "dangerous business to tamper with, or adulterate, so enduring and central a moral emotion" as "repugnance against starving people to death."

It might be argued that the similarities among all acts of providing nutrition and hydration are so great as to make it impossible to differentiate their *recipients* (e.g., dying patients) or their *methods* (e.g., a gastrostomy). Callahan cannot, however, take this line of argument because he admits that it is morally licit to discontinue medical nutrition under some circumstances. His argument can perhaps best be described as *symbol* or *sentiment utilitarianism*, by analogy with what is usually called rule utilitarianism. According to rule utilitarianism, certain *acts* should not be performed even when they appear to maximize welfare in particular circumstances, because over time and a wide range of circumstances the *rule* that prohibits those acts will maximize welfare. A rule utilitarian, for example, might defend a rule requiring that we provide nutrition and hydration by any means any time they will prolong life. If this rule would maximize welfare over time, it should be followed even if its violation might appear to maximize welfare in particular circumstances. (It may be difficult to imagine a utilitarian defense of this rule without any qualifications about the patient's interests and preferences, but the rule against "mercy killing" has been defended on utilitarian grounds.)[15]

Callahan does not "rule out" acts of withholding or withdrawing medical nutrition in order to produce good consequences or to avoid bad consequences. Nevertheless, in order to avert "social disaster," it is essential that professionals and others respect the symbol of feeding even in their decisions not to provide medical nutrition and hydration. How can this symbol be respected? Callahan's answer is peculiarly Protestant for a philosopher in the Catholic tradition: Respect for the symbol appears in the experience of revulsion even when not feeding is morally licit. Respect thus is a matter of emotions and sentiments of repugnance and repulsion. It is possible to avoid "social disaster" if there is widespread and "deep-seated revulsion at the stopping of feeding even under legitimate circumstances."

There are several weak links in Callahan's argument, including his image of "starvation," which, as I have suggested, may not be appropriate for the cases that concern us. However, I want to concentrate on his sentiment or symbol utilitarianism, particularly his claim that we can avoid "social disaster" by not educating people out of their emotion of revulsion, even in legitimate cases. Although we can morally differentiate acts, Callahan claims we cannot avoid "social disaster" if we try to differentiate the emotions regarding those acts because of the centrality of the symbol of feeding.

Callahan fears a twofold disaster. First, subversion of the commitment to feed the hungry, for example, poor people who are hungry. Callahan may be right that repugnance against letting people starve to death is a "necessary social instinct." But he provides no evidence for supposing that the social commitment to feed the poor will decline or disappear if the society tolerates withholding or withdrawing medical nutrition without revulsion. Nor does he offer any evidence that caregivers will become callous to the hunger and malnutrition of the poor if they do not experience revulsion in stopping intravenous nutrition when it is morally licit to do so. The widespread phenomenon of compartmentalization of the moral life, which is frequently inappropriate and often has bad consequences, suggests that it is easier than Callahan supposes to direct our sentiments and emotions selectively. He underestimates what Joel Feinberg calls "human emotional flexibility"[16]—our capacity for selective expression of sentiments and emotions, for example, when it is obligatory to feed the poor who are hungry and also permissible to discontinue IV lines for some dying patients. Callahan also assumes that there is great interdependence among our emotions and sentiments; although he uses the language of a "cluster of sentiments and emotions," his argument presupposes another image, perhaps a web or a fabric. Finally, as Jonathan Glover notes about what he calls "utilitarian moral conservatism," Callahan assumes that "the reasons we have for the policies we choose are relatively unimportant to the reverberations of those policies in other parts of our system of attitudes."[17] These assumptions are all very questionable.

If there is a moral presumption in favor of life-sustaining medical treatment, including nutrition and hydration, as in accord with the patient's interests and preferences, and if this moral presumption must be rebutted before treatment may be withheld or withdrawn, there is no need for the strong sentiments of revulsion, repugnance, and repulsion that Callahan affirms. Moral caution, hesitancy, and anxiety should be sufficient. In the early 1950s, the distinguished Catholic moral theologian, Gerald Kelly, S.J., used an actual case to clarify the concepts and criteria of ordinary and extraordinary means of treatment, including intravenous nutrition: "A patient almost ninety years of age, suffering from cardiorenal disease, had been in a coma for two weeks, during which time he received an intravenous solution of glucose and some digitalis preparation. The coma was apparently terminal. A member of the family asked that the medication and intravenous feeding be discontinued. With the approval of a priest, the doctor and Sisters acceded to the request, but they did so with some disquietude and they continued to be disturbed for some time after the patient's death."[18] Their disquietude and disturbance were sufficient, despite Callahan's insistence on repugnance, revulsion, and repulsion.

Callahan's second major fear is the "social disaster" of moving from a right to a duty. He worries that in our society acceptance of a right to discontinue medical nutrition and hydration will eventually, if not immediately, become a duty to discontinue them. He worries about the slippery slope—the slide from "may" to "must" or from permissible to required, especially under the pressures of economic rationality, which is instrumental rationality rather than expressive rationality and which thus tends to ignore the sorts of symbols and sentiments that Callahan rightly emphasizes. According to the logic of hard economic rationality, if feeding does not benefit an irreversibly comatose patient, stopping it is not only permissible but obligatory. "The only impediment to the enactment of that kind of policy," Callahan claims, "is a cluster of sentiments and emotions that is repelled by the idea of starving someone to death, even in those cases where it might be for the patient's own good." In our article, to which Callahan responds, Joanne Lynn and I focused mainly on the moral rightness, justifiability, or permissibility of discontinuing medical nutrition and hydration in some cases, indicating that agents have a right to discontinue these and other medical treatments under some circumstances.[19] Before turning to the question of economic rationality—and broader questions of justice in the allocation of resources—I want to explore circumstances under which withholding or withdrawing medical nutrition and hydration might be obligatory.

Neither the traditional distinction between ordinary and extraordinary means of treatment nor my distinction between obligatory and optional treatments is sufficient, because they do not envision circumstances in which it may be wrong to treat and obligatory not to treat. The following chart indicates the range of judgments:

Categories of Moral Judgment about Treatment

I. Obligatory	II. Optional	III. Obligatory Not to Provide
	A. Indifferent	
	B. Heroic and Praiseworthy	

Most ethical reflection has focused on I and II particularly in relation to incompetent patients; III has been developed mainly in relation to competent or previously competent patients. Even within the category of optional treatments, little attention has been paid to the different possibilities. A treatment might be optional in the sense that it is morally indifferent whether the agent provides it. However, it might be optional in the sense that providing it would be praiseworthy as a heroic act, though not providing it would not be morally blameworthy. Often the language of "extraordinary" and "heroic" suggests extra time, effort, and energy when there is limited chance of success.

Is it ever obligatory not to provide some treatments? If the principles of beneficence and respect for persons set the presumption in favor of treatment to prolong life and also the conditions for rebutting that presumption, the same principles may also dictate the conclusion that it is wrong to provide treatment in some cases and even obligatory not to do so. Perhaps the clearest example is the forcible or deceptive use of life-sustaining treatment against the unambivalent refusal of a competent patient, or previously competent patient, when that decision does not violate anyone else's rights. Even if such treatment expresses benevolence, it is a form of disrespect and insult to the patient as a person. Hence, it is appropriate to say that it is morally obligatory not to provide life-sustaining treatment in such cases. Even if there is no record of a previously competent patient's preferences or even if the patient has never been competent, life-sustaining treatment may sometimes violate that patient's interests. The treatment may be against those interests as well as not being required by them, for example, when the pain is so severe and the restraints are so great as to outweigh the limited anticipated benefits. It will often, perhaps even usually, be difficult to determine the balance of benefits and burdens to the incompetent patient, particularly where the patient has never had a life as a competent person expressing values; but in principle, the burdens could conceivably so outweigh the benefits to the patient as to make the treatment not only optional but also wrong and thus obligatory not to provide.

At this point we glimpse another potential danger of the argument for providing medical nutrition and hydration because of their symbolic value. Such an argument may imply that it is permissible or even obligatory to provide medical nutrition and hydration against a patient's wishes and interests in order to maintain the symbol and its associated emotions and sentiments. Parallel to rule utilitarianism, sentiment or symbol utilitarianism may justify overriding the interests and preferences of individuals in particular circumstances because of the overall value of undifferentiated sentiments and the symbol of feeding. It could justify forcing people, such as Elizabeth Bouvia and Ross Henninger, to receive nutrition even when they do not accept it or when its burdens outweigh its benefits to them.[20] It may nourish the symbol at the expense of actual patients.

In general, then, it is appropriate to hold that it is morally wrong to provide—and also morally obligatory not to provide—medical nutrition and hydration when it violates the principles of beneficence and respect for persons. However, to hold that there may be an obligation or duty (these terms are here used interchangeably) not to treat in some circumstances is not to repudiate obligations or duties to patients. It is important to distinguish an obligation to a person from a particular obligation to do X, Y, or Z to or for that person. The physician may have an obligation to a patient to care for him or her, but the content of that obligation of care should be shaped by the patient's needs and preferences. Within the context of the obligation of care, medical treatments to sustain life are sometimes obligatory, sometimes optional (in both senses), and sometimes wrong. Nevertheless, even if it is appropriate to say that caregivers act wrongly when they provide life-sustaining treatment in some circumstances, it may not be morally appropriate for society to enforce that judgment, at least in certain ways.

Callahan is worried not only about the societal enforcement of a moral obligation not to provide treatment but also about the grounds of that putative obligation, particularly the appeal to external factors, such as economic rationality, rather than to beneficence and respect for the patient. However, it is appropriate to say with the utmost caution that, in some circumstances, where a treatment is optional for the patient and also violates the interests of others, there may be a moral obligation or duty to stop (e.g., in order to avoid wasting resources that could benefit others). This is a matter of justice, which cannot be reduced to mere economic rationality. Earlier, it was important to protect the preferences and interests of patients against overtreatment; now it is important to protect them against undertreatment because of external factors, including cost containment. Thus it is necessary to be as clear as possible about the nature and the centrality of the care of patients as defined by their interests and preferences. Then it may be possible to assess the justice of various policies of allocation of resources in health care. I agree with Callahan that we need strong barriers to prevent the triumph of a narrow and hard economic rationality, but the barriers themselves will not hold if we erect them hastily and in the wrong places.

DISTANCING DEVICES AND SELF-DECEPTION

Erecting barriers, drawing lines, or setting limits is a major issue in debates about withholding and withdrawing medical treatments. Obviously there would be few debates if physicians and other health care professionals, as well as family members, drew lines and set limits in the same places in their decisions about the use of various treatments, including medical nutrition and hydration. For example, some people feel comfortable withholding treatment they have not yet started but do not feel comfortable withdrawing treatment they have already started.[21] Thus, they would not feel obligated to restart an IV if it became infiltrated and surgery was required to provide access, but they would feel obligated to continue an IV once it was started.

On one occasion an attending physician asked me, as an ethicist, to discuss a controversial case with the staff because of their disputes and discomfort about drawing lines. This case involved an elderly man who had numerous problems and could not be expected to survive long even with the most vigorous medical efforts. It was easy for all of the staff to accept an order not to resuscitate the patient in the event of cardiac arrest. Some of the staff also felt that the IV line providing fluids and antibiotics should be discontinued because it offered no reasonable chance of success, but others felt that the IV should be continued for the reasons given above. Yet many (but not all) in the latter group felt comfortable in not restarting the IV when it infiltrated.

My task as an ethics consultant was not to provide the right answer, or to tell the caregivers what they ought to do, but to analyze and interpret categories and distinctions that have evolved in religious and humanistic traditions of ethical reflection to illuminate such controversies. The distinction between withholding (not starting) and withdrawing (stopping) is difficult to defend and may even be pernicious if it leads to decisions not to begin treatments that may be useful in order to reduce the risk of being locked into a treatment that cannot be terminated. And yet the psychological and symbolic significance of the distinction cannot be denied. Similarly, agents may find it easier to withhold or withdraw some treatments than others, usually for psychological and symbolic reasons, as was evident in my discussion of natural and artificial treatments.

Where people draw lines and set limits often depends, at least in part, on their sense of temporal, spatial, and moral distance. Sometimes they can easily disassociate themselves from their acts or from the consequences of their acts because of their sense of distance. Such disassociation appears when people discount remote future consequences. For example, a few professionals and commentators favor withdrawing the IV rather than the respirator because of the *immediacy* of death from the withdrawal of the respirator. They note that, the respirator "creates an *immediate consequence of death* for which we must take responsibility."[22] It is not clear, however, why the authors suppose that they can avoid moral responsibility for discontinuing the IV just because dehydration develops slowly as a delayed consequence of their actions.

Several of our categories create moral distance, often by invoking temporal or spatial distance. Clear examples include the rule of double effect, which contrasts direct and indirect courses of action, and the distinction between killing and letting die. While such distinctions may be important and defensible, their function as psychological defense mechanisms should not be ignored. Thus, we should make sure that they are really rationally defensible, rather than matters of self-deception.

A clear but not uncommon instance of self-deception is to continue the IV but at a rate that will result in dehydration over time.[23] This action maintains the gesture and symbol of feeding but at a rate that will not sustain the patient's life. It could conceivably serve as a compromise for those who want to respect the symbol and yet also act in accord with the patient's needs and wishes. However, it necessarily involves self-deception, because an agent can carry it out only by failing to acknowledge that the patient will become malnourished and dehydrated while the IV line maintains the fiction and expresses the symbol of feeding. Otherwise the agent would have to take responsibility for the outcome. But moral responsibility requires that we face our actions and their consequences, looking clearly and critically at the lines we draw and the limits we set, particularly in decisions to withhold or to withdraw life-sustaining medical treatment, including medical nutrition and hydration.[24]

POSTSCRIPT: THE LIMITS OF MEDICAL
FUTILITY (1996)

In chapter 8, with Joanne Lynn, and in chapter 9, I argued that it is sometimes justifiable to withhold or withdraw artificial nutrition and hydration, parallel to other interventions that are clearly recognized as medical, and I developed the conditions under which such actions are justifiable, stressing that they are the same ones that apply to various medical interventions. Then in chapter 9 I further argued that it can sometimes even be wrong to provide—and thus obligatory not to provide—artificial nutrition and hydration. The standards used in both chapters include (1) when the artificial nutrition and hydration would be futile, (2) when they would provide no benefit, and (3) when their burdens would outweigh their benefits.

In chapter 8 Lynn and I used the language of futility in both narrower and broader senses. We first used it to mark interventions that could not be performed or would have little or no physiological effect if performed (e.g., little nutrition would be absorbed). However, we also noted that interventions that are disproportionate (the burdens outweigh the benefits) would be "futile in a broader sense." By implication, the second category of intervention (no possibility of benefit) could also be termed futile.

The debate about medical futility has become widespread and vigorous over the last several years. In its wake I would now argue for restricting the term more narrowly than in these chapters, in part because appeals to futility have become ways to restore a kind of medical paternalism, to reinstate medical authority over against patient and familial decision making, to mask value-laden judgments as value free and objective, to disguise rationing decisions, and so forth. Appeals to medical futility serve to stop conversation rather than to invite open discourse about values in treatment and nontreatment decisions, whether in caring for particular patients or in rationing care.

The case of Mrs. Helga Wanglie, which is frequently invoked in such debates, illustrates some of the ambiguities and limits in appeals to medical futility.[25] After emergency treatment in January 1990 for difficulties in breathing that resulted from chronic bronchiectasis, and months of follow-up treatment, Mrs. Wanglie, an eighty-five year old resident of a nursing home, was determined to be in a persistent vegetative state as a result of severe anoxic encephalopathy. Her care included maintenance on the respirator, with tube feedings, repeated courses of antibiotics, frequent airway suctioning, an air flotation bed, and biochemical monitoring. Weeks after this diagnosis, the physicians indicated that the life-sustaining treatment was not benefiting the patient and should be withdrawn; but Mrs. Wanglie's husband, daughter, and son insisted that it be continued. It was never clear whether the family's request reflected Mrs. Wanglie's previously expressed preferences or only their own preferences, but the family emphatically stated that "physicians should not play God, that the patient would not be better off dead, that removing her life support showed moral decay in our civilization, and that a miracle could occur." By contrast, in October 1990, a new attending physician, after consultation with specialists, concluded that Mrs. Wanglie was "at the end of her life and that the respirator was `non-beneficial,' in that it could not heal her lungs, palliate her suffering, or enable this unconscious and permanently respirator-dependent woman to experience the benefit of the life afforded by respirator support." Similar points also applied to artificial nutrition and hydration and other procedures.

The hospital sought to have Mr. Wanglie disqualified as decision maker for his wife, rather than trying directly to resolve the issues of putative medical futility. But the court appointed Mr. Wanglie, at his request, as his wife's conservator, and he requested that treatment be continued. Upon Mrs. Wanglie's death in early July 1991, the family described her care as excellent, but the daughter added that "we just had a disagreement on ethics." As Mr. Wanglie put it, "We felt that when she was ready to go that the good Lord would call her, and I would say that's what happened." The medical costs for the initial emergency treatment and subsequent care reached approximately $800,000; they were fully covered by Medicare and a supplemental private insurance policy.

Professional and public moral discourse frequently suffers, in my judgment, when such cases are discussed in the language of medical futility rather than in the language of proportionality (or even the language of utility). While many on the hospital staff and outside construed this case as a debate about medical futility, the family viewed it as a "disagreement on ethics." The language of medical futility has wide appeal in part because it appears to remove some decisions about treatment and nontreatment from ethical discourse. It appears to offer a more objective and even value-free way to make treatment decisions (however much concerns about health care costs fuel the debates). Nevertheless, only in the most limited set of circumstances can physicians legitimately claim that the treatment is futile in the strictest sense; that is, it cannot be performed or will not produce physiological effects, whatever ends or objectives are sought. In most other circumstances, there are debates about the value of the ends or objectives that are sought and about whether the procedures in question can realize those ends or objectives without disproportionate burdens, harms, and costs to the patient.

Even though judgments of medical futility in the strictest sense hold in virtual independence of the ends or objectives sought, most judgments of medical futility presuppose some judgments about the ends or objectives of treatment—for example, that it is not worthwhile for a patient to live in a persistent vegetative state. Only if there is agreement that such a life is not worth living can the procedure, whether a respirator or artificial nutrition and hydration, be deemed "non-beneficial." Hence, value judgments about the end state generally cannot be avoided.

Predictions of various outcomes also involve value judgments because they are usually probabilistic rather than certain. And the task of setting a statistical threshold is at least partly evaluative—where the line is drawn will reflect the line-drawer's values. For instance, while some clinicians view a treatment as futile only if there is no chance it will work, others view it as futile if its chance of success is thirteen percent or less.[26]

In addition to value judgments about end states and about probabilities, there are also value judgments about the symbolic significance of certain acts, even when they appear to achieve no medical benefit for the patient beyond the maintenance of mere biological life. As John Lantos and colleagues argue, "feeding patients in a persistent vegetative state may be futile if the goal is to restore cognition, but it may provide emotional and symbolic benefits to the patient's family or to society. These goals may be relevant to futility determinations and should not be automatically excluded."[27]

In short, claims of medical futility often presuppose value judgments about goals and means. Those who make claims of medical futility in cases where a procedure is deemed "non-beneficial," even though it can be performed and physiological effects produced, should not ignore or obscure the values involved. Insofar as the language of medical futility neglects or hides these value debates, it is faulty and should be replaced.

It is best, I believe, to refrain altogether from using the term *futility* for cases of failed *utility*. Such cases involve balancing the benefits that can be produced against the burdens, harms, and costs to the patient. The procedure in question can be performed, will usually produce a physiological effect, and will probably produce benefits for the patient, but its probable benefits are outweighed by the probable burdens, harms, or costs to the patient. These are judgments not of medical futility but of medical utility (a category that will be examined in chapter 11).

I do not claim that physicians must always present for patient and family consideration and provide all treatments that might possibly have some effect and benefit and might be proportionate. It is quite appropriate for the medical profession, using the best scientific and technical data, to develop, in public and with public input, guidelines for disclosure and provision of medical procedures. All I oppose is legitimating such judgments by appeals to medical futility, except in a narrow range of circumstances.

The last two chapters brought artificial nutrition and hydration under the moral categories that govern decisions about other medical treatments. In so doing, they implicitly operated within the framework that decisions to withhold or withdraw artificial nutrition and hydration could be construed as letting die rather than killing and thus could avoid the strictures against either assisted suicide or active euthanasia. However, the line between letting die and killing is not a hard and fast one, and it is difficult to deny the agent's intention to bring about death in withholding or withdrawing a procedure, such as artificial nutrition and hydration, that now delays or prevents death. I have not denied this intention, but have argued instead that the line between acts that our social, professional, and legal rules prohibit as killing and acts that they permit as letting die—that is, letting nature take its course, letting the disease kill the patient, letting God's will be done, etc.—reflects extrinsic moral considerations, which concern the dangers of abuse, the risks of the slippery slope, and so forth. While the weight of such extrinsic moral considerations may well shift over time, I continue to have serious reservations about the now rapid movement to alter the social, professional, and legal rules against assisted suicide and active euthanasia.[28]

If, despite protests, this movement continues, then it is crucial to institute procedural and other safeguards, particularly to protect the weak and vulnerable. One indispensable protection is a political-legal right to an adequate level of health care. Of course, such a right is important for the right to die—hence, I have limited my analysis of burdens, harms, and costs, including financial costs, to those falling on the patient himself or herself rather than including those that fall on other parties—but it is even more important for a right to assisted suicide or active euthanasia. It is morally indefensible for our society to recognize a right to physician-assisted suicide without a concomitant right to an adequate level of health care. Such a policy trumpets respect for autonomy over just care. The next four chapters address issues of just care in an attempt to determine fair and equitable policies of access to health care.

Part Four

■

ALLOCATION OF
HEALTH CARE

CHAPTER TEN

□

Who Shall Live When Not All Can Live?

Who shall live when not all can live? Although this question has been urgently forced upon us by the dramatic use of artificial internal organs and organ transplantation, it is hardly new. George Bernard Shaw dealt with it in *The Doctor's Dilemma:*

> SIR PATRICK. Well, Mr. Savior of Lives: which is it to be? that honest decent man Blenkinsop, or that rotten blackguard of an artist, eh?
> RIDGEON. It's not an easy case to judge, is it? Blenkinsop's an honest decent man; but is he any use? Dubedat's a rotten blackguard; but he's a genuine source of pretty and pleasant and good things.
> SIR PATRICK. What will he be a source of for that poor innocent wife of his, when she finds him out?
> RIDGEON. Thats true. Her life will be a hell.
> SIR PATRICK. And tell me this. Suppose you had this choice put before you: either to go through life and find all the pictures bad but all the men and women good, or go through life and find all the pictures good and all the men and women rotten. Which would you choose?[1]

A significant example of the distribution of scarce medical resources is seen in the use of penicillin shortly after its discovery. Military officers had to determine which soldiers would be treated—those with venereal disease or those wounded in combat.[2] In many respects such decisions have become routine in medical circles. Day after day physicians and others make judgments and decisions "about allocations of medical care to various segments of our population, to various types of hospitalized patients, and to specific individuals,"[3] for example, whether mental illness or cancer will receive the higher proportion of available funds. Nevertheless, the dramatic forms of Scarce Lifesaving Medical Resources (hereafter abbreviated as SLMR) such as hemodialysis and kidney and heart transplants have compelled us to examine the moral questions that have been concealed in many routine decisions. I do not attempt in this paper to show how a resolution of SLMR cases can help us in the more routine ones which do not involve a conflict of life with life. Rather, I develop an argument for a particular method of determining who shall live when not all can live. No conclusions are implied about criteria and procedures for determining who shall receive medical resources that are not directly related to the preservation of life (e.g., corneal transplants) or about standards for allocating money and time for studying and treating certain diseases.

Just as current SLMR decisions are not totally discontinuous with other medical decisions, so we must ask whether some other cases might, at least by analogy, help us develop the needed criteria and procedures. Some have looked at the principles at work in our responses to abortion, euthanasia, and artificial insemination.[4] Usually they have concluded that these cases do not cast light on the selection of patients for artificial and transplanted organs. The reason is evident: In abortion, euthanasia, and artificial insemination, there is no conflict of life with life for limited but indispensable resources (with the possible exception of therapeutic abortion). In current SLMR decisions, such a conflict is inescapable, and it makes those decisions morally perplexing and fascinating. If analogous cases are to be found, I think that we shall locate them in moral conflict situations.

ANALOGOUS CONFLICT SITUATIONS

An especially interesting and pertinent one is *U.S. v. Holmes.*[5] In 1841 an American ship, the *William Brown,* which was near Newfoundland on a trip from Liverpool to Philadelphia, struck an iceberg. The crew and half the passengers were able to escape in the two available vessels. One of these, a longboat carrying too many passengers and leaking seriously, began to founder in the turbulent sea after about twenty-four hours. In a desperate attempt to keep it from sinking, the crew threw overboard fourteen men. Two sisters of one of the men either jumped overboard to join their brother in death or instructed the crew to throw them over. The criteria for determining who should live were "not to part man and wife, and not to throw over any women."

Several hours later the others were rescued. Returning to Philadelphia, most of the crew disappeared, but one, Holmes, who had acted upon orders from the mate, was indicted, tried, and convicted on the charge of "unlawful homicide."

We are interested in this case from a moral rather than a legal standpoint, and there are several possible responses to and judgments about it. Without attempting to be exhaustive I shall sketch a few of these. The judge contended that lots should have been cast, for in such conflict situations, there is no other procedure "so consonant both to humanity and to justice." Counsel for Holmes, on the other hand, maintained that the "sailors adopted the only principle of selection which was possible in an emergency like theirs . . . a principle more humane than lots."

Another version of selection might extend and systematize the maxims of the sailors in the direction of "utility": Those are saved who will contribute to the greatest good for the greatest number. Yet another possible option is defended by Edmond Cahn in *The Moral Decision*. He argues that in this case we encounter the "morals of the last day." By this phrase he indicates that an apocalyptic crisis renders totally irrelevant the normal differences between individuals. He continues:

> In a strait of this extremity, all men are reduced—or raised, as one may choose to denominate it—to members of the genus, mere congeners and nothing else. Truly and literally, all were "in the same boat," and thus none could be saved separately from the others. I am driven to conclude that otherwise—that is, if none sacrifice themselves of free will to spare the others—they must all wait and die together. For where all have become congeners, pure and simple, no one can save himself by killing another.[6]

Cahn's answer to the question "who shall live when not all can live" is "none" unless the voluntary sacrifice by some persons permits it.

Few would deny the importance of Cahn's approach, although many, including this writer, would suggest that it is relevant mainly as an affirmation of an elevated and, indeed, heroic or saintly morality which one hopes would find expression in the voluntary action of many persons trapped in "border-line" situations involving a conflict of life with life. It is a maximal demand which some moral principles impose on the individual in the recognition that self-preservation is not a good which is to be defended at all costs. The absence of this saintly or heroic morality should not mean, however, that everyone perishes. Without making survival an absolute value and without justifying all means to achieve it, we can maintain that simply letting everyone die is irresponsible. This charge can be supported from several different standpoints, including society at large as well as the individuals involved. Among a group of self-interested individuals, none of whom volunteers to relinquish his life, there may be better and worse ways of determining who shall survive. One task of social ethics, whether religious or philosophical, is to propose relatively just institutional arrangements within which self-interested and biased people can live. The question then becomes: Which set of arrangements—which criteria and procedures of selection—is most satisfactory in view of the human condition (man's limited altruism and inclination to seek his own good) and the conflicting values that are to be realized?

There are several significant differences between the Holmes and SLMR cases, a major one being that the former involves *direct* killing of another person, while the latter involves only *permitting* a person to die when it is not possible to save all. Furthermore, in extreme situations such as that in *Holmes,* the restraints of civilization have been stripped away and something approximating a state of nature prevails, in which life is "solitary, poor, nasty, brutish and short." The state of nature does not mean that moral standards are irrelevant and that might should prevail, but it does suggest that much of the matrix that normally supports morality has been removed. Also, the necessary but unfortunate decisions about who shall live and die are made by men who are existentially and personally involved in the outcome. Their survival too is at stake. Even though the institutional role of sailors seems to require greater sacrificial actions, there is obviously no assurance that they will adequately assess the number of sailors required to man the vessel or that they will impartially and objectively weigh the common good at stake. As the judge insisted in his defense of casting lots in the Holmes case: "In no other than this [casting lots] or some like way are those having equal rights put upon an equal footing, and in no other way is it possible to guard against partiality and oppression, violence, and conflict." This difference should not be exaggerated since self-interest, professional pride, and the like obviously affect the outcome of many medical decisions. Nor do the remaining differences cancel *Holmes's* instructiveness.

CRITERIA OF SELECTION FOR SLMR

Which set of arrangements should be adopted for SLMR? Two questions are involved: Which standards and criteria should be used? And who should make the decision? The first question is basic, since the debate about implementation—for example, whether by a lay committee or physician—makes little progress until the criteria are determined.

We need two sets of criteria which will be applied at two different stages in the selection of recipients of SLMR. First, medical criteria should be used to exclude those who are not "medically acceptable." Second, from this group of "medically acceptable" applicants, the final selection can be made. Occasionally in current American medical practice, the first stage is omitted, but such an omission is unwarranted. Ethical and social responsibility would seem to require distributing these SLMR only to those who have some reasonable prospect of responding to the treatment. Furthermore, in transplants, such medical tests as tissue and blood typing are necessary, though they are hardly fully developed.

"Medical acceptability" is not as easily determined as many nonphysicians assume, since there is considerable debate in medical circles about the relevant factors (e.g., age and complicating diseases). Although ethicists can contribute little or nothing to this debate, two proposals may be in order. First, "medical acceptability" should be used only to determine the group from which the final selection will be made, and the attempt to establish fine degrees of prospective response to treatment should be avoided. Medical criteria, then, would exclude some applicants but would not serve as a basis of comparison between those who pass the first stage. For example, if two applicants for dialysis were medically acceptable, the physicians would not choose the one with the *better* medical prospects. Final selection would be made on other grounds. Second, psychological and environmental factors should be kept to an absolute minimum and should be considered only when they are without doubt critically related to medical acceptability (e.g., the inability to cope with the requirements of dialysis, which might lead to suicide).

The most significant moral questions emerge when we turn to the final selection. Once the pool of medically acceptable applicants has been defined and still the number is larger than the resources, what other criteria should be used? How should the final selection be made? First, I shall examine some of the difficulties that stem from efforts to make the final selection in terms of social value; these difficulties raise serious doubts about the feasibility and justifiability of the utilitarian approach. Then I shall consider the possible justification for random selection or chance.

Occasionally criteria of social worth focus on past contributions, but most often they are primarily future-oriented. The patient's potential and probable contribution to the society is stressed, although this obviously cannot be abstracted from his present web of relationships (e.g., dependents) and occupational activities (e.g., nuclear physicist). Indeed, the magnitude of his contribution to society (as an abstraction) is measured in terms of these social roles, relations, and functions. Enough has already been said to suggest the tremendous range of factors that affect social value or worth.[7] Here we encounter the first major difficulty of this approach: How do we determine the relevant criteria of social value?

How does one qualify and compare the needs of the spirit (e.g., education, art, religion), political life, economic activity, technological development? Joseph Fletcher suggests that "some day we may learn how to 'quantify,' or 'mathematicate,' or 'computerize' the value problem in selection, in the same careful and thorough way that diagnosis has been."[8] I am not convinced that we can ever quantify values, or that we should attempt to do so. But even if the various social and human needs, in principle, could be quantified, how do we determine how much weight we will give to each one? Which will have priority in case of conflict? Or even more basically, in the light of which values and principles do we recognize social "needs"?

One possible way of determining the values that should be emphasized in selection has been proposed by Leo Shatin.[9] He insists that our medical decisions about allocating resources are already based on an unconscious scale of values (usually dominated by material worth). Since there is really no way of escaping this, we should be self-conscious and critical about it. How should we proceed? He recommends that we discover the values that most people in our society hold and then use them as criteria for distributing SLMR. These values can be discovered by attitude or opinion surveys. Presumably, if 51 percent in this testing period put a greater premium on military needs than technological development, military men would have a greater claim on our SLMR than experimental researchers. But valuations of what is significant change, and the student revolutionary who was denied SLMR in 1970 might be celebrated in 1990 as the greatest American hero since George Washington.

Shatin is presumably seeking criteria that could be applied nationally; but at present, regional and local as well as individual prejudices tincture the criteria of social value that are used in selection. Nowhere is this more evident than in the deliberations and decisions of the anonymous selection committee of the Seattle Artificial Kidney Center, where such factors as church membership and Scout leadership have been deemed significant for determining who shall live.[10] As two critics conclude after examining these criteria and procedures, they rule out "creative nonconformists, who rub the bourgeoisie the wrong way but who historically have contributed so much to the making of America. The Pacific Northwest is no place for a Henry David Thoreau with bad kidneys."[11]

Closely connected to this first problem of determining social values is a second one. Not only is it difficult if not impossible to reach agreement on social values, it is also rarely easy to predict what our needs will be in a few years and what the consequences of present actions will be. Furthermore, it is difficult to predict which persons will fulfill their potential function in society. Admissions committees in colleges and universities experience the frustrations of predicting realization of potential. For these reasons, as someone has indicated, God might be a utilitarian, but we cannot be. We simply lack the capacity to predict very accurately the consequences which we then must evaluate. Our incapacity is never more evident than when we think in societal terms.

Other difficulties make us even less confident that such an approach to SLMR is advisable. Many critics raise the spectre of abuse, but this should not be overemphasized. The fundamental difficulty appears on another level: The utilitarian approach would in effect reduce the person to his social role, relations, and functions. Ultimately it dulls and perhaps even eliminates the sense of the person's transcendence, his dignity as a person, which cannot be reduced to his past or future contribution to society. It is not at all clear that we are willing to live with these implications of utilitarian selection. Willem Kolff, who invented the artificial kidney, has asked: "Do we really subscribe to the principle that social standing should determine selection? Do we allow patients to be treated with dialysis only when they are married, go to church, have children, have a job, a good income and give to the Community Chest?"[12]

The German theologian Helmut Thielicke contends that any search for "objective criteria" for selection is already a capitulation to the utilitarian point of view, which violates man's dignity.[13] The solution is not to let all die, but to recognize that SLMR cases are "borderline situations" which inevitably involve guilt. The agent, however, can have courage and freedom (which, for Thielicke, come from justification by faith) and

> can go ahead anyway and seek for criteria for deciding the question of life or death in the matter of the artificial kidney. Since these criteria are . . . questionable, necessarily alien to the meaning of human existence, the decision to which they lead can be little more than that arrived at by casting lots.

The resulting criteria, he suggests, will probably be very similar to those already employed in American medical practice.[14]

He is most concerned to preserve a certain *attitude* or *disposition* in SLMR—the sense of guilt which arises when man's dignity is violated. With this sense of guilt, the agent remains "sound and healthy where it really counts."[15] Thielicke uses man's dignity only as a judgmental, critical, and negative standard. It only tells us how all selection criteria and procedures (and even the refusal to act) implicate us in the ambiguity of the human condition and its metaphysical guilt. This approach is consistent with his view of the task of theological ethics: "to teach us how to understand and endure—not 'solve'—the borderline situation."[16] But ethics, I would contend, can help us discern the factors and norms in whose light relative, discriminate judgments can be made. Even if all actions in SLMR should involve guilt, some may preserve human dignity to a greater extent than others. Thielicke recognizes that a decision based on any criteria is "little more than that arrived at by casting lots." But perhaps selection by chance would come the closest to embodying the moral and nonmoral values that we are trying to maintain (including a sense of human dignity).

THE VALUES OF RANDOM SELECTION

My proposal is that we use some form of randomness or chance (either natural, such as "first come, first served," or artificial, such as a lottery) to determine who shall be saved. Many reject randomness as a surrender to nonrationality when responsible and rational judgments can and must be made. Edmond Cahn criticizes "Holmes's judge" who recommended the casting of lots because, as Cahn puts it, "the crisis involves stakes too high for gambling and responsibilities too deep for destiny."[17] Similarly, other critics see randomness as a surrender to "nonhuman" forces that necessarily vitiates human values (e.g., it is important to have persons rather than impersonal forces determining who shall live). Sometimes they are identified with the outcome of the process (e.g., the features such as creativity and fullness of being that make human life what it is are to be considered and respected in the decision). Regarding the former, it must be admitted that the use of chance seems cold and impersonal.

But presumably the defenders of utilitarian criteria in SLMR want to make their application as objective and impersonal as possible so that subjective bias does not determine who shall live.

Such criticism, however, ignores the moral and nonmoral values that might be supported by selection by randomness or chance. A more important criticism is that the procedure that I develop draws the relevant moral context too narrowly. That context, so the argument might run, includes the society and its future and not merely the individual with his illness and claim upon SLMR. But my contention is that the values and principles at work in the narrower context may well take precedence over those operative in the broader context, both because of their weight and significance and because of the weaknesses of selection in terms of social worth. As Paul Freund rightly insists, "The more nearly total is the estimate to be made of an individual, and the more nearly the consequence determines life and death, the more unfit the judgment becomes for human reckoning. . . . Randomness as a moral principle deserves serious study."[18] Serious study would, I think, point toward its implementation in certain conflict situations, primarily because it preserves a significant degree of *personal dignity* by providing *equality* of opportunity. Thus it cannot be dismissed as a "nonrational" and "nonhuman" procedure without an inquiry into the reasons, including human values, which might justify it. Paul Ramsey stresses this point about the Holmes case:

> Instead of fixing our attention upon "gambling" as the solution—with all the frivolous and often corrupt associations the word raises in our minds—we should think rather of equality of opportunity as the ethical substance of the relations of those individuals to one another that might have been guarded and expressed by casting lots.[19]

The individual's personal and transcendent dignity, which on the utilitarian approach would be submerged in his social role and function, can be protected and witnessed to by a recognition of his equal right to be saved. Such a right is best preserved by procedures that establish equality of opportunity. Thus selection by chance more closely approximates the requirements established by human dignity than does utilitarian calculation. It is not infallibly just, but it is preferable to the alternatives of letting all die or saving only those who have the greatest social responsibilities and potential contribution.

This argument can be extended by examining values other than individual dignity and equality of opportunity. Another basic value in the medical sphere is the relationship of trust between physician and patient. Which selection criteria are most in accord with this relationship of trust? Which will maintain, extend, and deepen it? My contention is that selection by randomness or chance is preferable from this standpoint, too.

Trust, which is inextricably bound to respect for human dignity, is an attitude of expectation about another. It is not simply the expectation that another individual will perform a particular act but that the individual will respect him as a person. As Charles Fried writes:

> Although trust has to do with reliance on a disposition of another person, it is reliance on a disposition of a special sort: the disposition to act morally, to deal fairly with others, to live up to one's undertakings, and so on. Thus to trust another is first of all to expect him to accept the principle of morality in his dealings with you, to respect your status as a person, your personality.[20]

This trust cannot be preserved in life-and-death situations when a person expects decisions about him to be made in terms of his social worth, for such decisions violate his status as a person. An applicant rejected on grounds of inadequacy in social value or virtue would have reason for feeling that his "trust" had been betrayed. Indeed, the sense that one is being viewed not as an end in himself but as a means in medical progress or the achievement of a greater social good is incompatible with attitudes and relationships of trust. We recognize this in the billboard that was erected after the first heart transplants: "Drive Carefully. Christiaan Barnard Is Watching You." The relationship of trust between the physician and patient is not only an instrumental value in the sense of being an important factor in the patient's treatment. It is also to be endorsed because of its intrinsic worth as a relationship.

Thus, the related values of individual dignity and trust are best maintained in selection by chance. But other factors also buttress the argument for this approach. Which criteria and procedures would people agree upon? We have to suppose a hypothetical situation in which several agents are going to determine for themselves and their families the criteria and procedures by which they would want to be admitted to and excluded from SLMR if the need arose.[21] We need to assume two restrictions and then ask which set of criteria and procedures would be chosen as the most rational and, indeed, the fairest. The restrictions are these: (1) That agents are *self-interested*. They are interested in their own welfare (and that of members of their families), and this, of course, includes survival. Basically, they are not motivated by altruism. (2) Furthermore, they are ignorant of their own talents, abilities, potential, and probable contribution to the social good. They do not know how they would fare in a competitive situation, for example, the competition for SLMR in terms of social contribution. Under these conditions, which institution would be chosen—letting all die, utilitarian selection, or the use of chance? Which would seem the most rational? the fairest? By which set of criteria would they want to be included in or excluded from the list of those who will be saved? The rational choice in this setting (assuming self-interest and ignorance of one's competitive success) would be random selection or chance, since this alone provides equality of opportunity. A possible response is that one would prefer to take a "risk" and therefore choose the utilitarian approach. But I think not, especially since I added that the participants in this hypothetical situation are choosing for their children as well as for themselves; random selection or chance could be more easily justified to the children. It would make more sense for people who are self-interested but uncertain about their relative contribution to society to elect a set of criteria that would build in equality of opportunity. They would consider selection by chance as relatively just and fair.[22]

An important psychological point supplements earlier arguments for using chance or random selection. The psychological stress and strain among those who are rejected would be greater if the rejection were based on insufficient social worth rather than on chance. Obviously stress and strain cannot be eliminated in these borderline situations, but they would almost certainly be increased by the opprobrium of being judged relatively "unfit" by society's agents using society's values. Nicholas Rescher makes this point very effectively:

> A recourse to chance would doubtless make matters easier for the rejected patient and those who have a specific interest in him. It would surely be quite hard for them to accept his exclusion by relatively mechanical application of objective criteria in whose implementation subjective judgment is involved. But the circumstances of life have conditioned us to accept the workings of chance and to tolerate the element of luck (good or bad): human life is an inherently contingent process. Nobody, after all, has an absolute right to ELT [Exotic Lifesaving Therapy]—but most of us would feel that we have "every bit as much right" to it as anyone else in significantly similar circumstances.[23]

Although it is seldom recognized as such, selection by chance is already in operation in practically every dialysis unit. I am not aware of any unit that has removed some of its patients from kidney machines in order to make room for later applicants who are better qualified in terms of social worth. Furthermore, very few people would recommend it. Indeed, few would even consider removing a person from a kidney machine on the grounds that a person better qualified medically had just applied. In a discussion of the treatment of chronic renal failure by dialysis at the University of Virginia Hospital Renal Unit from November 15, 1965, to November 15, 1966, Dr. Harry S. Abram writes: "Thirteen patients sought treatment but were not considered because the program had reached its limit of nine patients."[24] Thus, in practice and theory, natural chance is accepted, at least within certain limits.

My proposal is that we extend this principle (first come, first served) to determine who among medically acceptable patients shall live, or that we utilize a form of artificial chance such as a lottery or randomness. "First come, first served" would be more feasible than a lottery since applicants would make their claims over a period of time rather than as a group at one time. This procedure would be in accord with at least one principle in our present practices and with our sense of individual dignity, trust, and fairness. Its significance in relation to these values can be underlined by asking how the decision can be justified to the rejected applicant. Of course, one easy way of avoiding this task is to maintain the traditional cloak of secrecy, which works to a great extent because patients are often not aware that they are being considered for SLMR in addition to the usual treatment. But whether public justification is instituted or not is not the significant question; it is rather what reasons for rejection would be most acceptable to the unsuccessful applicant. My contention is that rejection can be accepted more readily if equality of opportunity, fairness, and trust are preserved, and that they are best preserved by selection by randomness or chance.

This proposal has yet another advantage since it would eliminate the need for a committee to examine applicants in terms of their social value. This onerous responsibility can be avoided.

Finally, there is a possible indirect consequence of widespread use of random selection that is interesting to ponder, although I do *not* adduce it as a good reason for adopting random selection. It can be argued, as Professor Mason Willrich of the University of Virginia Law School has suggested, that SLMR cases would practically disappear if these scarce resources were distributed randomly rather than on social worth grounds. Scarcity would no longer be a problem because the holders of economic and political power would make certain that they would not be excluded by a random selection procedure; hence they would help to redirect public priorities or establish private funding so that lifesaving medical treatment would be widely and perhaps universally available.

In the framework that I have delineated, are the decrees of chance to be taken without exception? If we recognize exceptions, would we not open Pandora's box again just after we had succeeded in getting it closed? The direction of my argument has been against any exceptions, and I would defend this as the proper way to go. But let me indicate one possible way of admitting exceptions while at the same time circumscribing them so narrowly that they would be very rare indeed.

An obvious advantage of the utilitarian approach is that occasionally circumstances arise that make it necessary to say that one person is practically indispensable for a society in view of a particular set of problems it faces (e.g., the President when the nation is waging a war for survival). Certainly the argument to this point has stressed that the burden of proof would fall on those who think that the social danger in this instance is so great that they simply cannot abide by the outcome of a lottery or a first come, first served policy. Also, the reason must be negative rather than positive; that is, we depart from chance in this instance not because we want to take advantage of this person's potential contribution to the improvement of our society, but because his or her immediate loss would possibly (even probably) be disastrous. Finally, social value (in the negative sense) should be used as a standard of exception in dialysis only if it would, for example, provide a reason strong enough to warrant removing another person from a kidney machine if all machines were taken. Assuming this strong reluctance to remove anyone once the commitment has been made, we would be willing to put this patient ahead of another applicant for a vacant machine only if we would be willing (in circumstances in which all machines are being used) to vacate a machine by removing someone from it. These restrictions would make an exception almost impossible.

While I do not recommend this procedure of recognizing exceptions, I think that one can defend it while accepting my general thesis about selection by randomness or chance. If it is used, a lay committee (perhaps advisory, perhaps even stronger) would be called upon to deal with the alleged exceptions since the doctors or others would in effect be appealing the outcome of chance (either natural or artificial). This lay committee would determine whether this patient was so indispensable at this time and place that he or she had to be saved even by sacrificing the values preserved by random selection. It would make it quite clear that exception is warranted, if at all, only as the "lesser of two evils." Such a defense would be recognized only rarely, if ever—primarily because chance and randomness preserve so many important moral and nonmoral values in SLMR cases.[25]

POSTSCRIPT (1995)

Background. Twenty-five years after these arguments first appeared in print, I want to reflect personally on the debates about patient selection over the years, the kinds of issues that concerned us, the various approaches we took (particularly the content of the relevant principles and the role of analogical reasoning about cases), and the limitations of several earlier approaches, including my own.[26]

The preceding essay represents my first serious foray into biomedical ethics. During my first year of teaching at the University of Virginia (1968–69), I was asked to lecture and participate in a discussion in a law school course on jurisprudence. The instructor had prepared a case involving the usual cast of exotic characters needing kidney dialysis. Its formulation had no doubt been influenced by Shana Alexander's article in *Life* and by the numerous other discussions that followed in its wake. In my session I drew heavily on analogous cases, particularly the law's handling of lifeboat cases, which had been discussed by Paul Ramsey in his book, *Nine Modern Moralists.*[27] Unbeknownst to me, Ramsey was already working specifically on patient selection, also drawing further on those analogous legal cases.

I thought little more about the problem of patient selection until the next academic year, 1969-70, when I somewhat reluctantly (because I was so busy) agreed to participate in a semester-long seminar involving faculty and students from medicine, law, and the humanities in a thorough exploration of the medical, social, and legal issues raised by artificial and transplanted organs. My seminar presentation focused on patient selection and was entitled (or at least was subsequently entitled) "Who Shall Live When Not All Can Live?" It provoked vigorous debate, particularly because of my argument for the use of a lottery or "first come, first served" in selecting patients for dialysis (and other scarce lifesaving medical resources), following a determination of their medical suitability. I soon submitted the essay to *Soundings.* In accepting it for publication, the editor asked whether I knew anyone who might write a critical response, and I recommended the head of the University of Virginia's renal division, who had offered sharp criticisms of my approach in the seminar.[28] Ramsey's book, *The Patient as Person,* with its major discussion of patient selection, appeared about the same time as my essay, and both largely represented deontological and egalitarian perspectives over against the utilitarian perspective that had dominated Nicholas Rescher's 1969 essay on "The Allocation of Exotic Medical Lifesaving Therapy."[29]

Principles and analogies. Both Ramsey and I paid attention to both principles and analogical reasoning in relation to paradigm cases, which casuists have recently stressed. The lifeboat cases were very important resources for analogical reasoning because they involved a conflict of life with life over a scarce lifesaving (though nonmedical) resource. As I put it later in my entry on "Rationing of Medical Treatment," in the *Encyclopedia of Bioethics,* most analyses

> have found the closest analogy in dramatic cases of killing survivors to eat
> their flesh or jettisoning fellow passengers after shipwrecks in order to save
> some lives when not all can survive. Nevertheless, there are several dissimi-
> larities: those involved in shipwrecks are deciding for themselves as well as
> for others, and often they must kill rather than merely let die. But according
> to Ramsey, there is no difference between these cases as far as the selection
> process is concerned.[30]

For centuries philosophers and religious authorities have debated such cases of microallocation or rationing. For instance, ancient Stoic philosophers, Jewish rabbis, and Christian theologians all debated how to distribute scarce lifesaving resources among different individuals. To take one example, there was debate about what ought to be done when two survivors of a shipwreck simultaneously reach a floating plank that can support only one of them. Some Stoics held that "one should give place to the other, but that other must be the one whose life is more valuable either for his own sake or for that of his country." If these considerations are of equal weight, "one will give place to the other, as if the point were decided by lot or at a game of odd and even."[31] In general, Christians tended to emphasize self-sacrifice out of love in such situations. According to Saint Ambrose, "Some ask whether a wise man ought in case of a shipwreck to take away a plank from an ignorant sailor? Although it seems better for the common good that a wise man rather than a fool should escape from shipwreck, yet I do not think that a Christian, a just and a wise man, ought to save his own life by the death of another, just as when he meets with an armed robber he cannot return his blows, lest in defending his life he should stain his love toward his neighbor."[32] Ambrose did, however, recognize the legitimacy of using force to defend innocent third parties. Several rabbis debated what should be done when two men who are on a journey through a desert have only one pitcher of water, enough to sustain one man but not both. One line of interpretation held that both should drink and die, but Rabbi Akiba held that "your life takes precedence over his life." In general, the Jewish tradition authorizes greater attention to self-preservation than some radical strands of the Christian tradition. Some Jewish texts also recognize *utilitas* ("utility"), but only when a person is choosing between two other human beings.[33]

These different positions suggest the variety in philosophical and theological arguments and conclusions about distributing lifesaving resources. In addition, the Hippocratic tradition offers little helpful guidance, because it focuses primarily on the individual patient, with little attention to possible conflicts of responsibilities. It directs physicians to benefit their patients, but fails to advise them on what to do when they cannot benefit all their patients who have equal needs.

Despite appeals to the lifeboat cases, it is important to note that legal opinions are not unanimous, and that, indeed, *Holmes*'s judge stands alone in several respects. First, as part of the background, nineteenth-century sailors had a fatalistic attitude toward shipwreck, and it was considered impossible to carry enough lifeboats to rescue all the passengers in case of a shipwreck. As a result, "when a passenger ship sank at sea, it was inevitable that some must be selected, either by circumstances or deliberate action, to die."[34] Usually, the crew took priority over women and children and male passengers who came last. For instance, in the sinking of the *William Brown*, at least thirty passengers were left on board, screaming for rescue, while the remaining thirty-five or so passengers and seventeen crew made it to the two lifeboats. Among the different stories of what happened, the mate said that he thought, in view of the danger from the leakage, it would be appropriate to throw overboard "those who were nearly dead." While some passengers later agreed that those thrown overboard were "nearly dead," at least one recent analyst of the case concludes that this was probably not accurate. Then all who remained on the longboat were rescued shortly after the last two men had been thrown overboard. Their sacrifice probably had no impact on overall survival. Finally, passengers were not told what was happening or given any chance to determine the procedure. (I'll return to the question of consent later.)

Some legal commentators note that while *Holmes* is a "leading case," it is not the "central case" in terms of precedent. I quote one commentary:

> The case is notable too as the only one in Anglo-American law that explicitly accepts the existence of a defense of necessity of homicide and as

the only case that explicitly accepts the priority, in appropriate circumstances, of selecting victims by lot. But the desultory judicial opinion affirming the charge, together with the inconclusive presentation of the problem as essentially a matter of jury decision rather than legal determination, deprived *U.S. v. Holmes* of the chance to become the central case.[35]

The legal history indicates that we cannot simply assume a consensus about lifeboat cases, a consensus that would mark either casting lots or making utilitarian judgments as paradigmatic. Rather we have to argue for them, and part of the argument will hinge on fundamental principles and values. The same holds for ethical assessments too.

Utilitarian and egalitarian principles. When I wrote "Who Shall Live?" I did not believe, and do not now believe, that we are forced to choose between principled reasoning and analogical reasoning about cases. The debate about principles in microallocation focused on utilitarian and egalitarian principles, both of which appeared in some practical decisions. Empirical studies of the allocation of kidney dialysis in the late 1960s identified the most important criterion in actual practice as the likelihood of success defined in medical and social terms (including job rehabilitation). According to two researchers, "Such factors as the congeniality of the patient as an individual, economic burdens of dependents if the patient wasn't selected, 'demonstrated social worth,' 'future social contribution,' . . . were considered of minor importance by a majority of centers, although one-fifth to one-third rated them important." Whatever institutions reported about their operative criteria, selected patients were predominantly male: 66 percent compared to 34 percent female (even when Veterans Administration hospitals were excluded). They were also predominantly white and high-school educated. These percentages were far out of line with the normal incidence of end-stage renal disease among these different groups and suggest that social-worth standards were actually operative. Several centers avoided social comparisons by using the criterion of "first come, first treated," and at least one center used a lottery.[36]

Among the arguments for utilitarian selection criteria, some focused on society's right to look for a return on its investment in the scarce lifesaving medical resource through patients' future contributions. Others focused on the rationality of the utilitarian approach in contrast to that of a lottery or "first come, first treated," which, utilitarian Joseph Fletcher argued, refuses to be rational and reduces us to the level of "things and blind chance."[37]

Arguments for the use of chance in final selection stressed the difficulty of determining substantive values for judgments of social worth, appealed to broad conceptions of rationality, and held that selection by chance can express and realize important individual and societal values. Both Ramsey and I sought to show that utilitarian selection criteria are difficult to determine and apply, particularly in a pluralistic society that lacks accepted criteria of social worth. It was also easy to criticize the values that emerged in actual judgments of social worth (as noted in the original essay).

The main positive reason for use of a lottery or "first come, first served" appealed to equal respect (what I am calling the egalitarian perspective). This principle involves equal respect for personal dignity through the avoidance of (at least total) comparative social judgments and the provision of equality of opportunity, apart from socially valued roles, for access to lifesaving medical resources. This perspective dominated most deontological alternatives to utilitarian criteria. In addition, its focus on symbolic, rather than merely instrumental, rationality is striking: It stresses that allocation procedures express and symbolize certain values, such as equal respect, and that value-expressive actions and policies are rationally defensible.

Paul Ramsey added a specifically theological analogy: "In allocating sparse medical resources among equally needy persons, an extension of God's indiscriminate care into human affairs requires random selection and forbids godlike judgments that one man is worth more than another."[38] (Some of these various reasons as well as reasons not then raised were later developed more systematically in excellent books by Gerald R. Winslow, *Triage and Justice* (1982), and John F. Kilner, *Who Lives? Who Dies? Ethical Criteria in Patient Selection* (1990).)[39]

Most of the positions I have identified recognized the possibility of legitimate exceptions to their primary criteria. According to utilitarian Rescher, if there are no "major disparities" in terms of utility—both medical and social—in the second stage of selection, then final selection could justifiably be random. He did not, however, attend to the independent values expressed and realized in random selection. By contrast, from an egalitarian perspective, I expressed my general opposition to exceptions to a lottery or to queuing. However, I also argued that if society found it necessary to make some exceptions on social utilitarian grounds, it should set a heavy burden of proof for an exception for an arguably desirable patient. Finally, Ramsey's exceptions appealed to the notion of a focused community, exemplified in triage after disaster as well as the lifeboat cases, as will be discussed in detail in the next chapter.

Major shifts. How has my mind changed since those early discussions of patient selection? I now believe that I oversimplified utilitarian selection, for instance, when I viewed the major alternatives at the stage of final selection as social-worth criteria, based on utility, and random selection, based on egalitarian premises. In particular, I came to appreciate the value and necessity of judgments of *medical utility,* as distinguished from *social utility,* in final selection.

More specifically, when I turned to the selection of recipients of organs for transplantation—in contrast to artificial organs, such as the kidney machine—I began to see a stronger role for what I would call judgments of medical utility—that is, not only patient need but also the probability of successful treatment, especially because organs that are rejected cannot be reused and thus are wasted. Furthermore, *medical utility* does not appear to me to be at odds with egalitarian concerns in the same way as *social utility.* And I reject proposals to count as egalitarian only those positions that focus exclusively on patient need without regard to probability of success. In essays that follow, I continue this argument for a role for medical utility in the context of triage and the allocation of transplantable organs. Furthermore, I now distinguish within social utility between broad and narrow conceptions, which will also be developed in the context of triage in the next chapter.

Another limitation of many early analyses was the relative lack of attention to the background conditions that determine the initial pool. Of course, we recognized that some medical and psychiatric criteria of inclusion/exclusion embodied unarticulated and undefended social judgments. We did not, however, pay enough attention to the often unjust social conditions, including unequal access to basic health care, that determine whether and when someone is referred for some forms of technological care. These background conditions are critical for determining the justice of particular microallocation procedures, such as "first come, first served." (See chapter 13.)

Furthermore, several of us concentrated on microallocation decisions of patient selection with only limited attention to macroallocation decisions. In an extreme statement, Paul Ramsey even insisted that macroallocation decisions are "almost, if not altogether, incorrigible to moral reasoning."[40] Thus, ethicists could not contribute anything significant to the societal debate about choosing among various social needs, including medical care, and choosing between preventive and rescue medicine. For Ramsey, these are all matters for the political process. Several ethicists soon rightly challenged the relative neglect of social ethics, including broad questions of access to health care.[41]

In practice, society found it difficult to accept the explicit determination of who would live and who would die according to ability to pay and other criteria of social worth, and thus decided through its representatives to eliminate the need for much patient selection by increasing the supply of dialysis by providing funds for virtually everyone with end-stage kidney disease. One consequentialist reason sometimes given for a lottery is that its bare equality will lead the wealthy and powerful to act to reduce scarcity and thereby reduce their chances of losing at a lottery—in my original essay I treated this as a side effect of, rather than a consequentialist reason for, a lottery. Indeed, based on our society's experience, "tragic choices" by whatever method—whether social utilitarian selection or random selection—may lead to efforts to avoid them. Still there are limits to the impact of microallocation decisions (who will receive a particular scarce good) on macroallocation decisions (how much of a particular good should be made available), as in the nineteenth-century acceptance of some shipwreck fatalities because of "tragic choices" necessitated by the inadequate supply of lifeboats.

Yet another limitation of many of our analyses appeared in our tendency to focus on life-and-death decisions—as in my question, "Who Shall Live When Not All Can Live?"—with less attention to more routine or less life-threatening decisions.

Response to recent criticism. Some objections to random selection confuse assessments of a lottery as a method (whether broadly or narrowly conceived) with assessments of the reasons for using it. Recall Shirley Jackson's powerful and haunting short story, "The Lottery," a fictional account of a town that gathers annually to draw lots to determine who will be stoned.[42] The death lottery was fair—it doesn't appear to have been rigged or otherwise unfairly conducted, even though the victim complained that her husband had to draw too fast—but we must ask why it was adopted and repeated again and again. The lack of a clear answer creates much of the horror of Jackson's story, for the community had largely forgotten why it conducted the death lottery every year, if it ever had a plausible rationale. Even if a lottery is fair, it may not be fair to use a lottery in particular circumstances.

Sometimes, however, the very blindness of the lottery is taken to represent arbitrariness, rather than impartiality, and hence, whatever its rationale, the lottery is rejected or at least considered to be inferior to some other approaches. Such an argument appears in a recent major criticism of "Who Shall Live?" Ralph P. Forsberg so carefully distinguishes the mechanism—the lottery— along with the outcomes it produces from its justification that the lottery can only be rejected as less than rational or even as nonrational and irrational.[43] He defines rationality broadly as "that which would be accepted by an impartial, objective observer as a justification for a position or conclusion" (p. 27). Such an observer need not find that justification the most cogent, but, in order to be rational, it must be "plausible" and "understandable." The impartial observer takes an unbiased perspective, that is, is disinterested in the argument's practical outcome, and is able to recognize nonrational arguments or positions by using such common logical tests as coherence and implausibility. Then Forsberg distinguishes *systematic reasons* from *particular reasons:* Systematic reasons are those given to defend the use of a particular system, while particular reasons are justifications "for a specific choice or decision reached regardless of the systematic procedure used to make it" (p. 28).

Starting from my argument that random selection "cannot be dismissed as a 'nonrational' and 'nonhuman' procedure without an inquiry into the reasons, including human values, which might justify it" (p. 176 above), Forsberg holds that, while both random selection and social-worth judgments are rational on the level of systematic reasons, random selection's outcomes are sometimes (at least once he writes "often") nonrational, unreasonable, and irrational. Or, stated differently, judgments of social worth "are more thoroughly rational" or, as he puts it at one point, "doubly rational," because they are "both systemati- cally and particularly rational" (p. 26). (While Forsberg only discusses social- worth judgments, as an alternative to random selection, he does not argue that social-worth judgments are rationally preferable to all other alternatives.) Everything finally hinges on his claim that it is logically indefensible to evaluate particular choices only by systematic choices—that is, to say that a particular choice is rational because of a rational system. It is not adequate to hold that particular choices derive their rationality from the rationality of the system that produces them: "Since we have conceded that both *systems* [random selection and social worth judgments] are rationally defensible in this sense, this does not weigh in favor of one or the other of the alternatives. At this point, then, it is the rationality of the particular choices that will determine which is the more reasonable procedure overall" (p. 30).

Forsberg also claims that there is an inconsistency between the values appealed to for random selection and those realized in particular cases. Several of his charges of inconsistency in my position appear to slip in particular substantive definitions, sometimes beyond the value itself and sometimes in a different sense than my proposal, even though he does not argue for or even acknowledge these definitions. He claims, for instance, that random selection can lead to ignoring "a basic tenet of equal treatment, namely, to treat relevant differences differently, by eliminating accomplishment or character from consideration as criteria. Eliminating such considerations is inconsistent with using them as justification for the very system that discards them" (pp. 30–31). His claim has the ring of plausibility only because of the different senses of equal treatment. Nothing in the language or norm of equal treatment requires that differences in accomplishment or character be accepted as "relevant differences" for allocating scarce lifesaving medical resources (or anything else for that matter). We may hold that they are relevant in the allocation of some goods but not others. And there is nothing inconsistent in drawing such distinctions. The argument for the relevance of particular characteristics in allocation must be made independently; it does not flow from the norm of equal treatment.

Forsberg claims that the procedure of random selection is irrational—provides no basis for the particular choice—because of at least three factors: the "arbitrariness" of the dice toss, the "denigration of individual worth implied by the use of such arbitrary means to determine who lives or dies," and the "inconsistency" of arguing for randomness as a way to preserve individual worth and then using the dice to toss that dignity away (p. 30). Two of these three points hinge on claims about individual dignity. Forsberg finds an inconsistency in my defense of a lottery by appeal to dignity mainly because he employs a different conception of dignity. My essay focused on our dignity qua human beings or persons, without regard to the dignity of our individual projects and the like, all of which he stresses as establishing our individual identities (p. 35). While there may be an inconsistency between *my* defense of random selection and *his* conception of dignity, there is no inconsistency between my defense of random selection and my conception of dignity. Furthermore, the relevance of individual projects, etc., to particular allocation choices requires defense beyond an appeal to formal rationality.

Forsberg insists that it is not rational to offer "his number came up" to justify a specific allocative decision. It is irrational to allow anyone to die on the basis of a throw of the dice: "No one deserves to die for such a trivial reason" (p. 33). However, I did not claim that anyone deserves to die. Instead, it is in part because such meritarian claims of desert, based on past or projected future contributions, are essentially contestable that we resort to random selection in the allocation of SLMR. He seems to assume, by contrast, that we have the necessary consensus about desert for particular choices about who might "deserve to be treated" and what might constitute their "just desert."

Forsberg attempts, unsuccessfully, I have suggested, to avoid morality in favor of formal rationality, except perhaps at the level of the system, which for him is not decisive. His attempt in this regard creates major problems in his interpretation and criticism of my position. First, he proposes that we look at the rationality of the particular choice apart from "the moral decision-making procedure" (p. 29). However, I do not suppose that random selection is a "*moral* decision-making procedure," as distinguished from others that are *nonmoral*; however, I do hold that it can be *morally justified* in some contexts. Nevertheless, Forsberg sometimes (not always) seems to put random selection on the level of moral systems, noting that it is "only one of many possible moral systems" (p. 32; contrast p. 33, where he uses the language of "morally defensible"). From my standpoint, both utilitarian and egalitarian moral reasons provide justifications for a system of random selection in some contexts.

Second, since Forsberg restricts morality to the systematic level, he is left with particular reasons that are nonmoral. Such a restriction might appear to offer the possibility of greater consensus, but that hope is not well-founded. Forsberg has to concede that the very categories he uses for particular choices, such as benefit, waste, etc., "still involve evaluation, but not moral judgment" (pp. 32–33). However, even these *nonmoral evaluative judgments* still presuppose substantial consensus about values at the level of particular choice. Yet I have argued that in a modern, pluralistic society we lack that consensus about values for use in judgments of social utility and thus must be suspicious of social-worth judgments that, at least in most cases, the allocation system must be designed to resolve.

"Who Shall Live?" rests on a different model of the cases the system of allocation will handle than Forsberg supposes. The basic question is whether the system should be designed for "hard" or "easy" cases, then which side should bear the burden of proof, and finally, how heavy should the burden be for those who argue for exceptions? For example, according to both societal values and philosophically defensible principles, there may be widespread agreement that a system of allocation should prefer former Surgeon General C. Everett Koop over "a hapless hobo who just drifts from place to place, holds no job, nor ever held other than menial jobs that were necessary for his immediate survival" (Forsberg's example, p. 31). If this easy case—or another easy case, such as a virtuous first-grade school teacher versus a convicted mass murderer—is taken as the model, the system is likely to stress social worth and place the burden of proof on exceptional procedures, such as the use of chance, when there are no substantial differences in social worth between the applicants. Thus, although Rescher included retrospective and prospective social worth in his allocation system, he indicated that the final selection should be made by random selection where there are "no really major disparities" within the pool of candidates.

If, on the other hand, "hard" cases are the model for the system, random selection may be rationally preferred. For example, apart from a social crisis following a disaster, where, in contrast to individual crises, there may be a "focused community" (Ramsey's language) with survival of the community as a goal, the microallocation system must choose between various people, most of whom are apparently dispensable from a functional standpoint and roughly comparable in past contributions and future promise. In such circumstances, how should the choice be made? By what criteria? By what certitude of social needs and a particular candidate's indispensability? If such "unfocused" settings and "hard" choices are normal, then social-worth judgments may seem inappropriate, and even impossible, and random selection even more plausible.

Forsberg does not adequately attend to the wide range of allocations, sometimes involving life and death, that are in fact and often quite reasonably based on casting lots, queuing, etc. He asks: "In anything other than a medical context would we be willing to allow a death based on a throw of the dice? It would seem not, unless the situation itself were irrational and thus not subject to rational analysis . . ." (p. 33). Two points are appropriate in response. First, more careful attention to real, even if not everyday, situations of choice might have led to a different perception. Consider, for instance, the solid justification for the use of random selection in determining who will serve in the military and who will thus bear substantially greater risks of death than others who have better numbers. Second, even tragic situations may be subject to rational (and moral) analysis, which can lead us to conclude that an impersonal, impartial system of random selection would be both rationally and morally justified.

Finally, the author attempts to construct from "Who Shall Live?" objections to his position, then offers his own rejoinders to them. I won't deal with all six—most hinge on what has already been discussed—but a few points are necessary. He seems to suppose that my use of the Rawlsian choice model commits me to holding that "selfishness is always rational." Nothing at all suggests that—indeed, it may be quite rational for an individual to yield his or her slot even if the random procedure assigns the medical good to him or her. Second, at a couple of points he dismisses some of my appeals as based on "psychological or emotional factors rather than rational reasons" (p. 38). But he fails to see how a social system of allocation in circumstances of tragic scarcity must be feasible, as well as (formally) rational (and, I would add, morally justifiable). Feasibility hinges in part on "psychological and emotional factors," which also play a role in public justification, which cannot, I would argue, be reduced to formal rationality.

As important as it is, formal rationality can take us only so far; under its cover, Forsberg, as I have shown, smuggles in substantive moral and nonmoral evaluative standards, without ever arguing for them. I have also shown that Forsberg's appeals to formal rationality to show inconsistencies in my position are mistaken. Furthermore, when we have to justify to each other our particular and general moral and nonmoral judgments for purposes of deciding how to allocate SLMR, formal rationality, while important, is not all-important. The psychological and emotional factors that Forsberg dismisses are also critical in the consent of various groups to the use of a lottery for allocation within those groups.

Group consent and recent examples. With his focus on formal rationality, Forsberg does not consider the hypothetical or actual consent of individuals to particular systems of allocation. He cannot admit consent to random selection as a procedure of allocation of SLMR as a sufficient reason for following it in particular circumstances because under virtually any other procedure there would be, in his view, some rational reason for the particular choices, beyond the systematic reasons. That a group, such as the passengers on the *William Brown* (who in fact were not informed and did not consent), actually consented to allocation by lots could not, for him, be a rational reason for the particular allocation that follows.

It has been suggested, however, that individuals in need or at risk will consent to egalitarian criteria, but will rarely if ever consent to (social) utilitarian criteria. While that suggestion is overstated and needs qualification, as I will argue in the next chapter, consent is important in the public justification of lotteries to allocate resources, risks, etc. As Barbara Goodwin observes, "quite unphilosophical human beings have resorted to the lottery to make tragic choices in the context of natural or man-made disasters for a long time. When a group of people must distribute some unavoidable evil between themselves, they will almost instinctively choose a lottery as the method even if in the early stages of the discussion each is concerned to assert why she herself should not receive the evil allocation."[44] She goes on to argue that the optimal context for the lottery, if it is to be just and viewed as just, is a "consensual group," particularly one with a common purpose. Once there is a common purpose of a group, what Ramsey calls a "focused community," it may be possible to make some allocative decisions on the basis of (limited) social utilitarian criteria (e.g., who is needed for the survival of the group, say, on the lifeboat). If, however, the common purpose is only that each individual in the group wants to survive—for example, patients awaiting an organ transplant or patients in the intensive care unit—a lottery or "first come, first served," which Ramsey calls an "ongoing lottery," may express equal respect, impartial treatment, and fairness in the competition, if matters cannot be resolved in terms of medical utility (to be developed in the next chapter). Throughout, I have stressed the social impasse in an unfocused, pluralistic society about other (social utilitarian) allocation criteria, as well as the fundamental values expressed and symbolized by a lottery. In some tragic circumstances people view the lottery as an acceptable way to allocate goods and burdens because of the limits of "mindful" choice, the blindness or impartiality of the lottery, and "the moral judgment that people should be treated as absolutely equal where basic life chances (chances of life or survival) are involved."[45]

Over the last few years lotteries have been used several times to determine who would gain access to new drugs that were available only in limited supply, either because they had only recently been approved or because they were still experimental but promising. And in some of these cases, the candidates for the scarce drugs either proposed or agreed to a lottery, within certain limits set by medical utility. Some of the lotteries involved AIDS patients, who wanted access to a promising new class of compounds (protease inhibitors). Since the demand for these drugs was much greater than the drug companies could handle (because the process of production is complicated and patients need large doses), lotteries were used. For instance, when Hoffman-LaRoche decided to offer Inviarase to patients with advanced AIDS (they must have CD4 counts less than 300 and be unable to benefit further from available drugs) outside current clinical trials, Dr. Alberto Avendano of the National Association of People with AIDS proposed a lottery, and both the drug manufacturer and the Food and Drug Administration agreed. (Sixty percent of the first 2,000 slots were set aside for patients with CD4 counts less than 50.) As Dr. Avendano stressed, "The lottery seemed the closest to the most fair way." In the lotteries by all the drug companies involved in the summer of 1994, over 18,500 patients competed for about 3,600 slots, which were expected to be almost doubled by the end of 1994. Although some argued that patients with AIDS who had participated in clinical trials should have priority, others stressed the symbolic value of the lotteries: "Lotteries say that after you meet medical criteria, all persons should have an equal shot at the good of society. Lotteries celebrate an understanding that all humans are endowed with equal dignity."[46]

Earlier, in the fall of 1993, Berlex Laboratories held a lottery to distribute Betaseron, a new genetically engineered drug that appears to slow the deterioration brought on by multiple sclerosis (MS). Sixty-seven thousand patients with MS entered the lottery, which was limited to patients whose doctors certified that they had the relapsing-remitting form of the disease and could still walk at least 100 yards without assistance (i.e., patients whose disease and symptoms matched the group for which it had proven effective in clinical trials). The company thought that it would take over a year to reach number 40,000 on the list. Regarding the method of allocation, Dr. Jeffrey Latts, Berlex's vice president for clinical research and development, said: "We talked to patient groups and doctors and distribution experts and the lottery seemed the fairest way. The patients were generally very supportive of the idea. Some of the doctors were less so, because they thought they should be able to choose which patients to put before which others." "Most of my patients are accepting the wait [for their numbers to come up] pretty well," observed another physician caring for MS patients, "because there is an element of fairness in a lottery."[47]

Whether a lottery, in a strict sense, or "first come, first served" as a natural form of random selection is preferable depends in part on which is more feasible in the circumstances. A lottery was more feasible in the allocation of these new drugs, in part because there was no prior waiting list, unless one could determine, say, when a particular medical condition developed. In other contexts, queuing may be preferable, for instance, in the allocation of organs for transplantation or space in an intensive care unit, apart from a community disaster when large numbers arrive all at once. The next chapter focuses on intensive care, and the following chapter on organ transplantation.

CHAPTER ELEVEN

□

Triage in Neonatal Intensive Care

THE POSSIBILITIES AND LIMITATIONS OF A METAPHOR

INTRODUCTION

Much of the moral content of medicine can be found in the notion of *care*. Our language suggests this: We talk about "medical care," about "health care professionals," and about "intensive care." At the very least, "care" means *responding to needs*. This response includes both attitudes and actions. Thus, it may be useful to distinguish "caring for" and "caring about" from "taking care of." "Caring for" and "caring about" refer more to attitudes, while "taking care of" refers more to conduct. For example, I may "take care of" my neighbor's cat while she is on vacation without "caring for" either my neighbor or her cat; I may be merely repaying a favor. Of course, actions of care may express attitudes of care, but "taking care of" others also involves skills. In the parable of the Good Samaritan, the Samaritan and the innkeeper are depicted (in the language of the Revised Standard Version) as "taking care of" the man who had been beaten by robbers and left for dead; their actions included feeding him and binding his wounds. Because they were not health care professionals, their skills may have been inadequate to meet the man's needs. However, in rendering care, they would have been expected to be careful, to avoid carelessness. This standard of care applies to everyone, professional or not. However, health care professionals, who present themselves as having certain knowledge and skills, are expected to render "due care," currently defined as that degree of care exercised by ordinary professionals in the community.[1] I will concentrate on "taking care of" rather than on "caring for" or "due care," though elements of both inevitably enter the discussion.[2]

Rendering care may take different forms, such as "ordinary care," "routine care," "extraordinary care," "palliative care," and "intensive care." Our focus is on "intensive care," that is, care that involves "the use of highly technological diagnostic and therapeutic techniques and sophisticated life-support systems."[3] "Neonatal intensive care" (NIC) covers the period from live birth through twenty-eight days of life. Prior to 1960, care of newborns was shared by several groups, but since the mid–1960s neonatology has developed as a subspecialty in pediatrics, and "neonatal intensive care units" (NICUs) have emerged. The *level* of neonatal intensive care varies greatly. In official classifications, there are three levels, depending on the intensity of care that can be provided by the available staff and equipment. Level I hospitals provide minimal newborn care for patients who are normal or have minor complications. Level II hospitals can provide intermediate care, which goes beyond Level I to treat certain types of neonatal illnesses such as mild respiratory distress syndrome (RDS). Such hospitals offer more sophisticated technologies and more specialized medical personnel. NICUs at level III hospitals provide care at all levels of intensity. Such units have the

> capacity for continuous and constant long-term intensive care and immediate availability of subspecialty consultants in fields such as cardiology and surgery. Level III services include continuous cardiopulmonary support and capability to treat those infants requiring long-term intravenous therapy, hyperalimentation, major surgery, and treatment of sepsis (wide-spread infection).[4]

Level III units also serve as referral centers for regions, arrange transport, etc.

I am interested in certain moral dilemmas that emerge in NICUs, even though they are not peculiar to NICUs. Technological developments, such as sophisticated diagnostic and therapeutic technologies, do not *create* these conflicts. While "neonatology has always had its ethical problems," as pediatrician Clement Smith notes,[5] its problems may nevertheless assume a different shape, present different nuances, become more frequent or common, or acquire a new urgency because of technological developments such as the equipment and skills that enable NICUs to salvage newborns who would have died just a few years ago. Examples are newborns who suffer from prematurity in general, from respiratory distress syndrome (RDS) in particular, or from such congenital anomalies as spina bifida and heart disease. (RDS is the cause of nearly 20 percent of all neonatal deaths, while congenital anomalies result in approximately 20 percent of all infant deaths.)[6] I want to explore some dilemmas in the distribution and allocation of intensive care to newborns.

METAPHORS AND ANALOGIES

The language we use structures, in part, the problems or dilemmas we encounter in NICUs and elsewhere. For example, the common language of "defective newborn" may be misleading because it seems to suggest a "defective product" that can and perhaps should be recalled. It may be more appropriate to use the language of "seriously ill newborn." Much of our language in this area is largely and fundamentally metaphorical. Prominent metaphors for hard and tragic choices in ICUs include allocation, rationing, and triage, the last two of which clearly reflect their development in wartime.

Other metaphors include "the commons"[7] (proposed by Garrett Hardin and adopted by Howard Hiatt and Raymond Duff) and the "lifeboat",[8] which has had a long history of discussion in allocation of health care. A lifeboat or ecological metaphor appears in our language of the "optimal carrying capacity" of an ICU.

Whether any metaphor, such as triage, is *adequate,* as I have suggested earlier in this volume, will depend in part on what it helps us see, that is, on whether it illuminates or distorts the reality of NIC and its distribution. However, metaphors *guide* us as well, and the adequacy of any metaphor to guide us in formulating policies for distributing lifesaving medical care under conditions of scarcity will depend, in part, on the *moral principles* it highlights and hides. In this discussion, I will emphasize the principles of utility—the greatest good for the greatest number—and equality—regarding each newborn as an equal. I will not argue for these principles here; rather I will assume them in order to test their implications for distribution of NIC.

TRIAGE: THE EVOLUTION OF A METAPHOR

Robert J. T. Joy contends that "triage is a technical term applied to situations of military or civilian catastrophe," and that it "is carried over into other situations, where it may become a metaphor for social, economic, or political decisions."[9] The use of "triage" as a metaphor has a long and complicated history.[10] In a sense, "triage" as applied to military or civilian catastrophe is already a metaphor, for the process of treating casualties in war or in disasters has been viewed as a process of sorting or grading, analogous to the classification of inanimate objects according to their quality. "Triage," a French word meaning sorting, picking, grading, or selecting according to quality, was first applied in English (as early as 1717) to the separation of wool according to its quality and later to the separation of coffee beans into three classes: "best quality," "middling," and "triage" coffee. The last class, consisting of bad or broken beans, was the lowest grade.

Although the sorting of casualties in war was developed by the French under Napoleon, it was not called "triage" until later. In contrast to earlier nonmedical usage, "triage" in war implied assigning priority to the worst off (within limits), rather than the best off. The systematic classification of wounded soldiers for purposes of treatment was not adopted in the Civil War in the United States; for the most part, treatment was provided in turn without regard to condition. In World War I, the United States army adopted the term "triage" and the idea of a sorting station from the French and the British armies; by World War II, the term "sorting" was more widespread than the term "triage." The current statement of U.S. military policy is that triage/sorting

> implies the evaluation and classification of casualties for purposes of treatment and evacuation. *It is based on the principle of accomplishing the greatest good for the greatest number of wounded and injured men* in the special circumstances of warfare at a particular time. The decision which must be made concerns the need for resuscitation, the need for emergency surgery and the futility of surgery because of the intrinsic lethality of the wound. Sorting also involves the establishment of priorities for treatment and evacuation.[11]

Similar formal policies have been adopted for civilian disasters, such as nuclear destruction and earthquakes. However, as we will see later, there are important variations.

The great increase in the use of emergency rooms after World War II led to the institution of triage systems in hospitals in the early sixties:

> Modern civilian triage involves more than identifying the most urgent and most salvageable cases for evacuation from a site where casualties are massed. Rather, it signifies an integral process whereby emergency patients are introduced into the system of emergency care. Good triage is the initial lubricant for an orderly flow of patients through the emergency department. . . .[12]

In such systems, a triage officer quickly assesses patients who have needs as "immediate" (posing a threat of death or serious physical impairment if not treated immediately), "urgent" (requiring prompt but not immediate treatment), and "nonurgent." A more complex, five-category system might include the following designations: "life-threatening," "urgent," "semiurgent," "nonurgent," and "no need for care."[13]

In all three settings—military conflicts, civilian disasters, and emergency rooms—triage became more important as medical care improved. As medicine could do more and more to meet serious human needs effectively, the importance of the system of distribution increased. Demands for equal access to medical care are not pressed unless medical care is *perceived* as important for human needs.

This history suggests that "triage" refers to particular systems of allocation; hence, the term "triage" should not be used as equivalent to "distribution," or "allocation," or "rationing" in general. It is a system that sorts or grades patients according to their needs, with some indication of the probable outcomes of interventions; in its more formal developments, it classifies patients according to set categories. Furthermore, as applied to medical care, it often represents allocation or rationing under crisis or emergency circumstances, where a decision has to be made immediately about particular patients because some of them probably have life-threatening conditions.

PRINCIPLES OF UTILITY: THEIR RANGE AND SCOPE

Systems of triage, whether informal or formal, have all had an implicit or explicit *utilitarian* rationale. That is, they have all been designed to produce the greatest good for the greatest number, to serve the common good, or to meet human needs most effectively and efficiently under conditions of scarcity.[14] Such triage or sorting is required because scarcity prevents health care providers from meeting all of the needs of all of the patients at the same time and to the same degree.

A utilitarian rationale or justification for a system of triage need not be perverse. First, utility is one important principle alongside several others; it merits serious consideration at every point. There is a moral duty to produce the greatest good for the greatest number, subject to other moral limits and constraints.

Second, viewing the principle of utility as the rationale for a system of triage need not have the implications that some critics immediately suspect. Some critics apparently overlook the distinction between rule- or system-utilitarianism and act-utilitarianism.[15] In the former, the principle of utility is not applied to acts directly but to a rule or a system (i.e., a set of rules and procedures); it may, for example, justify some system over none, several systems over a few, or a particular system over all others. In any case, the rules and procedures of the system of allocation are selected because they, on the whole, would probably produce the greatest good for the greatest number, considering various needs and desires. Such an application of the principle of utility may also incorporate symbolic or expressive values. For example, people may be as interested in what a system symbolizes about their equality as in the physical needs it meets. A system may thus be structured to *express* certain values, such as equality, as well as to *realize* certain ends, such as survival.

Third, a system of triage justified by utilitarian considerations, broadly conceived, may not incorporate some explicitly utilitarian criteria, especially of the sort usually associated with act-utilitarianism. For example, a system of triage may not give priority to the most productive or most valuable persons. It may simply use technical, medical criteria to determine which patients have the most urgent needs that can probably be met by the available resources.

Finally, at any level, utilitarianism has a strong egalitarian strain insofar as it insists that each one is to count as one and only one.

A fundamental variation in systems of triage, particularly in the categories of classification they adopt, stems from two distinct applications of the principle of utility. These two applications differ in the identification of the group to which the principle of utility is applied. In the current statement of U.S. military policy quoted earlier, the principle of utility is limited to "wounded and injured men" (presumably "people," without regard to gender). They are the ones who are counted in the maximization of good; their needs and interests are relevant and decisive. From this perspective of utility—which I will call "medical utility" in contrast to "social utility"—a system of triage is primarily *medical* in that it focuses on medical needs and medical salvageability.[16]

Historically, medical utility has been the major impetus for and the major determinant of systems of triage, but it is not accurate to suggest, as Stuart W. Hinds does, that medical triage always "has meant the allocation of priority according to need, and on no other grounds, for goods and services in short supply."[17] Not only has prospect of benefit usually been considered along with need, but there is also another strand in the interpretation and implementation of triage. In this other strand, the principle of utility includes the military and even the whole society, not only the wounded and injured. In the famous example of the distribution of penicillin during World War II, this scarce resource was distributed to soldiers who had been wounded in the brothels and had contracted venereal disease rather than to soldiers who had been wounded and injured on the battlefields.[18] The explicit rationale was that the soldiers suffering from venereal disease could be returned to military action more quickly and effectively than wounded and injured soldiers. Such an allocation presumably could not have been justified by medical utility, since some of the wounded and injured doubtless died from infections that could have been prevented or cured by penicillin; however, it could have been justified by social utility because it was important to win the war as quickly as possible. Such a decision clearly expands the area of the greatest good for the greatest number.

A similar expansion occurs in civilian disasters, such as plans in San Francisco to cope with a devastating earthquake or national plans to cope with nuclear destruction.[19] Regarding nuclear destruction, Thomas O'Donnell, a moral theologian, argues that "those casualties whose immediate therapy offers most hope for the conservation of the common good should receive first priority." Furthermore, he continues, "unless the individual is considered as very important to the common good, and is salvable, it would seem unreasonable for the medical personnel to expend their efficiency on a few when it could be conserved for the greater number in more remediable need."[20] This proposal, he notes, is in accord with national policy regarding a nuclear disaster.

In triage systems, there is an attempt to determine which patients are "salvable" or "salvageable," that is, capable of being saved or salvaged. Either term acquires different meanings according to the breadth of utilitarian calculation. For example, "salvageability" is primarily *medical* in utilitarian calculations that are limited to the pool of the "wounded and injured." It involves the reduction of loss of life and limb. But "salvageability" is *both medical and social* when specific or general functions are introduced in calculations of social utility. For example, as Beebe and DeBakey note, "traditionally, the military value of surgery lies in the *salvage* of battle casualties. This is not merely a matter of saving life; it is *primarily* one of *returning the wounded to duty,* and the earlier the better."[21] *Medical utility* recognizes the equal value of life; *social utility* recognizes the differential value of specific or general functions. The former does not infringe the principle of equal regard for life, but the latter does. Nevertheless, social utility, at least when focused on specific and urgent functions, may be justified in some crises and emergencies, though the heavy presumption against it must also be recognized.

Some of the confusion about triage can be traced to a failure to distinguish and assess different senses of "salvageability." For example, Paul Ramsey notes that if there were a fire in the NICU and a nurse could save only three babies, "she would gather up the three infants who *least* needed intensive care and take them to safety." Ramsey argues that this case "highlights a crucial aspect of triage medicine. . . . In disaster triage medicine one treats first those least in need of it":

> In case of nuclear disaster, for example, which destroys the medical centers of a metropolitan area, leaving only the dog and cat hospitals on the outskirts, the remaining medical resources are first called upon to bind up a broken finger so that a patient can use a shovel to help bury the dead or fix a leg so that another patient can hobble along carrying a stretcher or assisting the remaining medical team as together they move on to try to save, next, the more seriously injured. Set aside for last treatment or no treatment are *all* patients who would require intensive care.[22]

Ramsey is right about the justifiability (perhaps even the obligatoriness) of the nurse's action in saving the three babies who least needed intensive care, but his analogy with nuclear disaster is mistaken. The rationale is different in the two cases. In the case of the fire in the NICU those least needing IC would be chosen because they would be the most likely to survive apart from the NICU which is being destroyed. In disaster triage, however, the rationale is that those least in need can be restored to assist others. The former rests on medical utility; the latter rests on social utility, defined in terms of specific functions. Clearly different senses of salvageability are involved in these cases.

Even strong opponents of utilitarian approaches to allocation, such as Ramsey, sometimes accept *social utility* in assigning priority to patients with certain kinds of medical problems when they can discharge specific functions of great value to a community in a crisis or emergency, such as an earthquake. Most often Ramsey envisions a situation in which a pluralistic community, with various values and goals, becomes *focused* as a result of a threat to its survival. This survival then becomes an overriding goal in the emergency; it is a condition for realizing other goals. But Ramsey sometimes suggests the possibility and justifiability of a community becoming focused on goals other than its survival and appealing to those goals to identify exceptional functions for which people should be saved.[23] At any rate, he would not appeal to individuals' general or overall instrumental value; at most only their specific and urgent instrumental value can be used to establish priorities in limited, focused circumstances.

Throughout this argument about the justifiability of appealing to medical utility in systems of triage, I have assumed that counting numbers is compatible with a principle of equal regard for human life. But the assumption that numbers count and also that counting numbers is compatible with equal regard has been challenged. For example, John M. Taurek has argued that numbers should not count in deciding whether to distribute a lifesaving drug (a) to save five persons or (b) to save one person when it is not possible to do both (a) and (b). Suppose that their respective medical conditions are such that the drug could save either the five or the one, but not all. Taurek suggests the following approach:

> Here are six human beings. I can empathize with each of them. I would not like to see any of them die. But I cannot save everyone. Why not give each person an equal chance to survive? Perhaps I would flip a coin. Heads, I give my drug to these five. Tails, I give it to this one. In this way I give each of the six persons a fifty-fifty chance of surviving. Where such an option is open to me it would seem to best express my equal concern and respect for each person.[24]

Excluding special moral relations, such as contracts with any of the individuals in the case, I would argue that numbers should count and that they can count without violating equal respect for life. Although Taurek's argument for what Derek Parfit calls "innumerate ethics" is unsatisfactory for several reasons, I will only consider the relation between medical utility and equal regard. It is not plausible to suggest, as Taurek does, that saving the many over the one is like saving the rich over the poor. Rather, as Parfit argues, "if we give the rich priority, we do not give equal weight to saving each. Why do we save the larger number? Because we *do* give equal weight to saving each. Each counts for one. That is why more count for more."[25]

Some philosophers have held that medical utility might be an *acceptable reason* for saving the many rather than the one but that it does not thereby *obligate* anyone to use that system of allocation.[26] I would contend, however, that such an allocation is morally required—apart from more stringent, special moral relations such as contracts—when it would not impose greater and unwarranted risks or costs on the caregiver.[27]

Such a contention focuses on the caregiver apart from roles and relations. Usually within medical care, there are other reasons for giving the many priority over the one in distributing lifesaving medical resources. In contrast to the case presented by Taurek in which the *owner* of the drug had to decide how to use it, citizens commonly think that they have as much right to medical care as anyone else: "It is assumed that each of those in jeopardy has a citizen's equal claim to the use or benefit of that resource." Recognizing that this conviction is present in "our moral culture," Taurek contends that it usually stems from a sense of prior agreement or policy about the use of a resource to which each contributed through taxation or some other arrangement, rather than from an impersonal, comparative evaluation of outcomes—the loss of the many or the loss of the one. He uses the example of the Coast Guard captain's duty to rescue the many rather than the one; the duty of the director of the NICU in a public hospital would be another example. In both cases, citizens would have an equal claim on the scarce resource.[28]

TRIAGE IN NEONATAL INTENSIVE CARE

Sometimes the demand for NIC exceeds the supply of beds, equipment, and personnel. When NIC is limited or scarce, how should selection be made? When access is limited, who should have access? Who should have priority? I will consider this question for both *entry* and *discharge*, later trying to determine whether there are significant moral differences between selection among *only* newcomers and selection among *both* newcomers and current holders of space in the NICU. I recognize that the latter selection is more common; it is also more controversial.

In *Ethics of Newborn Intensive Care*, the editors, Albert R. Jonsen and Michael J. Garland, propose "A Moral Policy for Life/Death Decisions in the Intensive Care Nursery."[29] Most of this policy focuses on decisions about the care of seriously ill newborns without regard to scarcity, but it also includes the "allocation of limited resources" because, as the editors note, this introduces a "new element, that is, the comparison of need between two individuals." For this competitive situation, Jonsen and Garland explicitly propose (p. 152) a "rule of triage," analogous to

> the traditional medico-moral rule of triage, whereby casualties in military and civil disasters are typically divided into three groups: (1) those who will not survive even if treated, (2) those who will survive without treatment, and (3) the priority group of those who need treatment in order to survive. A further triage among the priority group can give preference to those who can be reactivated quickly, or who hold crucial positions of responsibility. Considerations of the common good become relevant in such decisions.

They argue by *analogy* from triage in war and civil disasters to triage in NICUs:

> Similarly, in the selection of infants to receive various kinds of treat-ments, the interest of the state can be invoked as an ethical consideration since the state has an interest in the recognition of values, in fulfillment of responsibilities and duties, in the fair and efficient distribution of benefits,

and in the promotion of the population's health. These interests are directed toward a common good which, in a situation such as this, may be the predominant consideration. *Because it is impossible to treat all infants in need, preference should be given those with the greatest hope of surviving with maximal function.* (P. 152, italics added)

Viewing access to the NICU as triage, and drawing on the analogy between selection for NIC and military-civilian triage, reflects the formal principle of justice—treat similar cases similarly. For example, if we believe that military triage is justifiable and that broad utilitarian criteria are defensible for military triage, we can argue by analogy in NIC and for broad utilitarian criteria for such triage. This is precisely what Jonsen and Garland do. However, the dissimilarities are significant and perhaps undercut the moral analogy. They may even call into question Jonsen and Garland's claim that the "principle of neonatal triage" is "instructive in general."

First, there are minor objections to the Jonsen and Garland formulation of the "traditional medico-moral rule of triage" and to their proposed analogous rule for the NICU. However minor, these objections do affect the analogy. Their formulation of the traditional rule is not sufficiently probabilistic; it refers to "those who will not survive" and to "those who will survive." However, their interpretation of the situation in NIC is more probabilistic and hence more adequate; it refers to "those with the greatest hope of surviving. . . ."

Another shift in language is also noteworthy. In their formulation of the traditional rule of triage, Jonsen and Garland indicate that in some circumstances "considerations of the common good become *relevant*," but in the proposed analogous rule for neonatal triage, they indicate that "a common good which, in a situation such as this, may be the *predominant* consideration." Identifying a *relevant* consideration is not tantamount to making it the *predominant* consideration. Relevance and weight are not equivalent.[30]

Second, the Jonsen and Garland formulation of the traditional rule of triage regarding priorities fails to capture the relation between *medical utility* and *social utility* in theory and in practice. Their position assumes that military and disaster triage can be justified, that we can sort patients according to need and then give priority to group 3—"those who need treatment in order to survive," and that it is justifiable to develop further priorities for treatment within the latter group on grounds of the common good. However, in contrast to Jonsen and Garland's interpretation of triage, most philosophical-theological commentators and actual practices indicate that it is even justified in some circumstances to give priority to group 2—"those who will survive without treatment"—because they can contribute more quickly and effectively to the common good. The penicillin example from World War II is a good one: The soldiers with venereal disease were not in danger of death, but they could be salvaged for immediate service. Most policies or practices of military triage and disaster triage incorporate both medical utility and (limited) social utility; but they also recognize that social utility can sometimes override medical utility.

What would be the greatest good for the military effort or the community after a disaster might not be the greatest good for those with medical needs.

Third, Jonsen and Garland identify several "interests" of the state and the community that might indicate the common good for formulating rules of triage; these are "interests in the recognition of values, in fulfillment of responsibilities and duties, in the fair and efficient distribution of benefits, and in the promotion of the population's health." All of these interests "are directed toward a common good." The common good, as Jonsen and Garland admit, can often mask special interests. That problem of application aside, the common good includes, as they suggest, not only ends to be realized, such as promotion of health, but also values and principles, such as fairness, to be expressed and respected. Then it becomes necessary to indicate how much weight various ends, values, and principles should have in the common good. For example, how much weight should be accorded to the expression or representation of the equal value of lives or to the principle of equal concern and respect? This value or principle may be so fundamental that it should not be sacrificed short of the exigencies of military battle and civilian disaster, and even then only when the survival of many lives or even the community itself is at stake. Perhaps it should not be sacrificed in the practice of medicine in the ICU.

Fourth, any system or rule of triage that incorporates social utility must consider patients' *specific functions* and/or *general social worth*. In war and civil disaster, Jonsen and Garland assign priority to "those who can be reactivated quickly, or who hold crucial positions of responsibility." Presumably these are conjunctive rather than disjunctive considerations in most circumstances— usually it would not be helpful to reactivate someone who could not discharge important responsibilities. However, for triage in the NICU, Jonsen and Garland propose that priority "be given to those with the greatest hope of surviving with *maximal function*" (p. 152, italics added). Considering "function" makes sense within a military effort or within a community organizing itself to save lives and to preserve or restore itself after a disaster, such as an earthquake. It may be possible—as well as justifiable—to identify *specific and urgent functions* that can be fulfilled only by individuals with certain characteristics and then to give those individuals priority. This is not possible in NIC, however, for babies hold no positions of responsibility and can fulfill no specific and urgent functions. Nevertheless, Jonsen and Garland apparently find a surrogate in "maximal function," which may be presumed to include an overall assessment of quality of life and, by implication, social worth. I would argue, however, that the society's interest in the equal value of life should override its interest in the highest quality of life. Otherwise, we would find ourselves struggling with criteria of "maximal function" in overall comparisons of newborns. Jonsen and Garland properly note some of the hazards and risks of such an approach, but they do not adequately emphasize its intrinsic moral limits. One reason may be that they had already incorporated too much "quality of life" into their judgments about the treatment of particular newborns apart from scarcity; when they turned to scarce resources they applied the same standard.

Having identified several deficiencies of the Jonsen and Garland rule of triage, I now want to propose an alternative priority rule, or rule of selection, which offers a more acceptable compromise between utility and equality for cases of scarcity: *If it is not possible to provide intensive care for all newborns when they need it in order to survive and when survival and IC would be in their best interests, priority should be determined by randomization or queuing unless there are major differences in their probabilities of survival.* I will unpack several aspects of this rule and indicate some provisos.

First, I assume that all of the newborns in question *need* IC in order to survive and that it unfortunately cannot be provided to all of them. Furthermore, I assume that we can identify those who are irreversibly dying and exclude them from IC, even though we continue to provide them other forms of care, such as palliative care. Finally, I assume the legitimacy of making judgments about the *best interests* of all newborns: If there are some for whom continued life would not be in their best interests (e.g., because it will be a very short and painful life), IC is also not in their best interests. Such a judgment introduces "quality of life," but only in a very limited and constrained sense. It presupposes that we can balance the benefits (such as prolonged life) and the burdens (such as severe pain) for a newborn and decide in some cases that the burdens outweigh the benefits. The IC may create its own burdens that need consideration. And if we consider only the newborn's interests (rather than the interests of the family or of the society), it may be permissible in some circumstances to withhold or withdraw IC.[31]

Obviously this priority rule will not be very stringent if expansive notions of the newborn's interests also include the "quality of life" that can be expected in view of the *quality of aftercare.* For example, Anthony Shaw offers a formula of "quality of life": $QL = NE \times (H+S)$.[32] NE represents the patient's natural endowment (physical and intellectual), while (H) represents the contributions made to that individual by home or family, and (S) represents the contributions made by society. An assessment of anyone's anticipated quality of life may legitimately include all of these factors; but an assessment of anyone's anticipated quality of life *for purposes of allocating scarce lifesaving medical treatment* should in general prescind from family and social contributions. Otherwise priority would depend on others' desires and efforts. A newborn's value would be dependent rather than independent, and unequal rather than equal.

Second, to hold that IC is in the newborn's best interests is also to hold that there is a *possibility* that it will salvage the newborn's life from the conditions that threaten death. In the case of irreversibly dying newborns, IC would be pointless or useless.

Third, if demand still exceeds supply, priority should then be determined by randomization or queuing unless there are major differences in probabilities of survival. This rule expresses both the principle of equal concern (or equal value of life) and the principle of medical utility. There is a presumption in favor of randomization or queuing, but that presumption can be rebutted when the probabilities of survival differ significantly for the newborns in question.[33] Where the differences in probabilities are negligible, there is no reason to depart from randomization or queuing.

This priority rule does not put the director of the NICU in the position of having to compare various newborns according to their respective social value—a nonmedical decision. It directs attention mainly to predictions about survival (with only severely limited attention to quality of life). One objection to this position might be that it is not possible to predict medical outcomes with accuracy. Insofar as this is true—and perhaps especially true with regard to newborns—the arguments in favor of randomization or queuing become even stronger.

Assigning priority to newborns with the best chance of surviving does not violate the principle of equality for the sake of social utility; it does, however, incorporate medical utility in the sense of probably producing the greatest good for the greatest number of newborns. Evaluative considerations have already been incorporated into judgments about the newborn's best interests; but those evaluations focus on the newborn himself/herself. Broader evaluative questions are precluded as infringements of the principle of the equal value of life.

Fourth, a proviso of *conservation* (or maximization) should be attached to the priority rule. It would require that the selection try to *save as many lives* as possible through the available resources. Thus, in situations of scarcity, it would be possible to give lower priority to those who would require a disproportionate share of IC. Basically, this proviso expresses medical in contrast to social utility.[34]

Fifth, a very difficult question concerns the moral significance of prior admission to the NICU: Do newborns already admitted to the NICU have absolute priority over those arriving later so that their respective chances for survival cannot be considered? In general, "first come, first served"—or queuing—is an appropriate expression of a commitment to equality in such matters. At the very least, this rule demands presumptive respect because of equality and also because of fidelity. Commencement of treatment morally precludes abandonment. Nevertheless, the presumption in favor of continuing to treat patients in the NICU can sometimes be rebutted when there are major differences in prognosis regarding salvageability in the limited sense of survival.

A survey of twenty participants at the Sonoma Conference on Ethical Issues in Newborn Intensive Care included responses to this question: "Would it ever be right to *displace* poor prognosis infant A in order to provide intensive care to better prognosis infant B?" Eighteen of the participants answered "yes" (for various reasons), while two answered "no." One of the negative responses held that the word "right" in the question was problematic: Displacement might be a matter of practicality (some might even say "necessity"), but would not be a matter of morality, that is, of right or wrong. The other negative answer invoked rule-utilitarian considerations. A policy of selective displacement would be too easily abused.[35] By contrast, I would argue that displacement can be a matter of right or wrong, that it is presumptively wrong but may sometimes be right, and that a narrow, carefully circumscribed policy can greatly reduce the danger of abuse.

From the standpoint of health care professionals, some anxieties at this juncture might be related to the difference that many recognize between *withholding* and *withdrawing* treatment. I have argued elsewhere that this distinction is not morally significant, even though it is psychologically important for decisions about treatment/nontreatment.[36] The context of that argument was the patient's best interests. However, in the context of competition for scarce lifesaving IC, where IC is presumed to be in the best interests of all the newborns in question, there is a conflict of interests. Then the question is whether admission to the NICU gives a newborn priority over other newborns who came later, whatever their respective conditions. It is not a violation of the principle of equality to displace newborns from ICUs under some circumstances, particularly if we limit poor/good prognosis to the probability of successful treatment for survival through IC (within the limits indicated above). And the principle of medical utility as expressed in the rule of conservation may also permit it.

It is often necessary to commence NIC in order to make a better diagnosis and prognosis of the newborn's condition. A rule prohibiting displacement or eviction would discourage such important assessments in some cases. Such a rule would also fail to take account of the changing condition and prognosis of various newborns. Finally, displacement from the NICU should not always be construed as "abandonment." Other levels and forms of care should be provided even if IC is not provided.

Sixth, *who* should determine access to NICUs? It is clear that there is a close relation between proposed criteria and appropriate decision makers. Since the rules I have identified for overriding either randomization or queuing are fundamentally *medical*, the director of the NICU and other physicians and nurses should continue to be the "gatekeepers" to the system. For another set of priority rules, structured around the relative value of lives to the society, another set of decision makers—perhaps mixed professional/lay—would be required in order to represent society's values as fully and as completely as possible. I have given reasons for opposing that set of priority rules in NIC. The director of the NICU should allocate access to the unit according to rules derived from the principles of the equal value of life and medical utility. The principle of the equal value of life requires randomization or queuing (and for continued treatment of those who have been admitted to NICUs) *unless* there are major differences in the prognoses regarding survival.

CONCLUSION

In conclusion, despite the claims of Jonsen and Garland, there are good reasons for being suspicious of the metaphor of triage in the allocation or distribution of NIC.[37] The metaphor is inadequate because it distorts the reality of the NICU and decisions about access and because it evokes distinct strands of moral reflection that are not necessarily compatible. Distortion results because the metaphor of triage obscures fundamental differences between the NICU and military or disaster medicine. To be sure, the term "triage" is often used to describe disposition of cases in the emergency room, but in such a setting, as we have seen, triage is "the initial lubricant for an orderly flow of patients through the emergency department." Triage in the emergency department is not problematic because it is designed to provide care according to needs in an efficient way. However, this sort of triage is not offered by Jonsen and Garland, or by others, as the model for intensive care, in part because it does not address the conflicts between patients for particular resources.

A fundamental feature of triage in war and in disaster—the paradigm for Jonsen and Garland—is that there is a perceived scarcity of lifesaving resources that cannot be overcome in the immediate future. The situation in the NICU rarely, if ever, involves this sort of crisis. Health care professionals usually try to gain time so that space, equipment, or personnel can be made available. Viewing the crisis as temporary, they try to "make do." If they cannot handle the crisis in this way, they may have to make decisions about access, about entry and discharge. Then the distinct strands of the tradition of triage pose problems for the use of triage as a metaphor or analogue for the NICU. In particular, the systematic sorting of patients according to need and the probability of successful treatment reflects both equal concern and medical utility. But another strand of the tradition invokes a principle of social utility in terms of specific, urgent functions and/or overall social worth. These two strands may not be compatible, and the metaphor of triage cannot guide decisions unless there is clarity about which strand is being used.

To move from triage based on medical utility to triage based on social utility requires a tremendous leap. In some contexts, social utility, in the form of specific, urgent functions, may justifiably override medical utility. However, social utility in the form of general, overall social worth should not override medical utility. Since it is not possible to identify specific, urgent functions for which newborns might be selected, there is a temptation to find a surrogate, as Jonsen and Garland do, in "maximal function," which becomes a place holder for general, overall value to the society. Such a position should be rejected on the grounds of equal concern, which permits considerations of medical utility, which may sometimes be overridden by social utility in the form of specific, urgent functions in emergencies, but which should not be overridden by social utility in the form of general, overall value.

The Jonsen and Garland formulation of a rule of neonatal triage, based on the analogy with military and disaster triage, may also reflect an attempt to generalize from emergencies to principles for nonemergency settings. As Charles Fried writes,

> I would acknowledge that triage is an appropriate concept on the battlefield and in emergencies, without being willing to generalize from that to the proposition that the rights in medical care that we posit for the more

usual situation are no better than special cases of the same principles that
dictate triage in an emergency. . . . The concept of emergency is only a
tolerable moral concept if somehow we can truly think of it as exceptional,
if we can truly think of it as a circumstance that, far from defining our usual
moral universe, suspends it for a limited time and thus suspends usual moral
principles. It is when emergencies become usual that we are threatened with
moral disintegration, dehumanization.[38]

In the NICU there are no *social* emergencies comparable to the social
emergencies that might justify appeal to specific, urgent functions in war or
in disasters.

Because of the inadequacies of triage as a metaphor for decisions about
access to the NICU under conditions of scarcity, I have proposed an alternative
priority rule that can express both equal concern for all newborns and medical
utility, without incorporating social utility: If it is not possible to provide
intensive care for all newborns when they need it in order to survive and when
survival and IC would be in their best interests, priority should be given to
those newborns who have the highest probability of survival. However,
because randomization and queuing express the principle of equal concern,
these procedures should be followed unless there are major differences in
prognosis (limited to survivability).[39]

POSTSCRIPT (1996)

Some additional historical, conceptual, and normative points are needed,
particularly in light of recent discussions. First, even though I have argued for
the importance of randomization and queuing, Larrey actually developed the
formal practice of triage as a form of "methodical succor" to overcome
treatment according to place in queue, that is, according to whoever happened
to be found first by a surgeon, as well as treatment by rank.[40] However, this
historical observation does not in any way establish that queuing should play
no role in triage, since it can be another mechanism of equality added to
(medical) utility, which itself is not necessarily incompatible with equality. In
some contexts queuing (or some other form of randomization) should enter in
only where there are no very substantial differences in medical utility.[41]

Second, questions have been raised about the historical accuracy of an
important case that figures prominently in many discussions of triage, including Paul Ramsey's and my own. This is the case, drawn from Henry Beecher,
about the allocation of penicillin to troops in North Africa.[42] According to the
usual story, decision makers gave priority to those wounded in the brothels
over those wounded in battle because of military needs (rather than medical
needs and probability of successful treatment). No available evidence indicates

that this was ever a formal U.S. policy, and historian David J. Rothmann dismisses the story as a "widely repeated but apocryphal tale."[43] Nevertheless, British government policy in allocating penicillin in 1944 was apparently (social) utilitarian in restricting its use to pilots and bomber crews, again on military grounds, and there is at least one reported case of a breach of these established rules of allocation to save another enlisted person.[44]

Is it important whether Beecher's story is historically accurate? In one sense, it is not important. What is most important from a casuistical stand-point is the consensus, if there is such, that such an allocation scheme is ethically justified and perhaps can even serve as a paradigm for other discussions of social utilitarian allocation. In another sense, its historicity is impor-tant as a way to indicate different strands in the historical tradition (or even different traditions) of triage. Robert Baker and Gerald Strosberg, two recent critics of (what they call) directly utilitarian triage, trace this history in order to show that such a policy can survive only when "coercively imposed or covertly practiced and [that it] tends to collapse in the light of publicity," and furthermore, that it fails to express the "logic of triage."[45] In general, these critics charge that directly (social) utilitarian selection departs from the ideal or "canonical" form of triage and that it is, in addition, unstable as a social practice since it cannot gain the consent and cooperation of the parties involved. For instance, the allocation of space on dialysis at least in part on (social) utilitarian grounds by the Seattle "god squad" survived only as long as it was not public; once it was public, it collapsed. I concur with these critics about the long-term instability of (social) utilitarian selection, as well as its tensions with egalitarian approaches, even if narrow versions of social utilitar-ian triage can sometimes be ethically justified when particular persons are needed for specific functions, whether in military triage or civilian triage in a disaster. However, in the critics' haste to stress the tension between utilitarian and egalitarian approaches to triage, they fail to see that criteria of medical utility are at the core of triage in all of its forms and that these criteria are not, in principle, in opposition to equality.

According to Baker and Strosberg, both "history and logical structure . . . support an egalitarian [rather than utilitarian] interpretation of ER and ICU triage."[46] They do not deny that indirect or rule-utilitarianism can support some forms of triage, including forms that stress equality. The debate, however, focuses on whether what they call direct (and others might call "act") utilitarian approaches characterize and should characterize triage. It is, in part but not only, semantic. They deny that triage is properly utilitarian at all, partly because they consider utility only as social utility, not medical utility.

In arguing that policies and practices of ICU triage are not utilitarian, Baker and Strosberg stress that although ICUs

> allocate access based on the probability of certain "outcomes," they do so one-dimensionally—in terms of an individual patient's prospects of survival. Nowhere in these calculations is weight given to emotional, financial, and

opportunity costs to families or to the community at large. ICU triage
policies thus seem patently non-utilitarian because they are oblivious to the
"aggregate welfare of the group," the defining feature of utilitarianism.[47]

Their statement actually involves two distinct (though not unrelated)
claims about the defining characteristics of utilitarianism: the range of values,
and the boundaries of the group for purposes of calculating the "greatest good
for the greatest number." In fact, ICUs are not oblivious to the "aggregate
welfare of the group"; they simply limit the morally relevant group to those in
need of IC and make their calculations of utility within that group by
considering the prospect of successful treatment along with medical need.
Furthermore, utilitarianism is not only a theory of values but also—and in the
sense I stress—a theory of why actions or types of action are right and wrong.
Nothing in utilitarianism itself dictates that (only) a certain set or range of
values can or must be used in calculating consequences in order to determine
right and wrong action. The comparative analysis and assessment of patients'
prospects of survival, for purposes of microallocation, is, I would contend,
utilitarian. The principle of medical utility involves comparing the respective
needs and probabilities of benefit of different parties in order to do the greatest
good for the greatest number.

If triage is understood as a classification of patients in order to do the
greatest good for the greatest number, then whether priority should be assigned
to the best off or the worst off depends on technologies, personnel, etc. What
can be done for sick or injured persons will vary from context to context; the
battlefield is different from the ICU. It is not inconsistent with (medical)
utility, for instance, to restrict admissions and hasten discharges of less severely
ill patients since they can best survive outside the ICU. To try to "improve
everyone's chances of survival" through triage requires a judgment of medical
utility.[48] It also expresses the principle of equality.

While not uncontroversial, the distinction between medical utility and
social utility, which I also stressed in the original paper, has now become very
important—I do not know whether my original paper was the first to use it or
whether I unknowingly borrowed it from someone else (as one wag put it,
originality is forgetting where you got it!).[49] If it is accepted, then—in contrast
to Baker and Strosberg—my claim is sound that "systems of triage, whether
informal or formal, have all had an implicit or explicit *utilitarian* rationale."

This distinction between "medical utility" and "social utility" (stated as
"social worth") also appears in Robert Zussman's recent empirical study of
triage in ICU in two hospitals.[50] Even when principles of medical utility are
widely accepted in IC, as he indicates, another important question is how they
are used in actual practice. Zussman's study provides a partial answer, particularly by identifying the ways in which certain biases distort decisions about
medical utility in practice. These biases result mainly from the organizational
and sociocultural contexts of IC.

First, IC "gatekeepers" tend to exclude patients who are "not sick enough" and to admit and retain patients who are "too sick," as measured by criteria of medical utility. This tendency can be explained in part by the persistence of the distinction between withholding (not starting) and withdrawing (stopping) treatment, which is, I argue, morally insignificant and perhaps even morally pernicious. In addition, there is often a felt need to keep a respirator-dependent patient in the ICU because of scarcity of beds (and essential personnel) for respirator support outside the ICU. Hence, the bias in favor of those who are "too sick" and have very limited prospects for survival stems in part "from the unwillingness or inability of hospital administrators to provide adequate step-down units."[51]

Distortion in ICU triage also occurs because of advocacy, both by families and by other physicians, especially private physicians. ICU physicians frequently form significant relationships with families of their patients—their patients are usually not responsive enough for such relationships—and these relationships often influence decisions about admission and discharge and thus smuggle in a version of social utility, even when these decisions are at odds with medical utility. Concern for families, Zussman notes, "introduces a systematic bias toward those patients with families, those whose families are frequent visitors to the unit, and, perhaps, those whose families the physicians find appealing. Such a bias is no more than a light crack in the application of medical utility to triage, but it is a crack through which criteria of social worth have begun to sneak back in."[52] Furthermore, physicians' advocacy for their own patients' admission to and retention in the ICU may pose conflicts with other physicians' roles as "gatekeepers" to IC. Hippocratic individualism, as I observed in the previous chapter, failed to prepare physicians to serve as "gatekeepers"; it exhorted them to seek the welfare of their own patients but provided no guidance for microallocation or triage decisions in situations of conflict. In IC and in other settings where physicians serve as "gatekeepers," patients may receive priority, even in the context of a triage system based on medical utility, because of the effectiveness of their advocates, often physicians who have been in long-term relationships with them. For instance, in the two hospitals Zussman studied, private physicians, often because of their prestige or authority, significantly influenced ICU admission and discharge decisions because of their ongoing advocacy for particular patients, including following them and sometimes even playing an active role in their care after the ICU staff had assumed primary responsibility.[53]

An appeal to medical utility (though without the label) and to one model of triage appears in Robert Truog's recent discussion of the use of a particular technology in the NICU: extracorporeal membrane oxygenation (ECMO), which provides cardiopulmonary bypass, particularly to support newborns who have life-threatening respiratory failure.[54] It is genuinely scarce because only a few centers have it, and highly trained personnel must be continuously present when it is used. Although ECMO can support newborns for over a month if needed, over 99 percent of the newborns who will survive ECMO therapy can actually be weaned off it in less than two weeks. The problem of triage emerges, Truog notes, when parents and caregivers want to keep a particular newborn on ECMO even though his or her prognosis is very poor and when continued treatment "can (and does) adversely affect the health and survival of other ill newborns [since] infants who require ECMO frequently die if they cannot gain access to the therapy within several hours, and many are too ill to survive an extended transport to an alternate ECMO center."

Truog recommends what I would call medical utility criteria: It is important to "seek to maximize the benefits obtained from scarce resources by giving them to those most likely to respond to the therapy." For ECMO, this implies removing from therapy newborns who cannot be weaned within a reasonable period (generally two weeks) "if another child with a better prognosis needed it." Early arrivers should not have absolute rights to ECMO or the NICU over later arrivers also in need but with a much higher probability of benefit, even though it is necessary to withdraw (and not merely withhold) treatment in some cases.

For those needing a particular treatment, such as intensive care, Truog focuses on a spectrum from "too well" to "too sick" in the context of the anticipated duration of the scarcity. He indicates what is involved by analogy with the battlefield physician caring for a large number of casualties following an explosion of a landmine in the middle of his platoon. When resource limitations are only *temporary* (e.g., the battlefield physician can expect evacuation and care for the wounded in several hours, or the ICU can expect to have a full staff of nurses in a few hours), then it is justifiable to concentrate resources on those who are "too sick." When, however, resource limitations are *not temporary* (e.g., the battlefield physician cannot expect additional assistance and evacuation, or the ICU cannot provide ECMO to any more patients or transfer new patients to other centers), then it is important to move more toward the "middle of the spectrum." When the scarcity is only temporary, concentrating on the sicker end of the spectrum will actually maximize the number of lives saved, but not when the scarcity endures for a longer period.

Truog also draws an analogy to the allocation of scarce organs, where there is a permanent shortage, as I will discuss in the next chapter. The main difference is that in organ allocation there is a "pool" of candidates on a waiting list, while candidates for intensive care appear sequentially, one at a time. For both the standard of "first come, first served" remains significant (I would say "morally significant" while Truog might say only "practically significant"), even though its application differs in these two contexts. Organ transplanters can consider the time on the waiting list of all the candidates for a particular donated organ, while directors of ICUs must consider newcomers in relation to those who arrived earlier and are already in the ICU. Hence, Truog notes, directors of ICUs, in contrast to organ transplanters, "cannot exclusively rely on *withholding* care to assure delivery to those most likely to benefit." While recognizing the problems involved in assessing probable benefit, he recommends setting a reasonable threshold for the probability of benefit in order to guide allocation decisions in ICUs, that is, to determine when the standard of "first come, first served" should be overridden in order to maximize medical utility.

One final point should be made: My discussion of microallocation and triage presupposes a context of public justification. Even when they are primarily or exclusively set by physicians, contemporary criteria for triage and microallocation presuppose a public context. Secrecy is often impossible, in part because of the public nature of much health care, especially in hospitals. Physicians participate in teams of caregivers and at a minimum must justify their decisions to other members of their teams (and increasingly to various administrators and regulators). In noting the "public nature" of ICU triage decisions, William Knaus stresses that they are made "in the presence of and with the involvement of a wide variety of interested spectators, other physicians and nurses, along with family members and friends."[55] Their public nature requires some "public consensus" about these decisions or at least the criteria for making them. Hence, he calls for "public accountability" to ensure adequate representation of our pluralistic society. In addition to strong general reasons for public involvement in setting triage criteria in various settings, there are strong specific reasons for public participation in some contexts, such as the criteria for the fair allocation of publicly donated organs for transplantation, the subject of the next chapter.[56]

CHAPTER TWELVE

□

Fairness in the Allocation
and Delivery of Health Care

A CASE STUDY IN ORGAN TRANSPLANTATION

INTRODUCTION

Several cases in the last few years have stirred rumblings of distrust in the process of selecting recipients of scarce organs for transplantation. Mickey Mantle's rapid liver transplant, which came shortly after he was put on the waiting list in June 1995, because of end-stage liver failure secondary to cirrhosis and cancer, was only the most recent. In 1993 Governor Robert P. Casey of Pennsylvania, age sixty-one, received an experimental and risky heart-liver transplant at the University of Pittsburgh Medical Center because his hereditary disease, amyloidosis, had caused his liver to produce an abnormal protein that then accumulated in the walls of his heart. Even though his medical team had expected a wait of four to six weeks, the thirteen-hour operation commenced just hours after his name had been entered onto the waiting list at the United Network for Organ Sharing (UNOS), the national organ transplantation and procurement network.

A case study of heart transplantation provides the starting point for my examination of fairness in the allocation and delivery of health care, which will continue in the next chapter when I explore "Rights to Health Care in a

Democratic Society." This chapter focuses on fairness in the allocation of organs, with some attention at the end to societal funding of organ transplants; the following chapter will put the debate about the societal funding of organ transplants in its larger context—whether there should be a political-legal right to health care in the United States.

THE CASE OF HEART TRANSPLANTATION: ISSUES OF FAIRNESS

Because of the dismal early results of heart transplantation following Dr. Christiaan Barnard's pioneering transplant in 1967, there was a virtual moratorium on the procedure for several years. A few centers in the United States, most notably one at Stanford University, continued heart transplantation programs; by the early 1980s, Dr. Norman Shumway and his colleagues at Stanford had achieved good success rates—65 percent of their carefully selected heart transplant recipients could be expected to survive at least one year, and they had a better than 50 percent chance of surviving five years. Because of these results, proponents of cardiac transplantation argued that it should be made widely available, but critics were not convinced that the problem of tissue rejection had been sufficiently solved or that the benefits of cardiac transplantation outweighed its substantial costs.

On February 1, 1980, the twelve lay trustees of Massachusetts General Hospital announced their decision not to permit heart transplants at that institution "at the present time." Their explanatory statement noted that "to turn away even one potential cardiac transplantation patient is a very trying course to follow" but that "in an age where technology so pervades the medical community, there is a clear responsibility to evaluate new procedures in terms of the greatest good for the greatest number." In June 1980, Patricia Harris, then secretary of the Department of Health and Human Services (DHHS), withdrew an earlier tentative authorization for Medicare to cover heart transplants, holding that such a technology must be evaluated in terms of its "social consequences" as well as its safety, effectiveness, and acceptance by the medical community. Cost is one social consequence, and the specter of another expensive program like the one for end-stage renal disease, which covers both renal dialysis and transplantation, engendered bureaucratic and congressional caution. For example, the cost of heart transplants at Stanford then averaged more than one hundred thousand dollars per patient. If two thousand transplants were performed the cost would be more than two hundred million dollars. If thirty thousand transplants were performed the cost would thus be three billion dollars. Estimates vary, but some reports suggest that each year in the United States there may be as many as thirty-five thousand victims of heart disease whose condition is hopeless without cardiac transplantation and who could possibly benefit from cardiac transplantation (if we assume that it would be possible to locate enough hearts for transplantation).

DHHS was also concerned about Stanford's screening criteria, which included "a stable, rewarding family and/or vocational environment to return to post-transplant; a spouse, family member or companion able and willing to make the long-term commitment to provide emotional support before and after the transplant; financial resources to support travel to and from the transplant center accompanied by the family member for final evaluation." "Contra-indications" for admission to the waiting list at Stanford included "a history of alcoholism, job instability, antisocial behavior, or psychiatric illness." Critics conceded that some of these criteria could be medically relevant but worried that others incorporated unarticulated, undefended, and even indefensible criteria of "social worth," perhaps in violation of procedural and substantive standards of fairness.

As a result of the uncertainty surrounding cardiac transplantation, DHHS ordered a major National Heart Transplant Study, under a contract with the Battelle Institute, to assess its "social consequences" before deciding whether to provide funds to pay for the operation. This study, which was submitted in 1984, determined that cardiac transplantation's success rate in length and quality of life, including rehabilitation for work, warranted viewing the procedure as nonexperimental. In 1986, the federal Task Force on Organ Transplantation also stressed the success of heart transplantation, noting that cyclosporine had recently improved the one-year survival rate of heart transplant recipients to 75 to 85 percent. Because of such successes, Massachusetts General Hospital reconsidered its decision and joined a consortium of Boston hospitals to provide cardiac transplantation, and the Health Care Financing Administration of DHHS in 1987 agreed to provide funds for a few heart transplants for Medicare-eligible patients at selected centers on the grounds that heart transplants are a medically reasonable and necessary service.

Some individuals can afford the one hundred thousand dollars required for a heart transplant and the expensive follow-up care, and some insurance policies cover these procedures. However, many patients lack these resources and cannot meet the stringent Medicare criteria. As a result, the decision about whether society should pay for heart transplants fell to the states. The most publicized decision not to provide state Medicaid funds for heart transplants (and other transplants, except for corneas and kidneys) occurred in Oregon. The Oregon legislature voted in 1987 to discontinue its Medicaid organ transplant coverage for an estimated thirty-four recipients over two years at an estimated cost of 2.2 million dollars in order to provide basic health care for approximately fifteen hundred low-income children and pregnant women. The governor, Neil Goldschmidt, stated when he signed the bill into law: "We all hate it, but we can't walk away from this issue any more. It goes way beyond transplants. How can we spend every nickel in support of a few people when thousands never see a doctor or eat a decent meal?" The Oregon Senate president, John Kitzhaber, a physician, asked, "Is the human tragedy and the personal anguish of death from the lack of an organ transplant any greater than that of an infant dying in an intensive care unit from a preventable problem brought about by a lack of prenatal care?" Medicaid boards in other states, such as Virginia, also chose not to provide funds for some organ transplants in order to have more funds for more patients in need. While the effects of such trade-offs are often invisible in society at large, they are quite evident in closed budgets, such as state Medicaid budgets or hospital indigent care budgets.

In addition to these debates about the so-called "green screen"—the criterion of ability to pay—patient selection criteria are still controversial, whether for admission to the waiting list or for assignment of a particular organ. The supply of donated organs is limited: There were over 3,000 patients on the waiting list for a heart transplant in March 1995, and there were only 2,337 heart transplants in all of 1994; many patients die before receiving a transplant. Over the last few years UNOS has developed and further modified criteria for distributing hearts and other organs, considering such factors as urgency of medical need, probability of successful transplantation, and time on the waiting list. There is vigorous debate about how to make these factors operational (e.g., how many points to assign to each factor) and about the relevance of other factors, such as age, contribution of a person's lifestyle to the end-stage organ failure, and social network of support. Perhaps because of unfair admission to waiting lists, patients receiving heart transplants have historically tended to be white males over forty-five years of age.[1]

I have several reasons for choosing this case study to focus my presentation. First, organ transplantation is often taken as a paradigm and a test case in debates about fair access to health care.[2] Careful examination of the debates regarding the allocation and distribution of organs for transplantation, including the relevance of ability to pay, can illuminate other debates about fairness in health care, even though, as we will see, organ transplantation may raise some unique issues.

Second, much of my own experience in public policy in health care has centered on organ transplantation. I served for two years on the Board of Directors of the United Network for Organ Sharing (UNOS), which is the national organ procurement and transplantation network (OPTN), and I have been on its Ethics Committee for three different terms. In 1985–86 I served as vice-chairman of the federal Task Force on Organ Transplantation. Both the Task Force and UNOS have helped to shape many of the public policies that now govern the distribution and allocation, as well as the procurement, of organs for transplantation.[3] The Task Force had the special responsibility of making "recommendations for assuring equitable access by patients to organ transplantation and for assuring the equitable allocation of donated organs among transplant centers and among patients medically qualified for an organ transplant."[4] (In this context I will construe "equity" as equivalent to "fairness," and I will sometimes use the term "justice" as equivalent to both of them.)

Third, this case study indicates the complex mix of ethical, social, scientific, and medical factors that appear in hard choices on the allocation and delivery of health care. As my own experience has made abundantly clear to me, it is not always possible to distinguish or separate these factors and to identify one factor as *the* ethical factor.

SELECTION OF RECIPIENTS OF SCARCE ORGANS FOR TRANSPLANTATION

The scarcity of organs for transplantation will probably remain a problem for the indefinite future; indeed, it is probable that the demand will always exceed the supply, unless animal organs can be used or artificial organs can be developed. Under these circumstances, there will be difficult questions regarding the procedural and substantive standards for patient selection. Who should choose recipients of donated organs and by what criteria?

Donated organs as public resources. Why not simply let the physicians, nurses, and others involved in transplantation select patients? Why shouldn't selection be viewed as a medical decision to be made by the appropriate professionals? There are some important reasons for developing general public criteria of patient selection—criteria that are developed with input from the public and publicly stated and defended. Apart from special cases—for example, when living donors of kidneys designate a recipient or beneficiary—it can be argued that, from a moral standpoint, donated organs belong to the public, to the community. This fundamental conviction undergirded the Task Force's deliberations and recommendations regarding fair access to organ transplantation: Donated organs should be viewed as scarce public resources to be used for the welfare of the community. Organ procurement and transplant teams receive donated organs as trustees and stewards for the community. Their dispositional authority over those organs should be limited and constrained.[5] (See also the discussion at the end of chapter 15.)

There is increasing demand that the public participate in formulating the criteria for patient selection in order to ensure that they are fair. In general, the evidence presented to the Task Force and published since then indicates that organ procurement and transplantation teams usually make morally responsible decisions in allocating and distributing organs. However, some widely publicized exceptions have generated public controversy and perhaps even reduced organ donations—some of these appeared to the public to reflect favoritism to wealthy foreign nationals. The demand for public participation in the formulation of criteria for the allocation and distribution of organs stems in part from the nature of the organ procurement system—it depends on voluntary gifts by the public, that is, by individuals and their families, to the community. Indeed, there are important moral connections between policies of organ procurement and policies of organ distribution. On the one hand, it is obvious that the success of policies of organ procurement may reduce scarcity and hence obviate some of the difficulties of patient selection. On the other hand, distrust is a major reason for public reluctance to donate organs, and policies of procurement may be ineffective if policies of distribution are perceived by the public to be unfair and thus untrustworthy.[6] Hence public participation—for example, in UNOS—is important. "Organ allocation falls into the region of public decision-making," as Jeffrey M. Prottas insists, "not medical ethics and much less medical tradition."[7]

Two additional points are in order about community ownership and public participation. First, while noting that prior to 1986 organs donated for transplantation belonged to the surgeons who removed them, Prottas contends that the fundamental philosophical shift effected by the Task Force both changed matters and left them the same. On the one hand, professional dominance remains. On the other hand, it is now more circumscribed and publicly accountable. With organ distribution now in the public domain, in part because transplant professionals sought government assistance, there are now more participants, particularly public participants, and the terms of the debate are now different. As Prottas puts it, "Alternative allocation systems are now defended in public debate, and equity as well as efficiency must be considered and defined. Physicians dominate the debate, through knowledge as well as power, but they must justify their actions now as trustees of the public. The organs are no longer theirs."[8]

Second, some unresolved ambiguities about community ownership persist in the formulation of public policies of organ allocation and distribution. The federal Task Force's final report urged that "donated organs be considered a *national* resource to be used for the public good" (my emphasis).[9] This claim provokes the question whether donated organs belong in the first instance to the local community where they are retrieved or to the national community. Even though different answers may not lead to different policies, they certainly create different presumptions and pose different problems. If the relevant community of donation is the national community, then that community has the right to assign the organs; because of logistical problems as well as the importance of incentives for procurement and donation, it may allow the organs to be used in the local community or region if possible and require sharing only under special circumstances. An alternative approach would start from local (or state or regional) ownership of donated organs, even though the local community may have to share organs under some circumstances (e.g., a six-antigen match).

The Michigan Transplant Policy Center presented the case for a federal system in organ sharing against (what it considered to be) the Task Force's recommendation of a "purely national system." Its arguments for a federal system—that is, "state or regional distribution systems which are federally regulated and then linked by a national network for distributing highly sensitized, pediatric, or excess organs as opposed to a system that is, in all important respects, national"—focused on efficiency through increased participation, organizational lubrication, and fine tuning; on justice that may be threatened when state or regions that are more conscientious and efficient in organ procurement have to share with others; on flexibility and research; and on personal care. The Michigan report tended to downplay the importance of human leukocyte antigen (HLA) matching of donor to recipient, which the federal Task Force considered essential in its recommendation of a policy of mandatory sharing of six-antigen matches or zero-antigen mismatches.[10]

UNOS, which was established as the national Organ Procurement and Transplantation Network (OPTN) under the terms of a contract developed largely out of the federal Task Force report, has itself wavered between national and federal philosophies, more often than not leaning toward the latter. Indeed, lack of clarity about its guiding philosophy affects many of its debates about procedural and substantive policies, insofar as they concern the relations among local, regional, and national communities. (As we will also see later, questions about donated organs as public resources concern not only intranational boundaries but also international boundaries.)

Justice and morally relevant and irrelevant characteristics. Justice not only involves public participation—a matter of fair process—but also substantive standards. "Justice" may be defined as rendering each person his or her due, and it includes both formal and material criteria. The formal criterion of justice is similar treatment for similar cases, while material criteria specify relevant similarities and dissimilarities among patients and thus determine how particular benefits and burdens will be distributed.[11] There is debate about the *moral relevance* and *moral weight* of various material criteria, such as need, merit, societal contribution, status, and ability to pay. Different theories of justice tend to accent different material criteria; however, some criteria may be acceptable in some areas of life but not in others.

A fundamental issue for organ transplantation is determining which material criteria are justifiable for the allocation and distribution of donated organs. Standards of justice permit rationing under conditions of scarcity, but they rule out selection criteria that are based on morally irrelevant characteristics, such as race or gender. The major debates focus on which characteristics of patients are morally relevant and which are morally irrelevant in the two stages of selection for organ transplantation: (1) formation of a waiting list, and (2) distribution of available organs to patients on the waiting list.

Waiting lists: problems in admission. There is general agreement that the waiting list of candidates for transplantation should be set largely according to medical criteria, that is, the need for and the probability of benefiting from an organ transplant.[12] There is, of course, debate about whether these medical criteria should be defined broadly or narrowly (for example, how high should we set the standard for minimal efficacy?), about how to specify these criteria, about the relevance of several different factors to the determination of need and efficacy, and about which criteria should have priority in case of conflict.

Why are both need and probability of success important? They reflect *medical utility,* which requires the maximization of welfare among patients suffering from end-stage organ failure. Medical utility should not be confused with *social utility.*[13] While social utility focuses on the value of salvageable patients for society, medical utility requires that organs be used as effectively and as efficiently as possible to benefit as many patients as possible. For example, if there is no reasonable chance that a transplant will be successful for a particular patient, it could even be unethical to put the patient in line to receive a scarce organ.

Efforts are made through UNOS and elsewhere to develop fair policies for allocating and distributing organs to patients on waiting lists, but it is more difficult to ensure equitable access to waiting lists for organ transplants. There is evidence that women, minorities, and low-income patients do not receive transplants at the same rates as white men with high incomes.[14] For example, in one study, females were approximately 30 percent less likely than males to receive a kidney transplant, black dialysis patients were only 55 percent as likely as white dialysis patients to receive a cadaver transplant, and patients receiving dialysis in units in higher income areas had higher transplant rates.[15] A primary source of unequal access appears to be in the decisions about who will be admitted to the waiting list rather than in the decisions about who will receive donated organs (although the waiting times may be quite unequal). Serious questions were raised, for instance, about the admission of Mickey Mantle to the liver transplant waiting list, even though there is no reason to believe that his selection from the waiting list was unfair. However, more research will be required to determine the extent to which unequal access to kidney transplantation, for example, hinges on patient choices and legitimate medical factors rather than on physician sequestration of patients in dialysis units, physician failure to inform and refer some groups of patients, or physician bias in the selection of patients seeking admission to waiting lists.

In the absence of minimum criteria for admission to the waiting lists for organ transplants, patients' physicians have the sole authority to decide whether and when to register a patient. According to a recent report, "a patient can be registered at any time before the transplant is urgently needed and can electively turn down available organs while accumulating waiting time that may misrepresent the patient's medical condition."[16] Hence, in order to prevent the unnecessary and unfair expansion of the waiting list by premature listings, as well as inappropriate admissions, UNOS and the relevant transplant organizations are rightly seeking to establish some minimum standards for waiting list admission.

Waiting lists: problems in multiple listing. UNOS's several shifts in actual and proposed policies on multiple listing (patients registering on more than one local waiting list) in part reflect its uncertainties about its underlying philosophy of organ distribution—is it national or federal?—and about the ranking of liberty in relation to equality, both of which have been viewed as regulative principles of justice.[17]

Some argue that the UNOS decision to permit multiple listing stems from another UNOS policy that gives local use priority except in the limited circumstances where sharing is mandated. In such a context, patients who are knowledgeable and who can afford to travel may seek admission to more than one transplant waiting list: "If the donor organs weren't going to be shared, the patients would be."[18]

The tension between liberty and equality as standards of fair, just, or equitable access to organ transplants pervades the debate about multiple listing. The main argument for permitting patients to appear on more than one waiting list stresses maximum freedom of choice and access; subordinate reasons include the facts that some patients live in different parts of the country at different times and that local transplant teams may sometimes be too busy or too tired to perform the procedure when an organ becomes available.

The main argument against multiple listing centers on the unfair advantage patients gain when they wait on several lists. In an oversimplified example (assuming a purely local sharing system and equal medical factors), if there are two local lists (A and B), each with 100 patients, each patient on list A or B has a 1 percent chance of receiving any available organ. But if ten of the patients on list A are also on list B, then B has 110 patients and each person on that list has only a .9 percent chance of receiving any donated organ. The ten people on both lists have increased their chances of receiving an organ to 1.9 percent.[19] This hypothetical example has real-life parallels. According to a 1992 study, patients on multiple waiting lists for kidney transplantation had a median waiting time of seven months, compared to a median waiting time of 13.3 months for those on only one list.[20] Furthermore, permitting multiple listings favors wealthy and mobile patients over others. It may lead to less effective patient care, may result in variations in tissue typing, and may increase expenditures for testing and patient workup that may be borne by the taxpayer (through Medicare). Even under a prohibitive policy, patients could choose their transplant center and their physician and would not be restricted to their geographical residence.[21] Even though the arguments against multiple listing, on grounds of fairness and equality, are cogent, they have not yet triumphed over political realities.

UNOS point system for cadaveric kidneys. Although UNOS has developed computerized point systems for the allocation of hearts, livers, and kidneys, I will use the point system for kidneys as the primary example because it has received the most attention and has undergone major alterations in light of conflicting values. In October 1987, UNOS implemented a point system for cadaveric kidneys, based on a proposal by Thomas E. Starzl and colleagues.[22] This system required that cadaveric kidneys be offered to patients on the local waiting list (defined as either the individual transplant center recipient list or a shared list of recipients within a defined procurement area) in descending order, with the patient with the highest number of points receiving the highest priority. The original point system consisted of three major parameters: the degree of sensitization, reflected in panel reactive antibodies (PRA, ten points maximum), time on the waiting list (ten points maximum), and HLA matching (twelve points maximum), with some attention to logistics and urgency. Critics noted that the point values for time waiting and high PRA overrode all other point allocations so that the first patient to appear on the print-out had high PRA levels but poor HLA matches. And most of the requests for area variances involved PRA and antigen match.

After much discussion, UNOS in 1989 adopted a revised point system that stressed HLA matching because of evidence about its long-term impact on graft survival. This revised point system accorded less weight to sensitization and to time on the waiting list, as well as to logistics and urgency. According to the UNOS policy statement, "for the national pool, the new allocation system will ensure optimal use of every cadaver kidney offered, since it will identify very well matched recipients. Highly sensitized patients will be chosen when excellent matches emerge. Kidneys will be shipped to highly sensitized patients generally only when negative crossmatches had [sic] been obtained at the donor center. Within each match category fractions of a point acquired for waiting time will determine the order in which patients with the same match score would be listed."[23] Medical urgency status could be requested under some circumstances, but it is rare because dialysis is usually possible as a backup. The policy of mandatory sharing of zero antigen mismatches continued, and a payback policy was adopted for centers receiving organs that had to be shared. ABO blood-group matching remained the same—blood group "O" kidneys could be transplanted only into blood group "O" patients except in the case of kidneys that were mandatorily shared because of HLA match; otherwise "O" patients would be greatly disadvantaged because "O" organs are usable in other blood groups, whereas "O" patients can only use "O" organs.

Another major change occurred in 1995 when the point system for kidney transplants was revised to increase the number of kidneys allocated by waiting time alone, thereby reducing the reliance upon certain HLA matches. Research had indicated that zero ABDR mismatched kidneys have a graft survival as high as kidneys with a six-antigen match, while the graft survival of kidneys with zero AB and three BDR mismatch levels are not significantly better than unmatched grafts. Thus, the UNOS board decided to mandate sharing of all zero mismatched kidneys but to eliminate points at the zero AB and three BDR mismatch levels and to assign seven points for zero BDR mismatches, five points for one BDR mismatch, and two points for two BDR mismatches. As a result of this shift in the point system, UNOS projected that the number of kidneys allocated by waiting time alone would increase from approximately 10 percent to approximately 40 percent. The remainder would be allocated according to a mix of HLA matching, waiting time, PRA, and age (with additional points given for young people, in a policy that will be discussed below).[24]

Assessment of point systems for allocating organs. How are such point systems to be assessed? I will first consider the value of computerized point systems in general and then the value of particular point systems. Many of the supposed advantages and disadvantages of point systems for the allocation of organs hinge on their alleged *objectivity.* Even though a point system does not eliminate the individual physician's judgment—the art of medicine—regarding, for example, the final decision about the use of an organ for any particular patient, it does reduce the physician's discretion. For example, Thomas Starzl contends that "the effect of [his original] point system was to diminish judgmental factors in case selection, which in the past probably had operated to the disadvantage of 'undesirable' potential recipients, including older ones and possibly ethnic minorities."[25] Even though many concede that some decisions have certainly been affected by physicians' subjective biases—for example, in admission to waiting lists—many also stress that it is important for physicians to be able to practice the art of medicine in view of the individual features of particular cases, such as predicting efficacy for a particular patient. According to Daniel Wikler, a computerized point system can systematize decision making by focusing on a full range of data and "can convince patients and the public that a routine, sound plan is in place," perhaps enhancing the perception of fairness in distribution or at least stimulating public discussion.[26] However, in focusing on objectivity, we must not forget that the selection and assignment of weights (points) to these factors rest on values.

With the exception of time on the waiting list, the criteria used in the different point systems for kidneys are medical in the sense that they involve medical techniques used by medical personnel and arguably influence the likely success or failure of the transplant. However, while medical in these senses, these criteria are not value free or value neutral.[27] The vigorous debate about how much weight each criterion should have is only in part technical and scientific (e.g., the impact of HLA matching); it is to a great extent ethical. In kidney transplantation, some factors, such as quality of antigen match and logistical score, focus on the chance of a successful outcome; in different ways both medical urgency and panel-reactive antibody focus on patient need; and time on the waiting list introduces a nonmedical factor, even though it may overlap with panel-reactive antibody because sensitized patients tend to wait longer for transplants. The points assigned to these various factors thus reflect value judgments about the relative importance of patient need, probability of success, and time of waiting—all factors stressed by the federal Task Force on Organ Transplantation.

Medical utility in patient selection. Both patient need for a transplant and the probability of a successful transplant reflect medical utility. Medical utility is not necessarily at odds with fairness, even though they sometimes come into conflict. It is a fundamental mistake to suppose that "medical utility" and "fairness" are necessarily in tension so that if one is met the other is infringed, and it is a fundamental mistake to suppose that "fairness" always dictates priority to queuing or randomization over "medical utility." Indeed, in some contexts determination of "medical utility" may be required by the principle of fairness. It may be "unfortunate" when one patient receives an organ over another because of "medical utility," but it is not necessarily "unfair." Appeals to "medical utility" in the distribution of organs do not necessarily violate the principle of equal concern and respect; judgments based on "medical utility" do not necessarily show disrespect and contempt, which, by contrast, are inevitable in judgments based on patients' comparative "social utility."[28] Furthermore, acceptance of "medical utility" does not commit one to utilitarianism as a foundational or substantive moral doctrine; "medical utility" can (and should) be accepted in any defensible deontological framework as well. Holding that a lexical or serial order of these criteria is impossible also does not entail utilitarianism. In addition, using a Rawlsian contract metaphor, we can argue that in a fair set of decision-making circumstances behind the veil of ignorance, patients not knowing their own medical conditions would choose criteria of "medical utility."[29] Such a hypothetical contract allegedly makes the distribution fair to potential recipients. Finally, others also argue that fairness to donors requires that organs be used effectively and efficiently.[30]

Judgments about medical need and probability of success, as already noted, are value-laden. Consider, for example, the debates about what will count as *success*—such as length of graft survival, length of patient survival, quality of life, rehabilitation—and about which factors influence the *probability of success*. Some contraindications are well established, such as mismatched blood group or positive donor-recipient crossmatch. Over time the UNOS point system for kidneys has stressed, to varying degrees, tissue matching on medical utility grounds. It is also not unfair to use tissue matching, not only because of medical utility, but because tissue matching functions as a kind of natural lottery, which involves the randomness of the HLA match between available donors and recipients.[31] However, there is vigorous debate about the relative importance of tissue matching now that cyclosporine and other immunosuppressive medications are available, and this technical debate influences judgments about the conditions under which kidneys should be shared outside the location where they are retrieved. For example, since cyclosporine is nephrotoxic, a retrieved kidney needs to be transplanted sooner than usual in order to increase the chances of successful transplantation when that immunosuppressive medication is used. Furthermore, studies also indicate that the length of cold ischemic time prior to the kidney's transplantation has an impact on graft survival.[32]

Tissue matching needs ongoing scrutiny. First, in view of the scientific controversy, it is essential to see if certain levels of tissue match or mismatch really make significant differences in the outcome of transplantation over time. Second, it is morally imperative to monitor the operation of the point systems to make sure that tissue matching does not have unjustified discriminatory effects, for example, against blacks and other minorities. As noted, discrimination may already occur against blacks and other minorities in admission to waiting lists, and tissue matching may have discriminatory effects for some patients on the waiting list. For example, most organ donors are white, certain HLA phenotypes are different in white, black, and Hispanic populations, the identification of HLA phenotypes is less complete for blacks and Hispanics, nonwhites have a higher rate of end-stage renal disease, and nonwhite populations are disproportionately represented on dialysis lists.[33] In this context, Robert Veatch argues that "if organs are to be allocated on the basis of degree of tissue match, the policy is, de facto, a whites-first policy."[34] Monitoring the operation of each point system will provide evidence regarding discriminatory effects. If such discriminatory effects emerge, then it may be necessary to sacrifice some probability of success in order to take affirmative action to protect blacks and other minorities. Indeed, such considerations figured significantly in the UNOS decision to alter, in 1995, the point system for the allocation of cadaveric kidneys to eliminate points given from some levels of match and to increase the role of time on the waiting list.[35]

Sometimes there is a tension between urgency of need and probability of success. Robert M. Veatch contends that "a justice-based allocation . . . would demand that highest priority be given to medical need and length of time the patient has been in need."[36] Apparently some potential recipients would choose such criteria. For example, in determining who will receive a heart, members of a Canadian transplant team note, "it becomes a difficult ethical issue as to whether the patient with the better outcome or the individual with the greatest urgency should receive the heart. The patients themselves would opt for the patient with the greatest urgency and by and large that is the decision taken by the team. However, one is conscious of the fact that one may be affecting the overall success rate by making choices in favor of individual patient urgency rather than making them on the basis of success."[37]

Tensions between medical urgency and probability of success may vary greatly depending on the organ in question. For instance, there is debate in heart transplantation about the use of artificial hearts and other assist devices, in part because they have sometimes given patients priority for scarce donor hearts on the basis of medical need, even though their chances for success may have been minimal. Critics such as George Annas charge that using the total artificial heart as a temporary bridge to transplantation does not save lives; it only changes the identities of those receiving heart transplants by giving very sick patients priority.[38] And UNOS revised its criteria for the allocation of hearts so that patients on mechanical assist devices would no longer receive priority over all other candidates in their area; under the revised allocation system, patients who require inotropic agents and are in intensive care units would also appear in the top priority group. One goal of this revision was to remove any incentive for a physician to put a patient on an assist device in order to improve his or her chances of getting a heart transplant.[39]

In liver transplantation, to take another example, the dominant practice has been to give the sickest patient the highest priority, but "medical utility" (and some would include cost-effectiveness) would often dictate placing the liver in the fittest patient and realizing the greatest medical benefit (at the lowest cost). Another reason for priority to those with a higher probability of benefit is that "as time goes on . . . the fitter patients become increasingly ill, their survivability on the waiting list declines, and their operative risk soars."[40] Nevertheless, as Olga Jonasson notes, there is clearly one case in which the sickest of all patients awaiting liver transplants is also the best candidate for successful transplantation—the young, previously healthy patient with fulminant acute liver failure.[41]

The category of medical urgency may not be as important when an artificial organ can be used as a backup (for example, dialysis for end-stage renal failure). However, some argue that medical urgency should include not only the immediate threat of death but also the likelihood of not receiving another organ because of presensitization, particularly because sensitized patients now constitute a hard core of the waiting lists for kidney transplants. The Task Force recommended that a highly sensitized patient who is predicted on the basis of either a computer antibody analysis or an actual crossmatch to accept the transplant should be given priority over equivalently matched nonsensitized patients. And yet the success rates may be lower for sensitized patients than for nonsensitized patients.[42]

Another problem is that medical urgency is a manipulable category. It is reportedly abused at times by physicians eager to protect their patients by declaring them medically urgent in order to increase their chances for a transplant.[43] These reports are not implausible in light of studies indicating that physicians are willing to lie in order to promote their patients' welfare in the health care system, such as using a misleading category in order to enable the patient to have a diagnostic procedure covered by health insurance.[44]

In short, it is not at all clear that a general, a priori formulation of the appropriate relation between medical need and probability of success within medical utility is defensible, in part because of the variations in organ systems. Thus, the proposals may have to be organ specific, and variations can be expected in policies from one organ to another. Ongoing monitoring and assessment of current policies, with public input and with special attention to the proper use of the category of medical urgency, appears to be the most appropriate action. As the federal Task Force recommended,

> A decision on how to apply the criterion of urgency must be developed by a thoughtful and broadly representative group, which must struggle with the concept of [the] best use of organs in the context of compassion and humanitarianism. Because donated organs are a scarce resource, policies to resolve conflicts between equity and efficiency that arise in the distribution of organs should be determined by a broadly representative group that includes patient, community, and ethical perspectives, as well as those of the medical professionals involved.[45]

Time on the waiting list. Many including this writer have argued that randomization or time on the waiting list is a fair way to allocate scarce lifesaving resources under some circumstances. I developed my argument to this effect in 1970 when the debate was mainly about kidney dialysis.[46] (See chapter 10 above.) At that time I argued that once the pool of medically eligible candidates has been determined, it is then fair to make the final selection by randomization or queuing. However, I believe that matters are somewhat different when the scarce medical resource is an organ, which cannot be reused. Now I would argue that medical utility should also be used to determine which candidate should receive the organ, after the eligible candidates have been identified on grounds of medical utility. A major reason is not wasting the gift of life; the organ has been donated for effective use. Giving an organ to a patient who has a very limited chance of success, perhaps because of poor tissue match, increases the probability that he or she will then need another transplant for survival, further reducing the chances for others as well as for his or her own successful transplantation. (Retransplantation will be discussed below.)

Nevertheless, it is important to reject positions that rule out queuing or time on the waiting list as morally irrelevant or even morally pernicious. For example, Olga Jonasson argues that "length of time on the waiting list is the least fair, most easily manipulated, and most mindless of all methods of organ allocation," and Ruth Macklin argues that the principle of "first come, first served" is inapplicable and even inequitable in the allocation of scarce medical resources because it ignores different medical needs and prognoses.[47]

By contrast to those positions, if two or more patients are equally good candidates for a particular organ according to the medical criteria of need and probability of success, their time on the waiting list may be the fairest way to make the final selection. This approach is similar to that recommended by the federal Task Force.[48] The original UNOS point system for kidneys gave more weight to time on the waiting list and also to sensitization, but since highly sensitized patients are likely to spend more time on the waiting list they were, in effect, counted twice. Some argued that such double counting is justifiable because of their difficulty of obtaining organs, while others noted that such patients might then receive priority over much better matched patients. In the revised point system in 1989, time on the waiting list functioned more as a tiebreaker in the allocation of kidneys in the UNOS point system, while in the revised point system in 1995, it again received greater weight (the point value was raised from 0.5 to one point for each full year of waiting time).

Queuing is often favored because it appears to be objective and impersonal, but the justification of its use in patient selection depends on certain values (or principles), such as fair opportunity. And there are both ethical and practical problems. It is not always easy to determine when a patient entered the waiting list; one way is the accession time on the UNOS list. But, as Jonasson notes, it is easy to manipulate this criterion, for example, by putting patients on the list before they become dialysis dependent.[49] In addition, it is important to note that the fairness of queuing (as well as of randomization) depends in part on background conditions. For example, some people may not seek care early because of limited financial resources and insurance; others may receive inadequate medical advice about how early to seek transplantation; and so forth.

There are clear differences in orientation between two approaches to time on the waiting list—use time on the waiting list unless there are substantial differences in medical utility, or follow medical utility unless there are no substantial differences and then use time on the waiting list. However, they should, practically speaking, end up at the same place, *if* there is a consensus on what constitutes "substantial differences in medical utility."[50]

Number of organs needed: multi-organ transplants. At the outset I noted some of the problems raised by the multi-organ transplant required by former Pennsylvania Governor Casey, who needed a combined heart-liver transplant because, as a result of a hereditary disease, his liver produced an abnormal protein that had damaged his heart. A heart and a liver had become available from a thirty-four-year-old man who had been fatally beaten several days before and remained on a respirator in nearby Allegheny General Hospital. The median waiting time in 1993 was 67 days for patients seeking a new liver, and 198 days for patients seeking a new heart. At the time of Governor Casey's transplant, six patients were ahead of him on the local waiting list for hearts, two for livers. While UNOS did not have a clear priority rule for combined heart-liver transplants, in part because the procedure was so rare—only six had been done in the United States at that time—the University of Pittsburgh Medical Center had a rule of giving the patient needing a heart-liver combination priority over patients waiting for one or the other single organ. In addition, the director of adult cardiac transplantation at the University of Pittsburgh Medical Center stressed that "the governor's life was in imminent danger." He received the highest priority ranking of medical urgency on the heart list, and the second highest priority ranking on the liver list.

Others wondered, however, whether the governor, who had indicated that he would seek no special preference, had in fact been the beneficiary of unfair policies and actions. Spokespersons in the case indicated that the governor received the two organs so quickly in part because patients needing multiple transplants are placed on a separate waiting list from those needing a single organ, and the governor was the only patient needing a heart-liver transplant at that time. The coordinator of organ procurement in the Pittsburgh area noted that if the donor had had healthy lungs, Governor Casey would not have received his transplant, because at least two Pittsburgh patients were waiting for a heart-lung transplant and they would have had priority over him.

In the wake of the controversy about whether Governor Casey's transplant breached fair allocation standards, the U.S. Department of Health and Human Services (DHHS) requested that UNOS clarify its policy for multi-organ allocation. At its annual meeting a few weeks later, the UNOS board indicated that a patient who needs a multiple organ transplant must be ranked No. 1 on the local waiting list for at least one of the lifesaving organs (heart, liver, lung, or heart-lung). Upon reaching the top of the local waiting list, the patient will be offered both the organ for which he was ranked highest as well as any other organ he or she needs. According to the UNOS president, the clarified policy "takes into account the urgency facing patients needing more than one organ, yet ensures that the needs of these patients do not supersede those of others awaiting a transplant. The board's . . . action was a clarification of existing policy."

Nevertheless, there is room for significant ethical debate, particularly regarding the weight of medical utility and fair equality of opportunity. From a medical utility standpoint, it may be justifiable, even obligatory, to use two organs to save two lives rather than one life, particularly since multi-organ transplants often have lower chances of success. And it is not plausible to argue that medical urgency gives such patients priority because, presumably, other candidates for single organs are also facing death.[51]

Number of organ transplants needed: retransplantation. Retransplantation raises important issues of fairness and medical utility—for example, in cases of patients receiving three or four liver transplants. Between 1977 and 1981, 10,818 patients received 11,615 kidney transplants, with 10,063 patients receiving one transplant, 713 patients receiving two, and 42 patients receiving three or more.[52] The argument is sometimes made that fair access to scarce organs for transplantation should limit a patient to one transplant, but others resist this conclusion on the grounds that denying a transplant recipient another transplant is tantamount to abandonment. It is difficult to sustain the claim of abandonment if there are back-up or alternative treatments, such as dialysis for end-stage renal disease. And from the standpoint of medical utility, it appears that the chances for successful transplantation decline somewhat after each transplant.[53] Others argue that "past use of resources is irrelevant to present need,"[54] but questions still arise about how many opportunities should be provided for access to a scarce resource under the principle of fair opportunity, particularly when some individuals die without ever receiving a single transplant.

Other controversial criteria. Several other criteria have been proposed for patient selection, and it is important to examine them carefully in order to make sure that they reflect medical utility, rather than social utility, and that they are not otherwise unfair. For example, as the case study noted, in 1980 former DHHS Secretary Patricia Harris withdrew the earlier tentative authorization for Medicare to cover some heart transplants in part because of her concern that some criteria for patient selection were more social than medical. Because judgments of social utility may masquerade as judgments of medical utility, it is essential that the criteria of patient selection be publicly developed and articulated as well as fairly applied. The public process is important to ensure that the criteria are based on medical utility, rather than on social utility, and also reflect both fairness and respect for persons. Several possible criteria are very controversial because they may function as guides either to medical utility or to social utility—for example, age, social network of support, lifestyle and behavioral patterns.[55] One question is whether they can legitimately function as guides to medical utility without implying judgments of social utility and infringing principles of justice and respect for persons.

THE ACCESS OF FOREIGN NATIONALS TO ORGANS DONATED IN THE UNITED STATES

Another major question arises from the Task Force's recommendation that "donated organs be considered a national resource": Should foreign nationals have access to organs donated in the United States? The issue of geographical boundaries, or "accidents of geography," is international as well as intranational. Of the approximately 6,000 cadaveric kidneys transplanted in the United States in 1985, 300 went to nonresident aliens who had come to the United States for medical care, and 200 to 250 additional kidneys were shipped abroad for use in other countries.[56]

The debate about transplanting organs obtained in the United States into foreign nationals and about exporting those organs invokes various moral principles as well as diverse convictions about the ownership of donated organs. Some who propose physician discretion in the selection of patients insist that the ideal of medical humanitarianism precludes the use of criteria such as national residence. By contrast, critics of physician discretion contend that the distribution of donated organs is not merely or even primarily a matter of medical humanitarianism but of social humanitarianism. I agree that donated organs belong to the community, that procurement and transplantation teams only serve as trustees and stewards of donated organs, and that the debate essentially concerns social rather than medical humanitarianism.

It should not be surprising that the most vigorous and divisive debate on the federal Task Force centered on the access of foreign nationals to cadaveric organs donated in the United States. Members of the Task Force sought to balance principles of fairness, beneficence (expressed as compassion and generosity), and utility and efficiency, particularly in the avoidance of kidney wastage. There was little or no interest in excluding foreign nationals altogether, for instance, when organs would otherwise go to waste. Rather, the debate focused on whether to adopt (1) a policy of U.S. citizens and residents first—sometimes called "Americans first"—which would allow some nonresident aliens on waiting lists but would not allow them to receive any particular donated organ unless no U.S. citizen or resident could benefit from it, or (2) a policy that would set a ceiling on the number of nonresident aliens on the waiting list, but would accord equal treatment to everyone on the list, regardless of national residence. The Task Force recommended the first policy for hearts and livers and the second policy for kidneys, recommending a ceiling of 10 percent until the matter could be reviewed by the OPTN (which became UNOS).

The majority of the Task Force distinguished between these two policies on the grounds that kidneys are not as scarce as extrarenal organs, and dialysis is usually available and feasible as a backup or an alternative to transplantation in the treatment of end-stage renal failure. However, eight of the twenty-five members of the Task Force dissented from the recommendation for renal organs. These dissenters argued that it is unfair to members of the national community to deny or to delay their access to organs donated in the United States and unfair to use taxpayers' money to obtain kidneys that would be distributed to nonresident aliens.[57] Furthermore, critics charged, the policy recommended by the majority, which I joined, would cost taxpayers in other ways because, for instance, each kidney transplant performed on a Medicare beneficiary in 1985 would have saved the Medicare system an average of

$62,000 over a five-year period. Thus, if 275 of the estimated 300 cadaveric kidneys that went to foreign nationals in 1985 had gone to Medicare recipients, the system would have saved approximately $17 million over a five-year period.[58]

According to Dan W. Brock, the division on the Task Force suggests two approaches with "considerable plausibility," and a public process could adjudicate the conflict.[59] The UNOS board later made an effort to determine public opinion in several ways—by involving the public on the UNOS board and on the committee that analyzed the options and recommended a policy, through public comment on the proposed policy, and through an opinion poll. The opinion poll was important because of the claim that when members of the public "learn that foreign nationals receive organs instead of U.S. residents and that they generally do so in a shorter period of time, they begin to question the fairness of that system and may very well become less inclined to donate. Indeed, in some communities where there has been much publicity about foreign nationals receiving transplanted kidneys, there has been a subsequent reduction in donation levels."[60] A telephone poll, conducted in January 1987, indicated that there is no clear majority opinion about whether domestic patients should receive preferential treatment over nonresident aliens in the United States for cadaveric organ transplants.[61]

After reviewing various arguments, UNOS adopted a policy that established some limits and directions but relied mainly on a procedure of accountability in the transplantation of nonresident aliens. It required UNOS members to charge the same fees for nonresident aliens as for domestic patients, to treat all patients accepted on transplant waiting lists according to UNOS policies for the equitable distribution of donated organs, and to arrange any exportation of organs through UNOS and then only after no suitable recipient could be located in the United States or Canada (included because of sharing arrangements). Accountability was established in several ways. On the local level, centers that accept nonresident aliens on their waiting lists should establish a mechanism for community participation and review. On the national level, the UNOS committee on foreign relations has a right to audit all transplant center activities relating to nonresident aliens and will automatically review any center that has more than 10 percent of its transplant recipients from foreign nationals.[62]

ABILITY TO PAY

So far I have examined issues of patient selection for organs for transplantation—both formation of the waiting list and selection to receive a particular organ—apart from questions of costs. Yet organ transplants are notoriously very expensive. Our society has responded very differently to different organ transplant procedures. Through Medicare's End-Stage Renal Disease (ESRD) program, virtually everyone who needs a kidney transplant (or dialysis) is covered, while coverage for heart and liver transplantation is at best spotty.

Should ability to pay function as a criterion for admission to waiting lists? Should there be a "green screen" for access to waiting lists for organ transplantation? These questions cannot be directly addressed by UNOS; they emerge on other levels for other social institutions, particularly for the federal and state governments.

As part of its efforts to propose policies to ensure equitable access to organ transplantation, the federal Task Force on Organ Transplantation offered several arguments in favor of increasing societal funding for organ transplants—on the one hand, for immunosuppressive medications for organ transplants already funded (mainly kidneys), and, on the other hand, for extrarenal organ transplants not currently funded.

The Task Force concluded that coverage for immunosuppressive medications was important because, for example, wealth discrimination had reentered the ESRD program, which had been designed to eliminate distribution of artificial and transplanted kidneys according to ability to pay. Noting that approximately 25 percent of the transplant population (for all organs) lacked state or private coverage for immunosuppressive medications, especially cyclosporine, which was then estimated to cost approximately $5,000 to $7,000 a year, the Task Force "found evidence that inability to pay for immunosuppressive medications had been a factor in the initial selection of patients for transplantation" and that some transplant recipients had undergone nonmedically indicated—and potentially risky—changes in their medications because of the costs.[63] The Medicare coverage that was subsequently approved, in part in response to the Task Force report, was limited to one year after the transplant. Further study is needed to determine the extent to which limited coverage of immunosuppressive medications limits access to transplants. (The coverage is now being gradually extended to a full three-year period.)

Much of the Task Force's concern about fair access focused on extrarenal organs—hearts and livers—in view of the limited and uneven provision of funds for them. And, according to the Task Force, there are several arguments for a societal obligation, to be discharged by the federal government as a last resort, to provide funds for extrarenal transplants in order to ensure fair access. One argument focuses on the *continuity* between extrarenal organ transplants and other medical procedures that are already covered, such as kidney transplants and dialysis. Appealing to the principle of consistency or universalizability, this argument accepts the precedent value of prior and current policy decisions. Still another premise in the argument is empirical—extrarenal transplants are comparable in efficacy and costs to procedures that are routinely covered. In response to worries about cost containment, defenders of public funding for organ transplantation hold that it is unfair to impose the major burden of cost containment on patients with end-stage organ failure who need transplants. The burdens of cost containment should themselves be distributed equitably across categories of patients needing health care.[64]

A second argument focuses on the *distinctiveness* or *uniqueness* of organ transplantation, particularly the social practices of procurement that provide the organs for transplantation. This argument identifies an important moral connection between organ procurement, including organ donation, and organ distribution and allocation.[65] In its efforts to increase the supply of organs, our society requests donations of organs from people of all socioeconomic classes— for example, through presidential appeals for organ donations or through state "required request" and "routine inquiry" statues, which mandate that institutions inquire about an individual's or family's willingness to donate, or even request such a donation. However, it is unfair and even exploitative for society to ask people, rich and poor alike, to donate organs if access to donated organs will be determined by ability to pay rather than by medical need, probability of success, and time on the waiting list.[66]

A third and related argument builds on societal opposition to commercialization and commodification of human body parts, as expressed in various laws and policies. (See chapter 15 below.) Federal legislation—as well as legislation in some states—prohibits individuals from transferring organs over which they have dispositional authority for valuable consideration.[67] In addition, various professional organizations involved in organ transplantation have taken a stand against the sale of organs for transplantation. It is difficult, according to this third argument, to distinguish (1) buying an organ for transplantation and then hiring a surgeon to perform the procedure from (2) purchasing an organ transplantation procedure that includes a (donated) organ as well as the surgeon's (and others') services.

These last two principled arguments may be combined with consequentialist arguments. There are legitimate worries about the impact of unequal access to organ transplants, based on inability to pay, on the system of organ procurement that includes gifts of organs from individuals and their families. As I noted earlier, there is substantial evidence that attitudes of distrust limit organ donation; this distrust appears to be directed at both organ procurement (e.g., the fear the potential donors will be declared dead prematurely) and organ distribution (e.g., the concern that potential transplant recipients from higher socioeconomic classes will receive priority).[68] Thus, it is not at all surprising that after Oregon decided to stop providing Medicaid funds for most organ transplants, "a boycott of organ donations was organized by some low-income people."[69] And cynical comments about how rapidly famous people, such as Mickey Mantle, receive scarce organ transplants reflect public suspicion of organ allocation policies.

A final argument is closely related to the first one, but instead of building on what the society has already decided to do regarding other health care, it focuses on the federal government's obligation, at least as a last resort, to ensure fair access to health care, including organ transplantation, by removing financial barriers, if necessary. For example, this argument might appeal to what the President's Commission for the Study of Ethical Problems in Medicine and Biomedical and Behavioral Research construed as society's obligation to provide equitable access to an adequate level of health care without excessive burdens.[70] (Even though this argument may appear to be independent of social practices, it may—and should—nevertheless appeal to principles and values embedded in those practices.) Whatever the foundation of the obligation, there will still be vigorous debate about what counts as an adequate level of health care and whether organ transplants qualify. And this debate moves us to questions of *macroallocation*. It is now standard to distinguish microallocation from macroallocation. For example, Engelhardt uses the term "macroallocation" to refer to "allocations among general categories of expenditures," and the term "microallocation" to refer to "choices among particular individuals as to whether they will be recipients of resources and in what amount."[71]

The question of microallocation is *who* will receive a particular scarce good; the questions of macroallocation focus on *how much* of a good will be made available, where financial resources can alter the availability of that good, such as organ transplants or AZT for AIDS patients. It is important to note that while macroallocation and microallocation are analytically distinct, they are significantly related. Obviously macroallocation decisions determine the extent of scarcity and the difficulty of patient selection by in part determining how much of a good will be made available in a society. If a particular technology or mode of health care is in limited supply, as is often the case, then there may be difficult microallocation decisions about who will receive this particular scarce good. But problems in microallocation may also have an impact on macroallocation decisions. For instance, it has been argued that the federal government decided to provide virtually universal funding for treatments for end-stage renal disease in part to eliminate the problem of patient selection for kidney dialysis and transplantation, that is, the problem of having health care professionals and committees explicitly determine who would live and who would die.[72] After a few concluding observations, I will put macroallocation in the context of debates about a political-legal right to health care in a democratic society, the subject of the next chapter.

CONCLUSION

In conclusion, I want to draw together several points and mention a few others that grow out of this analysis and properly set the context for discussion of a political-legal right to health care. First, I have stressed the principle of fairness (equity or justice). This principle is not easy to specify or to apply, for there are vigorous disputes about its meaning (e.g., whether it excludes medical utility) and about its weight (e.g., when it comes into conflict with other principles).

Second, fairness is not the only principle; there are others, including utility and respect for personal autonomy. Fairness does not always oppose such principles as medical utility, as I argued in my proposals for a fair use of medical utility (medical need and probability of success) in the allocation and distribution of organs for transplantation. Even when there are conflicts it is not possible to indicate in advance exactly which principle should have priority.[73] Seeking to balance these various principles is a worthy process and goal, but sometimes trade-offs are inevitable in policies of allocation and distribution.[74]

Third, if we ask about the fairness of providing or not providing funds for extrarenal transplants, it may be difficult to answer that question in an unfair system. If, as Norman Daniels argues, "our system is, in general, unjust,"[75] then it may be difficult to determine whether it would be just or unjust to press for and obtain funds for organ transplants or AZT or some other treatment. Which policy of allocation would be more likely to lead to a more just system? And, if neither would, which should be adopted as the morally preferable—perhaps because more just—policy within an unjust system? I will provide a context for such judgments in the next chapter, when I consider a general political-legal right to health care, which is currently missing from U.S. health policy.

Fourth, not only do our reflections about policies of allocation and distribution occur within a particular sociopolitical context, but that context itself often changes over time. Thus, it may be appropriate to develop policies to reaffirm some principles or values that have been neglected, or even overridden, in order to maintain their significance for the society over time.[76] There is no reason to suppose that within the range of ethically acceptable policies only one should be implemented over time.

Finally, ethical theories, including theories of justice and fairness, may have only limited applicability, partly in view of conditions that limit their feasibility. The phrase "applied ethics"—the application of ethical principles, rules, and theories—is not the most appropriate, for the task for ethics is more that of illuminating the ethical presuppositions and implications of the choices we have to make in the real world, in response to such questions as how organs should be distributed and allocated and whether organ transplants should receive societal funds. Those questions arise in a complex mix of social, political, scientific, medical, and other factors. "Illumination" rather than "resolution" is the main contribution of ethical theory, and it properly takes the form of "practical ethics" rather than "applied ethics."

CHAPTER THIRTEEN

□

Rights to Health Care in a Democratic Society

INTRODUCTION: SETTING THE STAGE

It is hard to decide whether liver transplants and heart transplants and other expensive procedures should be covered in the U.S. health care system, in part because it is difficult to determine morally appropriate policies within a flawed system. And there is a consensus that our current system—or nonsystem—is flawed and needs to be reformed. This consensus has not, however, translated into concerted, unified action for a particular solution, largely because of different perceptions of the fundamental flaw(s). Some—and I am among them—hold that the system is *morally flawed* and *unjust* mainly because more than 37 million people are uninsured at any one time, 15 to 20 million more are uninsured some time during the year, and millions more are underinsured for health care that costs more and more. Others, by contrast, identify the uncontrolled escalation of costs as the basic flaw, with health care expenditures now consuming about 13 percent of the gross domestic product, a much higher figure than for any other country, including those providing universal access.

People thus disagree about possible solutions to the crisis in our health care system, not only because the choices are hard and even painful, but also because they have different perceptions of the problems and their causes. A major gap now appears between the perceptions of the public and the perceptions of experts (i.e., policymakers and health care professionals).[1] They agree in substance that rising costs and limited health care coverage constitute serious problems. However, they disagree, for instance, about why costs are rising so rapidly. According to the public, the problem is human and moral—greed, high salaries, corruption, waste, and unnecessary testing. Hence, radical solutions are unnecessary. All that we need is more vigilant discipline for these chronic human propensities (what theologians might call sin). By contrast, according to the experts, our crisis of rising costs stems from novel factors, including the substantial increase in the numbers of older people who are major consumers of health care, along with the rapidly expanding armamentarium of medical technologies. What is needed then is health care rationing, setting priorities, and the like, all of which the public views as unnecessary, particularly in light of the widely shared conviction that health care is a right. Although both sides view the cost of health care as a major problem, many in the public concentrate on its negative impact on *individuals* seeking care, while many experts concentrate on its negative impact on the *system* as a whole, particularly as a percentage of the gross domestic product. Debates about health care reform thus involve disputes about empirical and conceptual matters, as well as about normative principles and values, or at least about their implications and trade-offs.

Before concentrating on these normative principles and values, particularly as expressed in rights and obligations, one important conceptual matter needs attention—the language of rationing. Compare the following statements, which are not uncommon in our debates about health care reform as well as about policies regarding specific procedures, such as heart and liver transplants: (1) "I hope we will never see the day when we ration health care in the United States"; (2) "We ration health care all the time in the United States."

For the first speaker, "rationing" refers to deliberate, nonmarket, usually governmental actions. For instance, the U.S. government rationed sugar, gasoline, and meat during World War II so that even those who could afford to purchase such goods were not able to do so beyond their "ration." Rationing in this sense affects those who have the money to pay for goods, such as health care, as well as those who do not, because it involves some mechanism of distribution other than the market.

For the second speaker, "rationing" includes any restriction on access to health care, whether by the government, an institution, or the market. It is a generic term for various limits on or barriers to access. Hence, even market economies ration goods by price, so that those who cannot afford those goods lack access to them (except for charity). More specifically, the market rations access to health care, and millions who want and need but cannot afford health care may not receive it.[2]

Where health care is rationed, in either sense of the term, people lack access to some health care that could be expected to provide a net benefit to them.[3] It is impossible to legislate proper use of the term "rationing"—even though I favor its broad use—in part because its use often depends on which principle or value is invoked. Since rationing in either sense tends to connote a nonideal (though perhaps inevitable) state of affairs, what counts as rationing will depend in part on whether liberty or equality is infringed, as well as on the extent of scarcity and limits on access. For the libertarian, governmental encroachment on the market to allocate goods is problematic rationing, which can be corrected by letting the free market operate; for the egalitarian, societal toleration of inequalities in access to health care in the market is problematic rationing, which can perhaps be corrected by governmental intervention in the market, at least as a last resort. The egalitarian thus points to the problematic rationing of health care in the marketplace—for example, the millions who lack access to health care—to justify governmental intervention to establish or ensure a political-legal right to health care. However, that political-legal right to health care itself probably cannot be implemented without some rationing, whether by the government, by other institutions, or by the market for health care beyond a decent minimum. Without attempting to legislate proper use of the term "rationing," we can at least note variations in usage and their bearing on the contemporary debate.

In this chapter I want to consider whether, and on what grounds, the United States should do what it has not yet done, namely, establish or ensure a political-legal right to health care. I will analyze and critically assess some major arguments regarding rights to receive (and obligations to provide) health care in a modern democratic society.[4] First I will briefly analyze the language of rights, distinguishing moral from political-legal rights and rights to health care from rights in health care. Then I will consider several possible grounds for a right to health care, particularly the grounds provided by the principles of fairness and compassionate beneficence in conjunction with an interpretation of the distinctiveness of health needs over against other needs and desires. Even if several moral arguments support a political-legal right to health care, there are major moral disputes about the content and scope of such a right.[5] Problems of content and scope include determining criteria for allocating resources for and within health care, as well as the acceptability of denying health care to those who have voluntarily taken health risks. I will next consider the question of the bearer of the duty to provide health care and the role of physicians and the government in discharging this duty before concluding with some concerns about the current and continuing revolution in managed care.

While I believe that there are strong arguments for a moral right to health care within democratic societies, I will attempt to show that various moral and nonmoral arguments, apart from such a moral right, support establishing a political-legal right to health care within the United States.

RIGHTS AND OBLIGATIONS

Much of the dispute about a right to health care centers on the appropriateness of the language of rights, "rights" being understood as justified claims against others. It is important to distinguish rights *in* health care (e.g., the right to consent to or to refuse medical treatment) from rights *to* health care.[6] The former are less controversial than the latter, in part because rights in health care largely consist of negative rights (i.e., rights to others' forbearance) along with some positive rights (i.e., rights to others' positive actions) that largely stem from negative rights or from customary expectations and contractual agreements. For example, the right to information from a physician or health care professional is a right in medical care; it is a positive right that is indispensable for voluntary consent to or refusal of medical treatment. It is more difficult to establish positive rights requiring the society or other individuals, who are not in special moral relationships such as promise-making and keeping, to act in positive ways to promote an individual's welfare. My interest here lies in determining whether there should be a positive (political-legal) right to health care, and, if so, its grounds, content, and scope.

There is a strong and understandable temptation to avoid the language of rights in discussing access to health care. For example, in its widely discussed report, *Securing Access to Health Care,* the President's Commission for the Study of Ethical Problems in Medicine and Biomedical and Behavioral Research assiduously avoided the language of rights and argued instead that there is a societal *obligation* to provide equitable access to adequate health care without excessive burdens to all citizens. The Commission held that, while all moral rights imply corresponding obligations, not all moral obligations imply corresponding rights: "A person may have a moral obligation to help those in need, even though the needy cannot, strictly speaking, demand that person's aid as something they are due."[7]

What is so special and dangerous about the language of rights? A right, I suggested earlier, is a *justified claim* against others.[8] The holder of a right can justifiably claim actions from others, and this justifiable claim appears to be both rhetorically and morally stronger when stated as a right (X's right to Z from Y) than when stated as an obligation (Y's obligation to provide Z to X), even if the right and the obligation have the same content (as in a right to receive health care or an obligation to provide health care). But whatever rhetorical significance the language of rights may have ("*my* right" vs. "*your* obligation"), rights and obligations appear to have the same moral strength in much of the moral life. First, there is a "firm but untidy" correlation of rights and obligations: if X has a right, Y has an obligation (and vice versa).[9] Some "obligations," such as charity, may be exceptions. However, when the Commission argues for a strong societal obligation to provide equitable access to health care, this obligation appears to imply a right on the part of citizens to equitable access to health care. Second, if rights are justified claims against others, they are justified by appeals to principles and rules in the same way obligations and duties are justified. The critical arguments hinge on principles and rules, which justify both rights and obligations. Thus, whether we start with rights or obligations, the argument generally encompasses the other and ultimately appeals to principles and rules.

Suspicion of the language of rights appears to stem, at least in part, from uncertainty about the relative strength and scope of rights over against obligations; for example, a right to health care appears to be both stronger and broader than a societal obligation to provide health care. Nevertheless, problems about content and scope, as well as about weight and stringency, are just as serious for obligations as for rights. For example, a right to health care may be absolute, prima facie, or relative; it may have varying content, such as equal access or a decent minimum; and it may be broad or narrow. All of these points hold for a societal obligation, too. Hence, a societal obligation to provide health care, as proposed by the President's Commission, also requires specification in terms of its content, range, and scope, and requires the assignment of weight relative to other obligations.

The principles and rules by which both obligations and rights are justified, as well as specified and weighted, may be assumed for purposes of argument, perhaps having been discerned in a society's own practices, policies, and laws. Rather than analyzing a right to health care in terms of a general, systematic theory, such as John Rawls's *A Theory of Justice,* I will analyze it in terms of several principles and rules that may be, and in fact are, supported by various theories. While such an approach lacks the timelessness of a more theoretical approach, its proximity to current debates may make it more valuable.

In contrast to earlier efforts to provide more theoretical justifications for political-legal rights to health care, many philosophers and theologians have recently focused in similar ways on the implications of intermediate, socially embedded principles and values for health policy. For instance, in largely defending the Clinton administration's health care plan, Dan Brock and Norman Daniels identify fourteen principles and values that "are deeply anchored in the moral traditions we share as a nation, reflecting our long-standing commitment to equality, justice, liberty, and community. . . . There is a widespread consensus on their central role in defining our common community," even though different moral, religious, and cultural traditions may emphasize some over others and weight them differently in cases of conflict.[10]

In limiting the context of my arguments to the United States, I do not here address questions about a "human right" to health care that cuts across national boundaries.[11] I rather attempt to build up from socially accepted principles and values to strong arguments for a political-legal right to health care in the United States. I do not deny the possibility of cross-cultural criticism. And "moral geography" is not unimportant. Questions about "moral geography" are sometimes raised to oppose capital punishment in the United States by observing that most other countries that retain and use capital punishment actually place a low value on human life. Similar concerns of "moral geography" also arise in discussions of rights to health care, because the United States stands alone among the developed countries—unless South Africa is included—in allowing such a high percentage of its citizens and residents to languish outside the system of health care because of financial barriers to access.

Humphrey Taylor and Uwe E. Reinhardt examine a major public opinion poll, in relation to other such polls, to determine which of three main hypotheses might explain why the U.S. health care system differs from those of other developed Western countries in not providing access to affordable care for all of its citizens.[12] The first hypothesis is that Americans have a different social ethic and are less compassionate about the underprivileged because everyone should be responsible for his or her own welfare. However, according to Taylor and Reinhardt, since the evidence shows that "majorities (some of them very large) or pluralities of Americans, like those in other countries, support positions in favor of universal coverage, it is clear that the American health-care system does not reflect the ideology or social ethic of most Americans." Rather, they argue, there is a tension between the American health care system and the American social ethic. Nor does American suspicion of government's capacity to accomplish major ends explain "why this country has failed to satisfy the belief of more than four out of five Americans that the government should be responsible for providing (or seeing that others provide) health care for those who need it." The most plausible hypothesis for our failure to establish a right to health care is the set of checks and balances and the role of interest groups in the American political system, rather than our social ethic.

I am emphasizing the moral or ethical arguments, particularly for ways to think about a right to health care. However, both rights and obligations appear in various spheres of life where they rest on distinct (though not necessarily separate) principles and rules. Moral rights and obligations are those justified by moral principles and rules. Just as obligations and duties may be political, social, legal, or religious, as well as moral, so rights may also take various forms and may be justified by different kinds of principles and rules. The debate about a right to health care is frequently confusing because of the distinctions and relations between a moral right to health care and an institutional right to health care, what I am calling a *political-legal right*, authorized through political processes and made a matter of law. Some apparently believe that the only adequate moral foundation of a political-legal right to health care is itself a moral right to health care. Of course, a well-founded moral right to health care would provide strong support for a political-legal right to health care. However, adequate moral justification, not to mention adequate moral motivation, may also be found in societal compassion, charity, and shared humanity, apart from any claims about moral rights. An act-utilitarian who recognizes no moral rights may also argue for a political-legal right to health care. And such a right may find additional support in nonmoral considerations such as personal prudence.

In short, arguments for a political-legal right to health care do not stand or fall on whether there is a moral right to health care; the language of rights does not exhaust our moral resources for argument. It is also very important to note the overlap or convergence of several moral (and other) arguments for a political-legal right to health care: We *ought* to adopt and implement a political-legal right to health care, even if arguments for a moral right to health care are not fully convincing.

POSSIBLE GROUNDS FOR A POLITICAL-LEGAL RIGHT TO HEALTH CARE

The grounds of rights and obligations will be found, I have argued, in principles and rules. Several moral principles operate in debates about whether there is a moral right to health care and whether there should be a political-legal right to health care. I will not deal with general egalitarian arguments for the equal distribution of all goods, including health care. Such arguments tend to focus on the implications of a commitment to the "welfare state," a term apparently first used by William Temple to mean that the "community makes corporate provisions for its citizens by guaranteeing them a minimum standard of life as of right, below which they will not be allowed to fall. It is not a matter of merit or desert. It is likely to include a guaranteed subsistence in the case of unemployment, as well as access to education and health care, and some housing provision."[13] Thus, one possible approach is to argue for a political-legal right to health care as part of the package of welfare to which every citizen should have access as a matter of course. Apart from such general egalitarian arguments regarding individual welfare, most arguments for a political-legal right to health care, particularly in the United States, appeal to several moral principles conjoined with claims about the special nature and importance of both health needs and health care.

Central to debates about the distribution of health care, as H. Tristram Engelhardt, Jr., suggests, are different views of the "natural lottery."[14] The metaphor of a lottery suggests that various health needs result largely from impersonal factors, such as genetics, and are thus undeserved. However, even if these needs are largely undeserved, insofar as they result from chance, this fact alone does not determine how, or on what grounds, the society should respond to them; for as Engelhardt further notes, undeserved needs may be viewed as either *unfortunate* or *unfair*. If needs are unfortunate, they may be the object of compassionate beneficence, and other individuals, voluntary associations, and even the society may voluntarily try to meet those needs out of compassion. If, however, needs are unfair, there may be a duty of justice to try to meet them. There is less agreement about a general obligation of justice to blunt the negative health effects of the natural lottery, which might generate a *general* right to health care, than about specific obligations of justice to blunt the negative health effects of particular individual and societal actions, thereby creating some *specific* rights to health care.

Specific Rights to Health Care

Before turning to a general political-legal right to health care, then, it is helpful to consider some specific political-legal rights to health care that have emerged or should emerge. This "spot zoning" or, to change the metaphor, "patchwork," of specific rights may be fairly extensive, though still short of universal. For instance, it is morally warranted and perhaps even obligatory, based on fairness, to provide health care to veterans who serve their country in the military. A similar argument may support the provision of health care, as well as other compensation, to subjects who were injured in research undertaken on behalf of the society, often under governmental funding, sponsorship or mandate. It may also be morally warranted and perhaps even obligatory to provide coverage for individuals who have been injured or suffered disease as a result of industrial pollution, which the society failed to force industries to correct. Our societal provision of health care for American Indians may also be morally obligatory in view of our past treatment of Native Americans. And, at the end of the previous chapter, I presented the arguments of the federal Task Force on Organ Transplantation for the provision of funds for extrarenal transplants, all based on specific contextual factors. However far specific "spot zoning" or "patchworking" extends, it stops short of a *general* political-legal right to health care.[15]

General Rights to Health Care

Fairness and the natural lottery. Perhaps the major argument for some sort of general right to health care hinges on the principle of fairness applied to health care needs. For example, Gene Outka has argued that the peculiar nature of health crises is the primary reason why the society should use the distributive principle of need and provide equal access to health care: "One should emphatically not separate the questions of special features of health care from particular conceptions of social justice. It is precisely because of such features that certain conceptions apply more fittingly than others."[16] In general, according to Outka, health crises are undeserved, randomly distributed, unpredictable, and overridingly important when they appear. Because of these features of health needs, such conceptions of social justice as distribution according to merit, according to societal contribution, or according to the satisfaction of the free desires of others in the marketplace, are mistaken. Instead, Outka argues, the criterion of "equal access" or "to each according to his needs" is more acceptable because it fits convictions that needs are given rather than acquired and that patients in need are innocent in relation to the natural lottery. While a critic might reply that health needs are comparable to other basic needs such as food and shelter, Outka insists that health crises are distinctive because they are "peculiarly random and uncontrollable."[17]

Other appeals to the principle of fairness in relation to the nature and source of health care needs appear in the report of the President's Commission for the Study of Ethical Problems in Medicine and Biomedical and Behavioral

Research, as well as in various philosophical discussions. Thus, the President's Commission notes that "if persons generally deserved their health conditions or if the need for health care were fully within the individual's control, the fact that some lack adequate care would not be viewed as an inequity. But differences in health status, and hence differences in health care needs, are largely undeserved because they are, for the most part, not within the individual's control."[18] And philosopher Norman Daniels, drawing in part on John Rawls, contends that health needs represent departures from normal species functioning and thus deprive people of fair equality of opportunity. Why? Because "normal species functioning is an important component of the *opportunity range* open to individuals in a society." Thus, the moral principle of fair equality of opportunity requires the provision of health care to "maintain, restore, or compensate for the loss of normal functioning."[19]

Compassionate beneficence. A second argument for a political-legal right to health care focuses on compassion, benevolence, and sharing, rather than on fairness, and views health needs as unfortunate rather than unfair. It often appeals to charity rather than to justice. While this argument is frequently applied only to individuals or voluntary associations (and might be accepted in those contexts by libertarians), it can be extended to the society and to the state in various ways. One way is through *social* virtue and character, which ought to be expressed in social policies. "The values we hold," writes L. H. LaRue, "can be used for our own self-definitions—to give a particular shape and meaning to our life. Furthermore, they can be used to represent the type of society in which we wish to live."[20] If we can distinguish *expressing* a value from *realizing* a goal, this approach emphasizes symbolizing, expressing, and conveying values, such as the society's compassion and care and the victim's pricelessness. (See also the discussion of symbolic rationality in chapter 1.) Themes from this approach echo in the report of the President's Commission:

> Since all human beings are vulnerable to disease and all die, health care has a special interpersonal significance: it expresses and nurtures bonds of empathy and compassion. The depth of a society's concern about health care can be seen as a measure of its sense of solidarity in the face of suffering and death...a society's commitment to health care reflects some of its most basic attitudes about what it is to be a member of the human community.[21]

In the Netherlands, for example, solidarity is often viewed as a collective obligation to take care of citizens, including the obligation to provide health care.[22] And Larry R. Churchill has argued for a communitarian obligation to provide health care, based on an understanding of ourselves as "social creatures" rather than atomistic individuals.[23] This social understanding reconciles compassionate Samaritanism with the claims of justice, because we do not have to create a social life from calculating self-interest; rather, we find the basis of justice in our "social passions and social interests." In short, "Samaritanism is best understood as a sign of our social nature rather than a supererogatory overcoming of alienated and egoistic motives to condescend to help another. Justice is Samaritanism enlivened by an impartial imagination." Such social arguments can also find support in some feminist approaches to care/caring, in relation to justice, even though such approaches often focus more on communities of choice than on the society as a whole.[24]

Another way to extend this second type of argument, based on compassionate beneficence, to the state is through the need for coordination in the effective *realization* of individual and associational charitable goals. Complex, technological health care requires coordination of efforts and activities, including capital investment and medical research. With some exceptions, modern-day Good Samaritans cannot deliver effective health care without such coordination and joint effort. Thus, support for a political-legal right to health care may derive from an "enforced beneficence argument," based on the need for coordination:

> There is a basic moral obligation of charity or beneficence to those in need. In a society that has the resources and technical knowledge to improve health or at least to ameliorate important health defects, the application of this requirement of beneficence includes the provision of resources for at least some forms of health care. If we are sincere, we will be concerned with the efficacy of our charitable or beneficent impulses. . . . The difficulty is that in the absence of enforcement, individuals who strive to make their beneficence most effective will thereby fail to benefit the needy as much as they might.[25]

So, Allen E. Buchanan argues, "rationally beneficent individuals," while denying that those in need have a moral right to health care, may nevertheless "agree to establish a coercively-backed principle specifying certain health programs for the needy and requiring those who possess the needed resources to contribute to such programs."[26] This version of the second argument— enforced beneficence—would hold only where there is a need for government coordination for the effective realization of charitable goals; the other more communitarian version—expression of social virtue and character—would presumably hold in all contexts. Although the "enforced beneficence" version is colder than the other version's emphasis on the value of community and the expression of compassion, both versions converge in their support for a legal-political right to some level of health care on the grounds of compassionate beneficence in response to unfortunate health needs. Neither version presupposes that there is a moral right to health care or that justice requires a corrective social response to undeserved health needs. Both versions build on less controversial foundations.[27]

Similarity of health needs to other needs met by society. A third type of argument also focuses on the nature of health care needs—it stresses that they are relevantly similar to the needs the society already meets through governmental action. Threats to health are similar to military threats on the national level and to threats from fire and crime on the local level. Just as the government on different levels meets the threats from military aggression and fire and crime in order to protect its citizens, so it ought to meet health threats through the provision of some health care beyond public health measures. Thus, as Gary E. Jones argues, "Closer examination of the present role of the State implies that there is a right to at least a minimal amount of health care."[28] Whereas the first two types of argument concentrate on what is distinctive and special about health needs and health care, this argument emphasizes what is *common* between health needs and services and other needs and services already receiving governmental support as part of the state's commitment to protect its citizens. Of course, identification of important dissimilarities among the above goods would deflect a charge of inconsistency against a society that provided some goods without also providing health care.[29] For example, individual health is not a public good in the same sense as national defense.

The implications of other sociomoral practices regarding health needs and services. Another type of argument considers our sociomoral practices regarding health needs and services rather than the nature of health needs (and services) in general or in comparison with other needs (and services). It examines what our society has *made* of health needs and services, not their intrinsic nature. An important version of this argument appears in Michael Walzer's *Spheres of Justice,* which holds that it is necessary to determine the "social meaning" of medical care.[30] In our society, Walzer argues, there is a "distinctive logic of the practice of medicine": "care should be proportionate to illness and not to wealth." How is this logic discerned? By examining the society's practices regarding health care. Our society has invested in health care, a "socially recognized need," and deprivation of health care involves a double loss—to one's health and to one's social standing as a citizen. When society identifies a good, such as health care, as needed, it blocks or constrains free exchange. While Walzer concedes that a two-tiered system—decent minimum of care and then free enterprise—would not be unjust in principle, he contends that it would be unjust in the United States because the "common appreciation of the importance of medical care" has carried the American people "well beyond" such an arrangement: "So long as communal funds are spent, as they currently are, to finance research, build hospitals, and pay the fees of doctors in private practice, the services that these expenditures underwrite must be equally available." The problem with this argument, as Ronald Dworkin contends,[31] is that it fails to appreciate the diversity of the American tradition regarding health care and thus fails to argue for the superiority of one strand of the tradition over other strands. Recognition of diversity would require an argument about the general principles involved in each strand of the tradition.

Prudential choices. All of the previous arguments for a general political-legal right to health care appeal directly to moral principles and values. However, it is also possible to appeal to prudence, as a supplement to, as a way to provide content for, or as a substitute for the moral arguments. Prudential justifications take two different forms. The first form is an appeal to hypothetical, idealized prudence, and it has been used to argue both for and against a right to a decent minimum of health care.[32] Examples in support of a political-legal right to health care include Norman Daniels's "prudential lifespan account" and Ronald Dworkin's "ideal prudent insurers," both of which appeal to hypothetical and idealized choices by prudent agents under circumscribed conditions.[33] Hence, over against the rescue principle, Dworkin proposes that we try to imagine the "prudent insurance" ideal, which involves imagining health care under "a free and unsubsidized market," without the current deficiencies of markets in health care—that is, the markets are transformed by virtue of a fair distribution of wealth and income; full information about the benefits, costs, and risks of various medical procedures; and the absence of information about the likelihood that any particular person will experience morbidity, either life-threatening or non-life-threatening, from diseases and accidents. Under such ideal circumstances, whatever aggregate amount this "transformed community" spends on health care is just, as is its distribution pattern, whatever it is. Why? Because "a just distribution is one that well-informed people create for themselves by individual choices, provided that the economic system and the distribution of wealth in the community in which these choices are made are themselves just." Dworkin proposes taking these assumptions about prudent choices under fair conditions to determine what it is "a universal health-care system should make sure, in all justice, that everyone does have. . . ." If most prudent buyers with average means in a free market would buy a certain level of health coverage, then it is the "unfairness" of the current system that prevents them from doing so now. Hence, justice would require that coverage. Although sometimes offered more as a guide for allocating health care than for justifying political-legal rights to health care, these objectives are not totally distinct, as Dworkin's approach clearly indicates. And Dworkin holds that "the prudent insurance test helps to answer . . . questions of justice" about allocation of funds for and distribution of health care, rather than substituting for them. Even so, some critics contend that these conditions of prudential rationality are too indeterminate to warrant conclusions of the sort Dworkin draws.[34] While contending that his proposed thought experiment is "not . . . beyond the reach of the imagination," Dworkin also concedes that answers to allocation questions cannot be determined "with any precision" through this imaginative exercise, because different people applying the "prudent insurance test" may give different answers.

On the other hand, there is also an appeal to actual prudential individuals, who are asked to realize that they may sometime need health care. Building on David Hume's conception of enlarged self-interest, Larry Churchill contends that both solidarity and personal security (in the sense of not fearing lack of attention to one's basic health concerns or impoverishment from expenditures for health care), as two major goals in health care, are linked as "natural outgrowths of enlightened self-interest, rather than as expressions of benevolence or as communitarian virtues."[35] Churchill appeals to "rational self-interest" in the context of four main factors: the scarcity of health resources in relation to health problems, the uncontrolled increase in health care expenditures, the largely unpredictable nature of health care needs, and the precarious status of being "insured." Rather than following Dworkin and others in the direction of idealized prudential decisions, Churchill contends that each of us, as a prudential individual, needs "to wake up to the precariousness of our situation, to place the I in a We context, and to acknowledge a common self-interest in universal care." Such a juxtaposition of the prudential, self-interested agent and others presupposes reciprocal standard-setting in which something like the Golden Rule or principle of universalizability operates in establishing the level and limits of services. In many respects, Churchill's appeal to prudence provides a motivating rather than a justifying reason, at least of a moral sort, for instituting a political-legal right to health care, even though the argument is constrained by moral considerations.

These various arguments for a general political-legal right to health care should not be viewed as mutually exclusive; they often overlap and provide mutual support. So far, however, I have focused mainly on possible arguments for such a right without regard to its content and scope, even though some major differences in content and scope emerged in the sketch of different grounds. For instance, even if the provision of communal funds creates a societal obligation to provide health care, as Dworkin notes in response to Walzer's claims, there is still legitimate debate about the content and scope of the obligation (and correlative right) within the American tradition. It is certainly not clear that equal access to health care finds stronger support in the American tradition than a right to a decent minimum of health care. Similar points can be made about other types of argument. Some of the above arguments for a general political-legal right to health care target equal access, some target a decent minimum, and still others appear to offer resources for either standard of access. Thus, it is necessary to consider ways to set and determine the content and scope of a right to health care.

CONTENT AND SCOPE OF A POLITICAL-LEGAL RIGHT TO HEALTH CARE

Whatever the grounds for a political-legal right to health care, it is essential to specify its content and scope. Specification must be consistent with and guided by moral principles, including the ones that may justify it. Even a political-legal right established by "legislative grace," as in the exemption of conscientious objectors from military service, cannot have arbitrary and capricious limits. A privilege extended out of social benevolence must still respect such principles as fairness.

The President's Commission argued that a standard of "equitable access to health care requires that all citizens be able to secure an adequate level of care without excessive burdens."[36] If this standard is unobjectionable, it is because the term "equitable," as Daniel Wikler notes, is only "a place-holder for a substantive moral standard for access."[37] Libertarians believe that the free market and voluntary, private charity should determine access to health care. Egalitarians could accept this standard of "equitable access," but they would define "equity" as equality or equal access; and strong egalitarians would insist that what is available to one should be available to all. By contrast, their major opponents defend a right to a decent minimum of care, which accords fairly well with the specification of equitable access offered by the President's Commission.[38]

I will explore the content and scope of a right to health care by considering standards of (1) equal access to health care, and (2) access to a decent minimum of health care in relation to three variables that appear in the formulation of the President's Commission—access, level of care, and associated costs or burdens (such as travel).

The question of a right to health care becomes one of access to what level of care with what burdens. That these questions are analytically distinct can be seen in debates about rights of availability and rights of access. As Paul Starr notes, "the right to postal service is a right of availability, not of access."[39] The government guarantees to transmit mail in and out of communities, but it does not guarantee that individuals will have sufficient funds to gain access to the system. Nor does the government guarantee delivery to one's front door; a person whose house is a considerable distance from the main road may have to drive to the mail box. Furthermore, while everyone may have equal access to police and fire protection in a local community, it is possible to judge that the protection is inadequate for that community and for individuals with special needs within that community. Thus, the content and scope of a right to health care must be defined in part in terms of availability and burdens as well as access. Concern for access focuses on removing barriers standing between individuals and certain goods; but it is also necessary to determine how much of those goods should be made available and at what burdens to the society and to the individuals involved.

Questions about the Standard of Equal Access

Let us look first at the relations between access and availability in the two major competing positions: a right to equal access and a right to a decent minimum, both of which may be defended by principles of societal benevolence and fairness in conjunction with health care needs (as well as by the other arguments previously discussed). Defenders of equal access contend that "every person who shares the same type and degree of health needs must be given an equally effective chance of receiving appropriate treatment of equal quality as long as that treatment is available to anyone."[40] Such a principle, strictly speaking, implies a single class of medical care, and it would rule out the freedom to buy and sell treatment that could not be made available to everyone. Ability to pay would not be admitted as a criterion of distribution at any level of access to the good of medical care.

Critics of the strict standard of equal access charge that it would either lead the society to put more into health care than its citizens want relative to other goods, or lead the society to limit the quantity and quality of care available to anyone in order to avoid having to make it available to everyone (e.g., it might refuse to permit research into and development of an artificial heart). Regarding the first charge, arguments for equal access do not necessarily imply any level of allocation of funds for making health care available; they only require that citizens not face financial barriers in obtaining whatever health care is available. As players under the late professional football coach, Vince Lombardi, noted, "he treated all of his players equally—like dirt." Thus, the standard of equal access may not imply too much, as some critics fear; it may imply too little. It may imply access only to what is available in the society, even if what is available is inadequate for a decent minimum of health care. In practice, however, the same reasons that favor equal access, particularly the nature of health care needs, will also support the allocation of sufficient resources for adequate health care. Certain health needs are overwhelmingly significant when they appear. At least those needs should be met, if they can be met, and health care should be made available to provide services to meet those needs.

If the provision of equal access successfully avoids the bottomless pit of health needs (and desires), it tends, according to its critics, to limit availability in order to avoid having to limit access, which should be equal. Thus, the criticism continues, proponents of equal access will sacrifice the amount and quality of care in order to avoid rationing, which necessarily involves unequal access. Most western European nations, as Robert Baker notes, choose to ration availability rather than to ration access; they tend, therefore, to restrict, for example, the number of operating rooms or the number of hours they operate or the number of procedures performed, all these restrictions producing delays in and sometimes de facto denials of care.[41]

Proponents of equal access often view their principles as stating an "ideal" or "a prima facie authoritative goal" for the society; it is a critical standard for the assessment and direction of policies toward health care.[42] Thus, they frequently argue that society should *strive toward* the goal or *seek to approximate* the ideal of equal access to health care while conceding that compromises may be necessary under conditions of scarcity. One such compromise may involve rationing scarce resources, such as dialysis, artificial hearts, or heart transplants, when not all can be treated. Nevertheless, the unrealized goal or compromised ideal may still leave "moral traces" that dictate the mode of rationing.[43] The "really effective chance" that proponents of equal access demand cannot be determined by income or status, which would violate principles of equal respect. While rationing might violate the right to equal treatment, it need not violate the right to treatment as an equal—to use Ronald Dworkin's distinction.[44] The requirement that a person be treated as an equal entails at least that his/her interests should not be disregarded and arbitrarily violated. The right to treatment as an equal thus undergirds certain procedural guarantees that express equality and fairness. Therefore, under conditions of scarcity, proponents of equal access tend to favor certain modes of rationing, triage, or microallocation, such as queuing, first come/first served, or randomization, along with the determination of medical utility. (See the examples in the three previous chapters.)

Questions about the Decent Minimum

Such an argument for rationing holds most cogently for lifesaving medical treatments known to be efficacious for clearly identified and urgent needs. However, not all health needs are equally basic, important, and urgent, and, correlatively, health care is heterogeneous. Thus, proponents of a right to a decent minimum of health care, who can accept the view of rationing developed above for certain levels or kinds of health care, contend that distributing *some* health care services according to ability to pay is not morally inappropriate. They argue that it is morally permissible and even essential to distinguish levels or tiers of health care. They support a floor below which no one should fall, but reject a ceiling beyond which no one can go. Everyone should have access to that floor, or, to change the metaphor, everyone should be protected by a safety net. Nevertheless, there is widespread debate about where that lowest level should be set. And similar debates emerge in various contexts, such as efforts to develop basic or essential benefits packages for insurance programs.

Some defenders of a political-legal right to a decent minimum of health care prefer to define the minimum in terms of a fair share of society's resources and to distribute that fair share in cash so that citizens can make their own choices.[45] A counterargument in favor of services or vouchers, rather than money, need not be paternalistic—that is, it need not restrict freedom of choice merely for the welfare of those whose freedom is restricted. Fairness is also a relevant and decisive consideration. If the society provided fair shares of money to individuals to spend at their discretion, and if the society for reasons of compassionate beneficence regarding health needs (or other reasons) would not deny critical care to those who had to spend their money on other goods, then fairness would dictate that the society provide health care services or vouchers instead of money. Otherwise, some individuals would consume more than their fair share of resources.

The criterion of adequacy. The main criterion for what health care services should be made available as part of the decent minimum is *adequacy* of the amount, level, kind, and quality of health care.[46] Tests of adequacy will at the least include whether health care is responsive to certain needs, for example, basic, urgent, or life-threatening needs.

Since the main argument for some right to health care is that health needs are either unfair (justice) or unfortunate (compassion), it is not surprising that an analysis of the content and scope of a right to health care focuses first on health needs. Part of what is at stake is a definition of health and thus of health needs and services. In general, there is a rough correlation between views about a right to health care and views about the objectivity and subjectivity of definitions of health and disease.[47] Opponents of a right to health care often contend that definitions of health needs and of health care are subjective and thus depend on the values of individuals and communities. They stress the *continuity* between various needs, including health needs, and between various needs and desires. By contrast, defenders of a right to health care tend to accept a narrower, more objective definition of health and disease, emphasizing *discontinuity* or at least sharp distinctions between health needs and other needs and desires. Nevertheless, both defenders and opponents of a right to health care agree that arguments for a political-legal right to health care are not cogent if clear *limits* cannot be set on health needs and services. As I stressed, proponents of both equal access and a decent minimum concur that some limits are needed; both recognize a limited right to health care and thus some form of rationing.

For example, Daniels adopts a narrow, objective definition: Diseases are departures from normal species functioning, and broadly speaking, "health care needs" involve what is needed to "prevent, maintain, restore, or compensate for . . . departures from normal species functioning." Hence, for Daniels, the fair equality of opportunity principle requires access to health care that can meet these needs but not necessarily to health care services—such as some cosmetic surgery or some counseling—that satisfy other wants and preferences.[48]

For Daniels, as for many others who argue for access to a decent minimum of health care, there should be a firm commitment to equal initial access to the health care system for diagnosis, for prognosis, and for treatment if necessary, beneficial, and so forth. What that system should make available will depend on such matters as distinctions between needs and desires and priorities among health needs—for example, urgent, life-threatening needs, or basic needs versus other needs.

Benefits and costs. Beyond the availability of services to meet certain health needs, tests of adequacy will certainly be complex and perhaps not susceptible to full specification. Nevertheless, certain directions and limits can be indicated by two additional tests of adequacy: (1) whether the health care in question offers a reasonable chance of benefiting the patient, and (2) whether the probable benefits outweigh the costs. Although I cannot discuss either test in detail, a few comments and illustrations are in order.

First, the society has no obligation to provide, and the patient has no right to expect, health care that does not offer a reasonable chance of benefit. A right to health care does not include a right to ineffective treatment. Nevertheless, major debates occur about which treatments are effective. For example, it has been difficult for patients to get private or public funds to cover some liver transplants. The debates about whether some liver transplants should be covered hinge in part on whether the procedure is considered experimental and whether it offers a reasonable chance of benefit.

Second, formulating the tests of adequacy to consider whether the probable benefits outweigh the costs introduces the last variable—costs or burdens—along with such techniques of cost-benefit analysis (CBA) and cost-effectiveness analysis (CEA), frequently using such measurements as quality-adjusted-life years (QALYs). These techniques are often used even when they are not explicitly identified or formally developed. Although both (especially CBA) raise numerous issues, including moral issues, they probably cannot be avoided and should be used with full awareness of their limitations. Such analyses can help specify the content and scope of a right to health care, even though they do not offer a foundation for that right.[49]

The debate about the guaiac test, an inexpensive qualitative test for minute amounts of blood, provides an instructive example. While a positive stool guaiac may indicate many different problems, one major cause is colon cancer, which can be cured if it can be diagnosed early enough. Several years ago the American Cancer Society recommended six sequential stool guaiacs to screen for colorectal cancer because a single test detects only about 12 percent of the cancers. The first test costs $4, and each additional test would be warranted by a cost-effectiveness analysis used to determine what should be made available as part of adequate health care. One study has indicated that the marginal cost *per case diagnosed* climbs dramatically: $1,175 for one test, $5,492 for two tests, $49,150 for three tests, $469,534 for four tests, $4.7 million for five tests, and $47 million for six tests.[50] It is doubtful that all six tests would be viewed as essential to adequate health care since the marginal cost per cancer detected far exceeds usual figures about the value of life from the standpoint of either discounted future earnings or willingness to pay.[51]

Procedures of Public Participation

When it is difficult to resolve disputes about substantive standards such as adequacy of health care, there is a tendency to revert to procedures. This tendency is understandable and frequently warranted, especially in a democratic society, if the procedures themselves are morally acceptable. Thus, the President's Commission rightly concludes: "It is reasonable for a society to turn to fair, democratic political procedures to make a choice among just alternatives. Given the great imprecision in the notion of adequate health care, however, it is especially important that the procedures used to define the level be—and be perceived to be—fair."[52] Deliberations and decisions within those procedures can be aided by careful cost-effectiveness and cost-benefit analyses, as long as we recognize that much of the debate really concerns the values that determine what will count as benefits and costs and how much particular benefits and costs will count, as well as how uncertainty will be handled (as in the discussion of additional guaiac tests).[53] It is not easy to determine a fair procedure for public participation, as indicated by the extensive debate about Oregon's efforts to rank medical procedures for Medicaid coverage through CEA using QALYs based on values articulated with public input.[54]

Broader Considerations of Costs and Cost Containment

The third variable—costs or burdens—has already been introduced in connection with CBA for determining adequate health care, but some of its other components need brief attention. I have viewed access mainly as overcoming financial barriers between individuals and the good of health care, but another important question is how the financial costs of health care will be *distributed* throughout society. And there are costs and burdens in health care other than financial ones. Some of these burdens include distance traveled for care and time involved in receiving care. Even when the financial costs of access to care have been covered, policies need to address these other burdens.

One reason arguments for universal access to health care have not triumphed in the public policy arena is the persistent and deep concern about the costs of health care, which, as I noted at the outset, many identify as the fundamental flaw in our current system. The common cry for "cost containment" is often ambiguous and elusive. True cost containment, Eli Ginsberg argues, is simply "a reduced inflow of real resources into the health-care system without a diminution in useful output that would adversely affect the satisfaction of patients or their health status."[55] Often what passes for cost containment is not *containing* costs but *shifting* costs from one sector to another, for example, from the federal and state governments to the private sector, sometimes through a redefinition of eligibility rules. However, when the private sector is unable to absorb all of these costs, some people will be denied access to adequate medical care. Or what passes for cost containment will successfully control costs at a substantial reduction in "useful output."

Societal concerns about the rising costs of medical care have directed attention to lifestyles and behavioral patterns.[56] Prevention, especially through alterations in lifestyles and conduct, has sometimes been touted as a major way to control costs.[57] If this is true, then it might appear that the exclusion of voluntary risk-takers from medical care when they develop health problems would be justified on cost-benefit grounds and would perhaps also be consistent with the principle of fairness. However, matters are not so clear-cut. Some risk-taking may actually require less medical care because it results in earlier and quicker deaths. Furthermore, a cost-benefit analysis should not be limited to the system of health care but should also include other programs such as social security and retirement. From this broader perspective, the case against providing medical care for medical needs that result to a great extent from individual risk-taking could disappear altogether. "Over the long run," Howard M. Leichter argues, "under public or private health and retirement systems, one can expect an increase rather than decrease in social expenditures as a result of avoiding health risks."[58] It would be fair, nevertheless, to ask individuals who engage in risky actions that result in costly medical needs to pay more along the way to cover the costs of their medical care, even if those needs do not ultimately cost the society more money when all programs are considered. They might contribute more to insurance schemes or pay a tax on their risky conduct, for example, an increased tax on cigarettes. However, it is not feasible to tax much or even most risk-taking in order to spread the costs of the risks more fairly and equitably.

Apart from a broad cost-benefit analysis, does the society have the same obligation to provide health care to voluntary risk-takers as it does to genuine "victims" of the "natural lottery"? Or do voluntary risk-takers forfeit their general political-legal right to health care? As we have seen, arguments for a general political-legal right to health care often appeal to fairness and/or compassionate beneficence in response to health needs, which are viewed as largely determined by the "natural lottery." According to the fairness argument, it would not be unfair to exclude voluntary risk-takers from medical care for health needs brought on by their own actions.[59] However, fairness is not the only argument for a general political-legal right to health care, and even this principle fails to justify the exclusion of voluntary risk-takers unless their actions are truly voluntary (after all, their predisposition toward those actions may itself be the result of the natural lottery, or they may not be aware that their actions are risky), unless their actions are the primary cause of their health needs, and so forth. Furthermore, a societal policy to exclude voluntary risk-takers from care would probably require a "health police" and breaches of privacy in order to identify certain risk-takers. All in all, such a policy is a bad idea.[60]

These several examples indicate the necessity of considering simultaneously all three variables—access and level of care as well as costs. What appears to be fair and just in the determination of a right to health care from the standpoint of one variable may be unjust and unfair from the standpoint of another variable.[61] At the end of this chapter I will consider the turn to managed care as a way to contain the costs of health care, perhaps without sufficient attention to access, availability, and adequacy.

WHO IS OBLIGATED TO PROVIDE HEALTH CARE?

Which individuals, professions, and institutions have an obligation to provide health care? This is an important question, whether we affirm a right to health care (and hence a correlative obligation), or whether we affirm an obligation to provide health care apart from rights. Part of what is involved in determining whether there is an obligation and/or right to health care is determining *who* bears such an obligation, independent of or correlative to rights. Paul F. Camenisch argues that health care deliverers have a special and significant obligation in their larger social context. His thesis is that every citizen has a right to needed health care, which health care deliverers are morally obliged to recognize by performing the enabling duties correlative to that right. These duties are grounded in the deliverers' knowing and willing participation in the sociomoral practices of gifts, roles, and promises through which they consent to take on such duties. For example, by accepting the benefits of medical education, with considerable support from the society, health care deliverers have a special obligation to meet the right of all citizens to health care.[62]

Although special obligations result from accepting benefits and gifts, from making explicit and implicit promises, from assuming certain roles, and from fair play, the *content* of those obligations is frequently determined by the social setting. It is not possible to determine the content of the obligation of health care professionals by looking only at formal transactions. For instance, some defenders of a right to health care note that over half of the costs of medical education and approximately 40 percent of the costs of health services in the United States come from public funds; and they conclude that medical professionals have an obligation correlative to the right to health care, because "accepting large quantities of public funds implies a significant obligation to the society."[63] Where there is an explicit contract (e.g., the government might support a certain number of years of service in a rural area), there may be no question about the content of the obligation. Apart from such clear expectations, however, many different actions may *meet, fulfill, or discharge obligations without themselves being obligatory.*[64] Although this distinction between *obligation-meeting actions* and *obligatory actions* is sometimes overlooked, it is essential in order to make sense of obligations that allow their bearers considerable discretion in the choice of actions to discharge those obligations. For example, I may be obligated *to* my parents because of what they have done for me; these obligations may stem from gratitude and fair play. I may not, however, be obligated *to do* any (or many) specific acts; I may have considerable discretion about *how* I discharge my obligation to them without having discretion about *whether* I discharge it.[65]

Similarly, physicians may not be obligated (except as a matter of charity) to choose a medical specialty or a location for their practice or even to treat patients who cannot afford treatment and are not covered by other funds. We might praise them for doing so, and society might even legitimately offer incentives, in addition to praise, to get physicians and other health care professionals to deliver certain modes of care in certain locations or to certain patients in certain settings. Implementation of a political-legal right to health care need not be detrimental to the relationship between physicians (and other health care professionals) and patients, in contrast to what some physicians and others fear.[66] Some rights in and to health care are defensible, and a political-legal right to health care in the sense discussed in this chapter can be instituted without damaging the relationship. Physicians should still seek, within the limits of access, availability, and costs set by the society, to do what is in the best interests of their patients, as limited by other patient rights (such as the right to consent to and to refuse treatment).

There is an important distinction, as Camenisch recognizes, between *deliverer* and *provider* of health care: Physicians and other health care professionals deliver health care, but ultimately the society provides it. If the society bears the obligation to provide health care, all citizens have a duty to work for a just health care system and to act in various roles and relationships (e.g., in voluntary associations) to make sure that health care needs are met. Because of their debts of gratitude and fair play, physicians and other health care professionals may have *special* obligations to work for a just health care system and to deliver some free health care to the poor in some circumstances. These debts of gratitude and fair play do not, however, constitute an obligation to deliver health care correlative to a right to health care apart from changes in expectations in the social system and from clear articulations of mutual responsibilities. Nevertheless, recognition of such debts should make it easier for physicians and other health care professionals to work for—or at least not to oppose—alterations in the social system that would more effectively meet the health needs of the population, even at some cost to them (e.g., in lower income than they desire). And their private charity is not fully gratuitous when their capacity to deliver health care results in part from the actions and policies of the society and the state.[67]

The President's Commission contends that there is a "societal obligation" to provide equitable access to health care. By "societal obligation," it refers to "society in the broadest sense—the collective American community," which consists of "individuals, who are in turn members of many other, overlapping groups, both public and private; local, state, regional, and national units; professional and workplace organizations; religious, educational, and charitable organizations; and family, kinship, and ethnic groups." The Commission argued that "all these entities play a role in discharging societal obligations."[68] If we distinguish (without necessarily separating) the society from the state and distinguish levels of government within the state, there is still a question about the proper role of the federal government in discharging the societal obligation. Some conservatives could even accept a "societal obligation" to provide health care if it meant only that individuals and voluntary associations "ought" to deliver or cover the costs of health care as part of their charitable actions; but they would not admit a significant role for the government. By contrast, within its pluralistic approach the President's Commission admitted a significant role for the federal government as the institution of last resort: The "ultimate responsibility" rests with the federal government "for seeing that health care is available to all when the market, private charity, and government efforts at the state and local level are insufficient in achieving equity."[69]

Principles in several moral, social, and political traditions can support *priority* for institutions other than the state—for example, the value of voluntarism, the principle of subsidiarity in Catholic thought, and claims about the ineffectiveness and inefficiency of government. If, however, the moral reasons for a societal obligation to provide health care are solid, they also support a role for the federal government, at least in the last resort. Some moral reasons may also make the resort even earlier, especially if recipients of health care as charity (rather than as a political-legal right) view themselves as dependent on others and do not seek care because it is so demeaning to do so. Within our current context, I would argue that the federal government needs to do more to provide access to adequate health care with fewer burdens. Professional charity and voluntary contributions are not adequate to provide for the millions who currently fall between the cracks in the floor of health care or wander through the Medicaid–private insurance corridor.[70]

Another effort—less coordinated by the government, which is nevertheless involved—attempts to reform contemporary health care through "managed care." However, it defines the fundamental problem mainly as one of costs rather than access, and it puts physicians in a central role in controlling costs by putting them under proper incentives to do so. Having always served as gatekeepers to health care, physicians now play that role with a different script and in a different setting.

CONTROLLING COSTS WHILE RATIONING
ACCESS AND CONSTRAINING CHOICES:
THE DILEMMAS OF MANAGED CARE

The sharp turn to managed care[71] as a way to deliver health care is the latest effort to control national health care costs in the wake of regulatory and other failures to limit overutilization in an open-ended insurance reimbursement system. The evidence is still mixed about how successfully managed care plans can contain health care costs in the long run,[72] but they are, nevertheless, touted as the main vehicle of the 1990s for controlling costs, even though many believe that other structural changes will also be required to produce true cost containment.[73] In 1993 almost 80 percent of U.S. citizens received health care insurance through their employers, and 51 percent of those were enrolled in managed care programs, a substantial increase over the 29 percent enrolled in such programs in 1988.[74] Clearly, as noted in chapter 1, there has been a shift in the primary metaphors for health care, especially from warfare to business, with a concomitant shift in primary orientation and values.[75]

Yet the term "managed care" has been used so widely and loosely as to become almost meaningless. It embraces a diverse set of organizational and financial arrangements, from tightly bound group practice health mainte-nance organizations (HMOs), to looser affiliations of physicians and hospitals linked by payment formulas, to traditional forms of practice managed through third party restraints. Nevertheless, all types of managed care attempt to control costs through modifying the behavior of physicians and patients in order to promote a more rational (though not necessarily more just) use of resources. To accomplish this goal, they use various devices, such as case management and utilization control.

Managed care is clearly driven less by concerns about availability or access or even quality than by concerns about the costs of health care. It is commonly assumed that managed care can limit expensive unnecessary care without reducing appropriate treatment, but this assumption may not be warranted— it is simply too early to say. At the very least, a mechanism that primarily manages cost does not fit easily into a system that has traditionally sought to manage patient care through physicians serving as advocates for patients' medical needs. Because physicians are responsible for 75 percent of all health care expenditures, their judgments about appropriate care for their patients obviously have a major economic impact as well. In order to control these costs, managed care programs expect physicians to temper their health care recommendations based on their personal financial incentives as well as their appreciation of the limits on resources. The various ethical problems emerging in managed care include damaging the physician-patient relationship, adding financial incentives for the provider to limit care and thereby creating a serious conflict of interest, and limiting physician and patient autonomy, as well as rationing health services without the kind of public accountability that justice appears to require.[76]

It might be argued that serious conflicts of interest have always plagued medical care. However, the patient is in a very different position when the physician has incentives to *withhold needed treatment* than when the physician has incentives to *provide unnecessary treatment.* In the latter situation, represented by fee-for-service, the patient can obtain another opinion; in the former, the patient may be totally unaware of a needed treatment because no one has recommended it.[77] Conflicts of interest in managed care thus threaten individual patient choice, and it is unclear whether such mechanisms as the required disclosure of conflicts will be adequate, particularly in light of other major limits on patient choices.

Constraints on physicians' abilities to act on behalf of their patients pose serious ethical problems for these medical gatekeepers. On the one hand, even if their own financial interest is not decisive, physicians may sometimes be unable to act effectively on behalf of their patients without "gaming the system," but such actions may threaten, perhaps unfairly, the allocation pattern of the particular system, such as an HMO. On the other hand, physicians have a moral responsibility to participate in setting defensible priorities in the global budgets within which they practice, rather than leaving those decisions to administrators. (In global budgeting, a plan or unit receives a fixed amount of money to use in meeting the health needs of a particular population.) Beyond physician participation, there is also a strong case for public participation in setting those priorities. Rather than leaving the decision at different levels to clinicians (or administrators), "a responsible society must choose its own principles for allocating global healthcare budgets."[78]

This point holds for all global budgets, such as state Medicaid budgets, which can, for instance, sometimes cover liver transplants only by substantially reducing other care. In a closed system with a global budget, where resources cannot be expanded (at least immediately), trade-offs may be clear; and it may be entirely appropriate to make a careful cost-benefit analysis and decide not to provide funds for certain transplants, as has occurred in several states, including Oregon and Virginia. However, the arguments may be quite different in a more open system, such as the society as a whole, where it is not clear that withholding funds from extrarenal transplants will actually free those funds for basic care for pregnant women and children, to take one example.[79] Nevertheless, public participation is ethically indispensable in setting priorities in both closed and open contexts.

The fundamental social-ethical issue that arises from managed care is the tension between containing the costs of health care and providing just access to and broad consumer/patient choice in health care. In order to be ethically sound, managed care, with its firm dedication to cost control, must include— or the broader society must include—equitable access to health care, which could occur (1) through increasing the social responsibility of managed care organizations, beyond cost containment, or (2) through providing societal funds to ensure equitable access within a general political-legal right to health care. In principle, the cost savings from managed care could be used to provide wider access to health care, but in practice, this will almost certainly not occur without a fundamental societal commitment expressed in new policies to ensure equitable access to health care. For instance, in many managed care programs, there is a temptation to shift risks or to neglect the needs of chronically ill populations in order to limit costs.[80] And, regarding the other option, our society has not yet displayed the political will to provide anything close to fair access to health care.

In this context the illusion of choice—rather than real choice—prevails. One constraint is that individuals often cannot even choose the health plan they pay for. While Americans typically choose their own home, automobile, and life insurance plan, their choices about access to health care providers and services are largely determined by their place of employment. Seven out of ten Americans purchase health insurance at the workplace, and a growing number of those are enrolled in managed care plans that limit choice of doctors.[81] Ideally, in an open market, they would be able to choose from various options, but in reality, they face limited choices. Eighty-four percent of employers who offer health insurance to their employees offer only one plan,[82] and some of these plans offer a limited choice of providers. Such restrictions are allegedly justified by the need to reduce administrative and other costs, even though, as a result, employers have little say in their health care, which is, nevertheless, purchased through employee copayments, deductibles, out-of-pocket cost sharing, and foregone wages.[83]

The employer's need to control costs often conflicts with the employee's desire for certain services and providers. Particular plans change over time, and small businesses, in particular, often switch to less expensive health care coverage for their employees, further compromising continuity of care and constraining choice. In addition, the average worker changes jobs eight times over a lifetime, thus moving from one plan to another. Despite all the disadvantages of an employer-based health insurance system, it remains in place because of the tax-free status of the company-provided fringe benefit, Americans' largest tax break, worth approximately $90 billion in federal, state, and local taxes.[84] Under the illusion of choice, it actually undermines individual choice in two ways: by disadvantaging those who wish to purchase health insurance outside the employment arena (where there is no tax relief and nongroup rates are markedly more expensive) and by substituting corporate decisions for consumer choices.

Finally, more than 37 million people who are uninsured in this country, one quarter of whom are children, have little choice in the type of health care they receive or how they receive it, if they can gain access to it at all. Traditionally, uncompensated indigent care has been provided—however incompletely and inadequately—by private and public hospitals which shift the costs onto their paying clients (patients and insurers) in the form of higher charges or insurance premiums. However, managed care programs severely limit hospitals' ability to subsidize non-Medicaid indigent care because lower cost, capitated plans prevent cost-shifting to private payers, and already strained state budgets cannot easily pay for the care of the uninsured. In one study, 34 percent of the uninsured reported that over the previous year they had needed medical care but did not get it, and 71 percent stated that they had postponed medical care because they could not afford it.[85] Working Americans and their families without company-provided benefits often must go without care or purchase bare-bones policies with only catastrophic coverage. Either way, their access to health services is severely limited. By all accounts managed care arrangements will further reduce the availability of health care to those with inadequate or no insurance and will further limit their choice. Thus, our current de facto health policy—the corporatization of health care with strong pressures to downsize the system and impose tighter economic discipline on providers—appears to reduce substantially the choices about access for patients and prospective patients as well as for physicians. This new world of corporate managed care threatens choice in several interconnected ways—employers limit health plans; health plans limit physicians and hospitals; and employees are limited in their ability to protect their interests or find other employment since they usually lack any means of access to health care other than those provided by their health plans. These reductions of choice will continue without serious public accountability unless there are major changes in societal discourse and policies.

Our society's failure to recognize and attempt to correct the limitations of managed care reflects, in part, its ambivalence regarding the role of government. It also reflects our temptation, at least in times of economic exigency, to view the less fortunate as victims of their own failures rather than as victims of a "natural lottery" or "social lottery." Without a publicly defined social policy concerning health care, our national health policy is becoming, de facto, the sum of financial incentives and disincentives within a largely private system of care. Such managed care arrangements to control costs have their own human and ethical costs—including the further reduction of individual choice even as they maintain the illusion of choice—particularly when the society has chronically failed to address serious, persistent limits on access to health care resulting from individuals' limited financial resources and insurance.

In conclusion, managed care will perhaps (though not certainly) reduce the societal costs of health care. Such a reduction, if it actually occurs, will not be accompanied by increased access to health care because there is a lack of societal commitment to use any possible savings to bring others into health care. Managed care may thus address one problem of contemporary health care—its rapidly escalating costs—while failing to attend adequately to access or to availability and quality. Without a societal perception of and commitment to resolve the problem of access, a less costly system will still remain an unjust system. The fact that it is less costly in no way diminishes its injustice. A more just health care system awaits a vigorous moral, social, and political commitment—whose appearance is, sadly, nowhere in sight—to establish a general political-legal right to health care.

Part Five

■

OBTAINING ORGANS
AND TISSUES FOR
TRANSPLANTATION

CHAPTER FOURTEEN

□

Ethical Criteria for Policies to Obtain Organs

for Transplantation

LANGUAGE, CONCEPTS, AND MORAL NORMS

Several controversies in biomedical ethics revolve around transfers and uses of human body parts (HBPs). (An HBP is here defined as "an organ, tissue, eye, bone, artery, blood, fluid, or other portion of the human body.")[1] The language, concepts, and moral norms for cadaveric organ and tissue transplantation are often contested, in part because of complex beliefs, sentiments, symbols, and practices surrounding the dead body.

Examples of problematic language include "harvesting," "salvaging," "procurement," and "retrieval" of organs, each of which has ambiguous and even controversial implications."[2] The widely used term "donor" itself presents problems, because it is used to refer to the cadaveric source of the organs as well as to the decision maker about donation. However, the first use is inappropriate unless the source of the organs also made the decision to donate. Hence, it is inappropriate to describe as "donors" deceased individuals who never had the capacity to decide or never decided to donate. A deceased child or a decedent who never in any way indicated his or her wishes about donation cannot appropriately be described as a "donor" of the organs that are removed from his or her body and transplanted into others. Such organ providers are at most sources of organs, rather than donors. Whoever made the decision to donate the decedent's organs is properly described as the donor. Likewise, someone who sells an organ is a "vendor" or "seller" rather than a "donor." And descriptions of "paid donations" or "paid donors" mix the moral practice of making donations or giving gifts with the distinct moral practice of selling and buying.[3]

Developments in transplantation and in reproductive technologies pose significant questions about the moral rights and obligations of individuals, families, health care professionals, and society at large in the transfer and use of HBPs. These questions include: Who has dispositional authority over cadaveric HBPs—individuals while alive, their next of kin after their death, medical and health care professionals, or the society? And what moral and legal limits constrain their dispositional authority? What rights do those with dispositional authority have—the rights to possess, to use, to exclude others from use, to destroy, to transfer by various means? What are the philosophical presuppositions of judgments about dispositional authority, particularly regarding the nature of personhood and embodiment? Is it appropriate to construe this dispositional authority in terms of ownership and property rights? (This last question will be addressed in the next chapter, along with the more specific topic of transfer of organs and tissues by sale and purchase.)

Actual and proposed public policies (here defined as whatever governments choose to do or not to do), as well as social practices for obtaining organs for transplantation, frequently assume answers to these questions—often without facing, much less resolving them, in our pluralistic society. In addition to testing our language and concepts surrounding HBPs and their transfer and use, we also need to evaluate actual and proposed policies to obtain organs according to their ethical acceptability, preferability, and political feasibility. We need to determine which policies meet our shared ethical standards in a minimally acceptable way and then to determine which, among the policies that pass ethical screening, are actually ethically preferable. For instance, given two ethically acceptable policies, one might be ethically preferable because it promotes communal altruism. This distinction between "ethically acceptable" and "ethically preferable" is often overlooked. Some commentators on Paul Ramsey's *The Patient as Person* indicate that he ruled out a policy of presumed donation; but he did not, in fact, rule it out—rather, he gave express donation priority because of its ethical preferability.[4]

My assessment of ethical acceptability and ethical preferability will hinge on several moral principles or values that have become commonplace in biomedical ethics.[5] They have been widely accepted (with only slight variations), and they are embedded in various policies and practices, which will be my focus rather than any particular moral theology (e.g., divine revelation) or moral philosophy (e.g., the contractarianism of John Rawls). While losing some of the distance and timelessness provided by abstract ethical theories, my approach also gains from its appeal to principles that already operate in public policies and practices and can still be invoked to criticize and direct them. Embedded principles are not immune to criticism from the standpoint of ethical theory, and it is crucial to move dialectically between ethical theory and ordinary general and particular moral judgments. My approach to this embedded "common morality" is close to the approach of the common law. I will appeal to principles that can be discerned in our public policies and practices, recognizing that they may be inchoate, incomplete, and even inconsistent at times. This approach can be assessed in part according to its capacity to illuminate and direct our debates about policies to obtain HBPs.

The relevant moral principles are respect for persons, including their autonomous choices and actions; beneficence, including the obligation both to benefit others (positive beneficence) and to maximize good consequences—that is, to do the greatest good for the greatest number (utility); nonmaleficence, the obligation not to inflict harm; and justice, the principle of fair and equitable distribution of benefits and burdens. These principles are prima facie binding rather than absolute or merely relative—that is, they are binding insofar as their features are present. As prima facie binding, these principles should not be infringed, other things being equal. However, any act or policy may have other features (including the infringement of another principle), and these prima facie principles may thus come into conflict under some circumstances in which one may justifiably be overridden in order to protect another. In balancing these principles, intuition is not our only guide. It is important to follow a definite procedure of reasoning that shows that the infringement of one principle is necessary to protect another principle, that it would probably protect the principle which is considered stronger or weightier in the situation, and that the infringement is the least necessary to realize the objectives that are sought. (For this procedure see chapters 4 and 6 in this volume.)

After determining which policies are ethically acceptable and ethically preferable, we still have to ask which are politically feasible. Political feasibility certainly includes—but is not limited to—a policy's ability to survive legislative and judicial scrutiny. It also includes cooperation from various other parties—a policy to obtain organs for transplantation is not feasible unless it can elicit the necessary cooperation from institutions, professionals, and individuals and their families. And, as noted earlier, organ transplantation is a complex area because of the various beliefs, emotions, symbols, and practices evoked by the dead body. Any policy that ignores or neglects these complexities will probably fail to increase the supply of transplantable organs.

I will evaluate several possible modes of acquisition or transfer of HBPs, in part according to their effectiveness in realizing our societal goal of increasing the supply of organs for transplantation to save lives and enhance their quality. Rather than attempt to justify this societal goal, I will assume it; I will then seek to determine which policies to obtain organs are ethically acceptable and which are ethically preferable in light of the prima facie moral principles identified earlier and, furthermore, which are politically feasible, especially as suggested by public and professional attitudes and practices.

Finally, my analysis separates policies to obtain organs from policies to distribute them. However, their moral connections are significant.[6] It is obvious that policies to obtain organs have a significant impact on distributional problems. For example, successful policies to obtain organs will reduce scarcity and thus eliminate some problems of patient selection. However, it is less clearly recognized that policies of organ distribution may also have an impact on the success or failure of policies to obtain organs. For example, in a system that relies on donations, the public's perception that distributional policies are unfair will probably reduce organ donations. Depending as it does on public goodwill, the current system of organ donation is fragile in many respects. (Evidence for this claim will appear below.) This fragility stems in part from the very factors that lead to media and public fascination with organ transplantation: the saving of human life through the transfer and transplantation of HBPs from one (living or dead) person to another. Unless this fragility is recognized, proposed public policies to increase the supply of organs may actually do more harm than good—they may even be counterproductive as well as ineffective.

MAJOR METHODS OF ACQUIRING ORGANS FOR TRANSPLANTATION

There are six main ways to acquire (or transfer) HBPs: express donation, presumed donation, routine removal, expropriation, abandonment, and sale/purchase. All currently play some role in our society or other societies. For example, under our present laws and policies, express donation is virtually the only mode of transfer of solid organs; it also became the dominant mode for transferring blood following a period in which the market approach was used. However, in the acquisition of semen for artificial insemination, sales are common, and in the acquisition of tissues for the development of cell lines, abandonment is also frequent.[7] Expropriation is rare and more controversial; however, it is evident in legally required autopsies and tests for drugs. The laws in several states that permit the removal of corneas from bodies under the auspices of the medical examiner without the decedent's prior or the family's current permission may be characterized as either presumed donation or routine removal, depending on other contextual features. This chapter will concentrate on donation, with passing attention to other modes of transfer except for sale, which will be examined in the next chapter.

It is not clear that there should be a single method of transfer for all HBPs, partly because of the significant differences among the HBPs (e.g., renewability), and it is not clear that any HBP should be transferable in only one way. One fundamental question is whether solid cadaveric organs should continue to be transferred almost exclusively through some form of donation or whether an alternative mode of transfer could more effectively increase the supply of organs for transplantation without infringing important moral principles.

Express Donation

The legal framework for organ procurement in all fifty states and the District of Columbia is the Uniform Anatomical Gift Act (UAGA), which was rapidly adopted with some modifications in the late 1960s and early 1970s.[8] Within the legal framework of the UAGA, individuals may determine what will be done with their organs after their death; in the absence of a valid expression of the decedent's wishes, the family may decide whether to donate the organs. Thus the UAGA reversed the trend in several jurisdictions of viewing the cadaver as the "quasi-property" of the family rather than of the decedent, if there was a conflict between them. Some ethicists, such as Paul Ramsey, were ambivalent about the UAGA, partly because its emphasis on the individual's autonomous choices while alive appeared to threaten the legal tradition of the family's quasi-property right in the corpse.[9]

Many thought that the UAGA's individualistic framework would generate a large supply of organs because opinion polls indicated that people were willing to donate their own organs after their death. A Gallup poll taken in 1968 (around the time of the first heart transplant) determined that 70 percent of the U.S. public would be willing to donate organs for delivery after their death.[10] In practice, however, the family, not the decedent while alive, has emerged as the primary donor of cadaveric organs.[11]

Individuals as donors. Few individuals take advantage of the main mechanism for expressly donating their organs, the donor card, or what the UAGA calls the "document of gift." Indeed, there has been a significant decline in the stated willingness of individuals to donate their organs by signing donor cards. According to several polls over the last several years, only 40 to 50 percent of adults are very or somewhat likely to donate their organs after their death.[12] The reasons cited include lack of thought about it and reluctance to face the prospect of death; but some of the most significant reasons reflect distrust, or limited trust, of the system. Respondents worry that if they sign a donor card, physicians might take premature action to obtain their organs before they are really dead or might even hasten their death.[13] This concern increased as the relationship between patients and physicians became more adversarial during the 1970s (as reflected in part in the increase in malpractice suits). The stated willingness to become organ donors is even lower among groups (such as blacks) who perceive themselves to be on the margins of the system and to have even less reason to trust it.[14] In addition, organ procurement teams still regularly consult the family.[15] And if a conflict exists between the decedent's wishes and the family's wishes, the family tends to win, in part because of unwarranted fears of legal liability (the UAGA in fact grants blanket immunity for good faith actions on the basis of signed donor cards), because of legitimate fears of jeopardizing organ donation and transplantation as a result of negative publicity about the institution's and professionals' "aggressive" actions (the loss of many organs for the sake of a few organs from this source), and because of the desire not to inflict harm on a family already experiencing profound tragedy.

If we seek to increase express donation by individuals, which policies would be ethically acceptable, ethically preferable, and politically feasible? First, as an expression of the principle of respect for persons, the individual's wishes should be established as determinative in cases of conflict with familial wishes. This priority was intended in the original UAGA; the amended UAGA eliminated any uncertainty about the matter, stating that "an anatomical gift that is not revoked by the donor before death is irrevocable and does not require the consent or concurrence of any person after the donor's death."[16]

Second, as a way to further respect for the legal and moral priority of the decedent's wishes, the amended UAGA rightly imposes an affirmative obligation on the part of appropriate officials, such as law enforcement officers and firemen, to "make a reasonable search for a document of gift" or refusal by potential sources of organs and to notify the hospital of their findings.

Third, as the amended UAGA also provides, there should be a clear way for the individual to refuse to be a donor, because both the failure to sign and the revocation of a document of gift are ambiguous. Neither necessarily indicates an agent's intention to block the donation of his/her organs; each indicates only that the agent did not express, or actually revoked, the intention to donate, perhaps because of the fears noted earlier about signed documents of gift, while leaving open the legal possibility of familial donation after his or her death.

Fourth, in addition to providing definite ways for individuals to make or to refuse donations, and recognizing the priority of decedents' wishes to donate or not, the UAGA should be amended to allow individuals to designate decision makers on their behalf—that is, surrogates who can express the decedent's values in the circumstances of donation.[17] There is a parallel in some natural death acts, which now allow not only the specification of standards (e.g., "I don't want to be kept alive by X under Y") but also the designation of a decision maker who can presumably reflect the agent's values in the actual situation that develops (e.g., "I designate John or Mary to make decisions about my treatment if I am incompetent to do so"). Similarly, in the context of organ transplantation, individuals may be willing to designate a decision maker whom they trust rather than signing a document of gift that they fear might put them at risk at the hands of strangers in large, impersonal, bureaucratic institutions that are eager to gain organs.

Routine inquiry. Different settings of routine inquiry, aimed at individuals while alive and competent, need careful scrutiny. On the one hand, routine inquiry of individuals as they enter the hospital is problematic. Nevertheless, the amended UAGA suggests a routine inquiry to ascertain whether the patient being admitted to the hospital is a donor at the same time questions are asked about insurance. If the patient is not a donor, a hospital designee, with the consent of the attending physician, "shall discuss with the patient the option to make or refuse to make an anatomical gift." While expressing the principle of respect for personal autonomy, routine inquiry is not clearly appropriate at the time of hospital admission, and because of the reasons for which individuals are reluctant to sign documents of gift, it would probably lead most patients to avoid putting themselves on record as donors of organs. Indeed, they might even refuse donation rather than express no decision, thereby precluding the possibility of familial donation.

Routine inquiry directed at individuals in many other settings is not ethically problematic and may even be modestly effective—for example, when individuals obtain a driver's license. Nevertheless, routine inquiry still reflects excessive attention to individually signed documents of gift, as well as severely constrained interpretations of autonomy, problems that mar, even seriously, proposals for mandated choice.

Mandated choice. Both the misdirected orientation toward donor cards and the faulty conception of autonomy appear in common proposals for mandated choice. As an extension of routine inquiry, mandated choice would require individuals to state their preferences regarding organ donation in conjunction with some other state-mandated task, such as renewing a driver's license or filing income tax forms. It would force individuals to make a choice (yes, no, or undecided), and their choice could then be entered into a registry.

In its defense of mandated choice, the American Medical Association Council on Ethical and Judicial Affairs states: "Requiring a decision regarding donation would overcome a major obstacle to organ donation—the reluctance of individuals to contemplate their own deaths and the disposition of their bodies after death—and individual autonomy would be protected and even enhanced."[18] The AMA Council thus claims both rightness and effectiveness for a policy of mandated choice; that is, it would both respect individual autonomy and increase the supply of transplantable organs. However, the AMA Council's explanation for mandated choice reveals the proposal's inadequacy from the standpoint of effectiveness: "Under mandated choice, individuals who feel this reluctance [to contemplate their own deaths and the disposition of their bodies after death] would have to confront it, thereby removing it as a barrier to donation." This explanation addresses only one set of reasons for individuals' reluctance to sign donor cards—and not necessarily the most significant reasons at that: It does not address, for instance, important fears and attitudes of distrust. Forcing the choice, I have long argued, would decrease, rather than increase the supply of organs. Why? If forced to choose, many individuals would check "no," not because they necessarily oppose the donation of their organs after their death, but because they are afraid of being on record as donors of organs, with only the time of delivery of their gift to be determined. In saying no, they would block their family's possible decision to donate. Now there is some limited evidence along those lines: In Texas, a version of mandated choice has resulted in a refusal rate of 80 percent. And those refusals block familial decision making.[19]

Proponents of mandated choice could retreat to rightness over effectiveness, but their assessment of rightness in terms of respect for autonomy distorts that norm by narrowly focusing on, and attempting to institutionalize, a particular conception of autonomy, one that is very individualistic, rationalistic, formalistic, and legalistic.[20] They fail to see that people might exercise their autonomy about donation in various ways—for example, by delegating the decision to a family member or friend, or by refraining from blocking the family's subsequent decision. Furthermore, the AMA Council, quite surprisingly, not only seeks to respect or even to promote individual autonomy but also to foster what it describes as an "ethically preferable" exercise of autonomy: "The individual's interest in controlling the disposition of his or her own body and property after death suggests that it is *ethically preferable* for the individual, rather than the family, to decide to donate organs" (my emphasis).[21] It is unclear why it is ethically preferable for an individual to decide to donate his or her organs, rather than, for example, explicitly delegating or merely leaving that decision to someone else. The AMA Council thus invokes an *ideal* of personal autonomy, rather than a principle of *obligation* of respect for someone else's autonomy. And that ideal is difficult to defend and even hard to take seriously in the absence of any plausible defense.

Perhaps the AMA Council's response can be partially discerned in its claim that it is only articulating what the UAGA also holds: "The Uniform Anatomical Gift Act's emphasis on individual autonomy and individual decision making would be protected and enhanced by a system of mandated choice, in which the donation decision would have to be confronted before death and would have to be made by the individual donor, not by a surrogate."[22] In contrast to the AMA Council's interpretation, however, the UAGA only supports an individual's *right* to dispose of his/her body parts—the amended UAGA does not indicate or argue that it is "ethically preferable" for the individual rather than the family to make the decision. It is both ethically acceptable and ethically preferable to have a system in which individuals can make their own decisions and in which those decisions triumph over familial objections. The AMA Council report badly confuses an individual's right to make a decision with the ethical value—and putatively supreme value at that—of the individual making his or her own decision.

This confusion also mars the AMA Council's appeal to a survey in support of its recommendation of mandated choice as an effective option. As the AMA Council reports, 90 percent of the respondents in that survey indicated that they would support a *program* of mandated choice. However, there is no evidence that they stated that they would individually donate organs under such a program or that they would in fact do so, especially since they do not currently do so even under routine inquiry. Public acceptance of a program of mandated choice should not be confused with individuals' willingness to say "yes" to organ donation under that program. The AMA Council's conceptual confusion leads it to claim, quite mistakenly, that there is "empirical evidence that mandated choice would be acceptable to the public and therefore effective in increasing the organ supply."[23] There may be empirical evidence that the public would accept such a *program*, but there is no empirical evidence that it would be effective in increasing the organ supply. And as already indicated, all the available empirical evidence about what people say they would do, and what they actually do, is sufficient to rebut such a claim about effectiveness.

Families as donors. In practice, the family has become the main donor (i.e., decision maker about donation) of cadaveric organs for transplantation. Even though it is clear that family members (as specified in the UAGA) have the legal authority to donate in the absence of knowledge of the decedent's wishes, it is still appropriate to examine the family's moral authority to donate. If the decedent, while competent, asked the family to donate his or her organs, the family's action simply conveys the decedent's wishes, which he or she may have been reluctant to put on a donor card for various reasons. In such a case, the decedent may be viewed as the donor and the family as the instrument of that donation. If the decedent, while competent, did not make a decision or express a preference, the family may still decide to donate. Even if family members believe that they are doing what the decedent would have wanted (based on his or her values), they are the donors and the decedent is only the source of organs.

P. M. Quay has argued that "no one, including the state, has any right to make use of a person's cadaver or its parts for research, transplantation, or other purposes, if the deceased has not given his free consent to that use."[24] He construes the decedent's failure to donate as a refusal of donation, but the evidence presented above suggests that people do not sign documents of gift for various reasons, often preferring to leave the decision to the next of kin who can reflect their values and protect them against abuse in a situation of serious vulnerability. In this context, a decedent's refusal to block familial donation may even be considered altruistic, though it is not necessarily a decision to donate. Even though I would argue that morally the individual's wishes have priority—and should have legal priority in cases of conflict with familial wishes—the family still has moral authority to donate or not to donate in the absence of evidence about the decedent's wishes.

Opinion polls indicate that individuals are more willing to donate the organs of family members including their own children than they are to donate their own organs by signing donor cards. In a series of polls, over 80 percent of the respondents indicated that they were very or somewhat likely to donate the organs of family members.[25] A cynical interpretation of these results would be mistaken. When put in the context of the stated reasons for reluctance to sign donor cards, the opinion polls suggest instead that individuals who donate family members' organs can view themselves as buffers or barriers between the untrustworthy system and the potential source of organs. This point is supported by the fact that 63 percent of the respondents to the 1985 Gallup poll approved of the statement: "Even if I have never given anyone permission, I wouldn't mind if my organs were donated upon my death."[26] When there is a signed donor card, people cannot identify a protective buffer or barrier because the donation in principle, though not in practice, has already occurred unless the donor changes his or her mind before incompetence or death.

Required request. First proposed by Arthur Caplan in 1984, required request was subsequently enacted by most state legislatures, recommended by the revised UAGA, required by the Joint Commission on Hospital Accreditation, and mandated in federal legislation for all institutions receiving Medicare or Medicaid funds.[27] Another term might better stress the features that should appear in institutional protocols for approaching the family of a potential source of organs. For example, organ procurement teams should inquire about the decedent's wishes regarding organ donation; if the decedent's wishes are unknown, they could inform family members of their own right to donate and ask them whether they wish to do so or even ask them to donate.

The arguments for required request presupposed a *willingness* on the part of families to donate—as reflected in various opinion polls—and a concurrent *unwillingness* on the part of physicians and other health professionals in hospitals to notify them of their rights to donate and to ask them to donate. For supporters of required request, the main bottleneck appeared to be the failure of institutions and health care professionals to notify families of their rights and to invite them to donate a relative's HBPs under the appropriate circumstances.[28] In short, the problem was a shortage of askers, not a shortage of givers. Even though the reasons for failure to ask include concerns about harming the family, members of the family have a moral and legal right to know they can donate the decedent's organs, and they often report that such a donation was helpful rather than harmful. In these circumstances it is fair to require institutions to work out appropriate mechanisms of notification and request in accord with their own particular circumstances. (Some laws and policies require that institutional mechanisms be established, while others require that requests be made.)

After approximately a decade of implementation in some states, the effectiveness of required request laws and policies remains unclear. Even in the face of early physician resentment and resistance, there was evidence of a substantial increase (10–20 percent) in tissue donation but not in organ donation.[29] Arguments emerged for sanctions to ensure physician compliance, but opponents charged that such measures would produce a backlash.[30] Still other critics contended that "a policy of required request creates conflicts of interest at the clinical, psychological, and social/economic levels that produce considerable disvalue."[31] However, when carefully examined, those conflicts are avoidable through improved education of professionals or through appropriate procedures, or they are inherent in organ procurement itself, whether under required request or under some other policy.[32] Another criticism is that "required request laws represent a further step away from voluntary donations";[33] but this criticism either rests on a conceptual error about voluntariness or misconstrues the situation of request and donation.

Even though required request laws and policies may have contributed early on to some increases in tissue donation, they do not appear to have increased the supply of donated organs—the number of acts of donation of solid organs each year has remained relatively constant for several years, at 4,000 to 4,500, with most increases occurring as a result of changes in the criteria of donor eligibility (e.g., accepting donations from older persons). However, "slow increases in the number of donors, or having the number of donors hold steady given the background of a shrinking donor pool and more restrictive eligibility criteria, suggest that the present level of donation may be higher than it would be without the laws."[34] Required request may have had positive indirect effects in making requests for donation more routine in hospitals. Nevertheless, the reluctance of physicians and other health care professionals to make such requests may account for the modest direct effects. One proposal to overcome professional reluctance is "routine referral," or "required referral," which would require that institutions and health care professionals routinely notify a trained procurement team in circumstances where cadaveric organs and tissues might be available.[35]

Educational programs. Whatever legal and policy changes are enacted, public and professional education will remain crucial. In addition to professional education, public education is still obviously needed, even in the context of required request, since the familial refusal rate is so high. For instance, some studies of the effects of required request indicate that familial refusal rates may be as high as 60 to 70 percent.[36] However, public education needs to be drastically reconceived. Effective educational programs will have to address individuals as members of families, rather than merely as potential signers of donor cards, and will have to address confusion about brain death and attitudes of distrust that appear to pose obstacles to express donation.

Educational efforts are still too often directed at individuals as potential signers of donor cards—even the Mickey Mantle inspired campaign of 1995 promoted baseball "donor cards"—and now need to be redirected at *individuals as members of small communities, mainly their families.* The efforts should be targeted mainly at familial donors rather than at individual donors and at changing attitudes rather than at providing information. To recognize the legal and ethical priority of the individual's wishes to donate (or not) is not to determine the direction and focus of educational policies. Since individuals are more willing to donate organs of family members and also to have family members donate their organs (even if they have not signed documents of gift), efforts should be concentrated not on educating individuals to be donors of their own organs but on educating individuals to authorize family members to donate as well as to donate family members' organs. Hence it is important to secure intrafamilial communication of wishes rather than signatures on documents of gift. (However, as indicated above, it is important for individuals to be able to block any transfer of organs after their death.)

Educational efforts too often address supposed cognitive rather than attitudinal barriers. However, the available data indicate that potential donors know about organ transplantation, its successes, the need for more donated organs, etc. The major obstacles appear to be attitudinal, particularly attitudes of distrust. However, such attitudes are difficult to overcome and simply may not be amenable to educational efforts. Effective education of the public (and professionals) about brain death could help to reduce individual and familial distrust of the system of organ procurement. In addition, education about the system of organ distribution could help to reduce the suspicion that when a famous person such as former Pennsylvania Governor Casey or former baseball player Mickey Mantle obtains scarce organs quite quickly, the distribution criteria have been misapplied to give them priority. In the final analysis, however, many of the causes of distrust lie outside the system of organ procurement/distribution itself, stemming instead from perceived injustices in the wider health care system. And those can only be expected to worsen under the "reforms" brought about by managed care.

We should also not overlook the possibility that vigorous societal and professional actions to alter the criteria of death, which were so essential to the growth of organ transplantation, may be partly responsible for the public's distrust. Warnings appeared in the late 1960s and early 1970s that updating the criteria of death in order to gain access to additional organs for transplantation could lead to a backlash. For example, Ramsey contended that there is a parallel between the necessity on the practical level of differentiating professional roles in declaring death and transplanting organs (so that the physician involved in the transplant should not determine whether the potential source of organs is dead) and the necessity on the intellectual level of differentiating reasons for updating the criteria for determining death.[37] Criteria of death should be updated, he insisted, as part of the care of the dying patients themselves, not as part of the effort to increase the pool of potential donors of organs. Hence, great caution is needed before further altering the criteria of death, even to be able to obtain organs from anencephalic newborns to save the lives of potential pediatric recipients who have end-stage organ failure. As Alexander Capron argues, changing the criteria of brain death—or even creating an exception to brain death—to accommodate the anencephalic may further threaten a fragile system. In addition, it is necessary to educate both professionals and the public that there is only one concept of death, not one for potential sources of organs and another for other people.[38]

Presumed Donation

Legislation not uncommonly establishes presumed donation (or consent) or routine removal of HBPs for transplantation. Frequently called "routine salvaging" or "routine harvesting," such laws have been adopted in several states in the United States to obtain corneas, and in several countries—including Austria, Belgium, Finland, France, Norway, Portugal, and Singapore—to obtain organs and tissues.[39] These laws authorize the removal of organs or tissues, usually without notification, in the absence of the decedent's prior and/or the family's current explicit dissent. They thus involve "opting out" rather than "opting in"—that is, individuals and/or their relatives must act to stop the process of donation or removal rather than to initiate it. Even though countries that obtain organs through "opting out" policies still have waiting lists for renal transplants, they "seem to come closer to meeting their needs for transplant kidneys" than countries that rely on express consent.[40] But would one or both of these policies of "opting out"—presumed donation or routine removal—be ethically acceptable, ethically preferable, and politically feasible in the United States at this time?

Vigorous debate continues about how to construe and evaluate laws and policies of "opting out." "Presumed donation" and "routine removal" are often considered interchangeable, but their moral bases are different. "Presumed donation" assumes that the individual (and/or the family) owns and has dispositional authority over cadaveric organs and that the decedent's prior failure (or the family's current failure) to dissent constitutes consent. In contrast to some misguided criticisms, presumed donation, properly understood, merely shifts the presumption about an individual's and/or family's wishes in the absence of their express statement of those wishes. In contrast to routine removal, it does not presuppose societal ownership of cadaveric HBPs.

In order to determine whether presumed donation is ethically acceptable as a basis for the removal of organs and tissues from a cadaver, it is important to explore further the nature of its presumption. If donation is presumed on the basis of a general theory of human values or on the basis of what reasonable people would do, without any reference to this particular person's (or family's) actions, it is unwarranted. If it is based on what this person would have done if asked, it is more acceptable but still problematic. If, however, it is based on a construal of this person's (or family's) wishes as silently expressed, it can be defensible under some conditions. Perhaps the best way to explicate this approach is to view it as tacit consent rather than as presumed consent.

Where does presumed consent (donation) fit on the spectrum of varieties of consent? The paradigm case of consent is express consent, and it is most prominent in the doctrine of informed consent in therapy and research as well as in the UAGA framework of organ donation. Such express consent creates rights in others. Implied or implicit consent is not expressed but inferred from other actions, even though the consenter may not have understood or intended such an implication. By contrast, tacit consent, appealed to by John Locke, among other political philosophers, is consent that is expressed silently or passively by omissions or by failures to indicate or signify dissent. Under certain conditions, as A. John Simmons notes, the failure to dissent or to object constitutes tacit consent.[41] The potential consenter must be aware of what is going on and know that consent or refusal is appropriate, must have a reasonable period of time for objection, and must understand that expressions of dissent will not be allowed after this period ends. He or she must also understand the accepted means for expressing dissent, and these means must be reasonable and relatively easy to perform. Finally, the effects of dissent cannot be "extremely detrimental to the potential consenter." Some of these conditions ensure the consenter's understanding; others ensure the consent's voluntariness. When these conditions are met, the potential consenter's silence may be construed as tacit consent. Such consent may be ethically valid in some circumstances.

In principle, then, presumed donation, as tacit consent, could be an acceptable basis for organ removal. However, it must satisfy very rigorous standards in order to be ethically valid in practice, for silence may only indicate a lack of understanding of the means of dissent or of the proposed course of action. Hence, vigorous public education would be required, along with easy ways to register dissent. Uneasiness about what silence means may account for the similarity of practices in various societies, whatever their legal structure.[42] For example, in countries with presumed consent legislation, professionals still regularly ask the family.[43] Finally, ethically acceptable presumed donation may not be more cost-effective than enhanced and redirected educational efforts in the context of the UAGA, along with the other changes proposed above.

Another possibility is to approach the family on the basis of the decedent's presumed donation rather than to view it as the decedent's legally valid and sufficient donation. Health care professionals would probably find it easier to approach the relatives of a dead person with a presumption about the decedent's preferences, as established by tacit consent. For example, professionals could approach the family by saying, "Since John didn't dissent, we would like to remove his organs; do you have any objection?" In short, presumed donation, in the form of silent or tacit donation, could provide the basis for a valuable ritual in a difficult, often tragic set of circumstances.[44] However, there is no evidence to suggest that this psychological advantage for professionals would translate into significant changes in rates of organ donation.

In addition to stressing the stringent conditions that would have to be met for valid presumed donation, some argue for the ethical preferability of a system of express donation in order to symbolize and build a community of altruistic concern among strangers.[45] However, a policy of presumed donation also rests on altruism, that is, passive altruism, and does not preclude active altruism. Nor, as I have argued, does it presuppose that HBPs belong to the state. Final control still rests with the individual and/or the family.

Nevertheless, presumed donation of solid organs faces a serious practical problem. Although such a law could possibly have been adopted in the late 1960s, it is not politically feasible now. In one Gallup poll, 86.5 percent of the respondents indicated that they would be opposed to legislation that would give doctors "the power to remove organs from people who have died recently but have not signed an organ donor card without consulting the next of kin."[46] However, supporters of "opting out" over "opting in" stress that several states already have laws that authorize the removal of corneas, without notification of the next of kin or consent of the decedent while alive, by medical examiners or coroners who examine bodies following accidents or homicides. Such laws have survived constitutional challenges in some states, even when they do not require notification of or consent from the next of kin. And they have been effective: For example, substantial increases occurred in cornea transplants in the few years after such legislation was passed in Georgia (from 25 in 1978 to over 1,000 in 1984).[47] It is possible that these laws endure without major vocal opposition because of different social views about different HBPs (corneas versus solid organs), because most solid organs are obtained from brain-dead cadavers (and thus involve all the uncertainties and confusion that surround brain death), and because the public is largely unaware of these laws, even in several states where they exist. If the last is the case, as I suspect it is, presumed donation is not ethically valid, in the current circumstances, because of a lack of understanding on the part of the "donors" who are allegedly "donating" by their silence. Under such circumstances the policy is actually one of routine salvaging or even expropriation or conscription masquerading as a form of consent or donation.

Finally, even if a policy of presumed donation for organs could be adopted at this time, it would probably reduce rather than increase the number of donated organs. Although this claim may appear at first to be counterintuitive, it rests on evidence about the distrust of the system that potential donors have registered in some polls as the major reason for their reluctance to sign donor cards. In a system of presumed donation, it is highly probable that such attitudes of distrust would lead individuals to take affirmative actions to remove themselves from the list of presumed donors. Such actions would prevent familial donations and, under current circumstances, reduce the number of donations.

Routine Removal

In contrast to "presumed donation," "routine removal" presupposes that the society has a right of access to cadaveric organs and tissues because it owns them or because individuals and families have an enforceable social duty to provide them. Such a communitarian approach to organ removal may, nevertheless, allow opting out for various reasons, such as conscientious objection.[48] To identify the operative moral basis of laws that authorize removal without express consent, it is necessary to look beyond the society's rhetoric to its practices. If its policy is "presumed donation," then the society should undertake vigorous efforts to ensure public understanding and voluntariness. Otherwise a decedent's prior or a family's current failure to dissent may only indicate a lack of understanding of the opportunity or means for dissent or of what will be done in the absence of dissent. If, however, the society's policy is "routine removal," public understanding and voluntariness are superfluous because social ownership or social obligation provides the moral foundation for the policy.

Virtually all the objections that stand in the way of "presumed donation" also extend to "routine removal" and need not be repeated. The major presupposition of "routine removal"—that the society has final dispositional authority over cadaveric body parts—provokes the most fundamental objection, based on what William F. May calls the "principle of extra-territoriality," which holds that even the deceased individual is not reducible to the needs of the social order.[49] This fundamental objection to "routine removal" may in part account for the prevalence of the rhetoric of "presumed donation" in various societies. And uneasiness about society's dispositional authority over cadavers—as well as about the meaning of an individual's or family's failure to dissent in presumed donation—may account for the similarity of practices in most countries, despite their different legal frameworks.[50] For example, in most countries with what is termed "presumed consent," professionals regularly ask the family or follow the family's known wishes, just as they do in countries with "express donation." Among European countries with either "presumed donation" or "routine removal," only Austria, with its long tradition of authorizing autopsies on all persons without requesting permission, does not, in practice, notify the next of kin.[51] Finally, in addition to various moral reservations, such a policy appears to be politically infeasible in the United States.

Expropriation

The public health justifies autopsies in some cases, even against the conscientious objections of adherents of religious groups. Strong communitarian positions extend the public health rationale to the conscription or expropriation of cadaveric organs.[52] Albert R. Jonsen has condoned salvaging organs even against the will of the decedent and the family, arguing that the primary and secondary purposes of consent are either irrelevant to or insignificant in the context of organ salvaging. The primary purpose of consent, according to Jonsen, is to protect the moral autonomy of persons, allowing them to govern their lives in terms of their own values and to protect themselves from harm and exploitation; but this purpose is "no longer relevant to the cadaver which has no autonomy and cannot be harmed." The secondary purposes of consent include respecting the beliefs that the decedent held while alive or observing cultural practices about burial; however, these secondary purposes "would seem to yield before the significant value of therapy for those suffering from serious illness. . . . The genuine possibility of significant benefits to others overrides any secondary purposes that consent and permission might have."[53]

While meriting careful consideration, Jonsen's position fails to see that "the principle of extraterritoriality" of the person extends symbolic protection to the cadaver, that actions can be symbolically rational (in respecting symbols, such as the cadaver as a symbol of the person who once existed) even when they appear to be instrumentally nonrational or irrational, that individuals can be wronged even when they are not harmed (e.g., by having their will thwarted after their death), and that sociocultural practices of disposal of the body remain very important for various communities, including families and religious communities.

Even if respect for (individual or familial) autonomy, like all other principles, is only prima facie binding, it cannot be justifiably overridden unless there is no acceptable alternative to protect other important principles, which are deemed more important in the circumstances. In an emergency (similar to a wartime policy of conscription) a justification could possibly emerge for the forced removal of HBPs for transplantation against the will of the decedent and his or her family. However, conditions for an emergency are also very stringent—the society must be in imminent danger and lack feasible alternatives.[54] It is not clear that expropriation of cadaveric HBPs can be justified to benefit other individuals by preventing death. And there are acceptable and preferable alternatives that need further experimentation and implementation. Furthermore, in view of other public responses, there is little reason to believe that this method of acquisition would be politically feasible.

Abandonment or Failure to Claim Bodies and HBPs

Another mode of transfer of tissues, fluids, and sometimes organs, is abandonment, that is, the failure to claim bodies and their parts. For example, patients simply abandon their excised organs, tissues, blood, urine, etc., which may then be used by researchers, at least under some conditions. This mode of transfer has been the subject of some dispute—for example, in the case in California in which a patient's spleen, which was removed with his permission in a therapeutic procedure, was subsequently used, without any notice to or consent from the patient, to develop a profitable cell line.[55] Nevertheless, this mode of acquiring organs appears to characterize the unclaimed body statutes or coroner's statutes that authorize the removal and use, including transplantation, of organs and tissues. In one case in California, controversy arose when a man's heart was transplanted after he died unexpectedly and no one identified him or claimed his body for over twenty-four hours.[56] In practice, this mode of acquisition of HBPs will be limited, and it will not significantly increase the supply of organs for transplantation. Furthermore, strict rules should be established to make sure the body is actually unclaimed—that is, abandoned.

The next chapter will examine the remaining major mode of transfer of cadaveric HBPs—sale/purchase—in terms of its ethical acceptability, ethical preferability, and political feasibility. It will consider arguments for and against conceiving HBPs as property, and arguments about the rightness and the effectiveness of moving from "quasi-property" rights in HBPs to full property rights, including the right to transfer by sale/purchase. In addition, it will consider whether various incentives, such as financial incentives, could enhance donation without crossing the boundary into sale.

CHAPTER FIFTEEN

□

Human Body Parts as Property

AN ASSESSMENT OF OWNERSHIP, SALES,
AND FINANCIAL INCENTIVES

INTRODUCTION: ONE STORY OF HOW WE
GOT TO THIS POINT

In the United States organs become available for transplantation through gifts or donations, whether express or presumed. The Uniform Anatomical Gift Act (UAGA) of 1968 clearly authorized express donation by individuals and then by families, if decedents had not expressed their wishes; presumed donation, as noted in the previous chapter, surfaced in several states for the removal of corneas (as well as in other countries for solid organs).[1] However, the UAGA did not rule out transfer of organs by sale—it simply did not deal with sales at all. According to the chair of the drafting committee of the UAGA, the drafters did not intend either to encourage or to discourage payment for organs: "It is possible, of course, that abuses may occur if payment could customarily be demanded, but every payment is not necessarily unethical. . . . Until the matter of payment becomes a problem of some dimensions, the matter should be left to the decency of intelligent human beings."[2] Even though sales had been illegal in some states prior to the adoption of the UAGA in all states in the late sixties and early seventies, most prohibitory statutes were repealed when the

UAGA was adopted; the legal status of the sale of organs for transplantation "remained uncertain."[3] Meanwhile, the sale of blood and sperm has remained legal.

In 1983 there were several hearings on organ transplantation before congressional committees and subcommittees. In April 1983 in hearings on Organ Transplants before then Congressman Albert Gore's Subcommittee on Investigations and Oversight of the House Committee on Science and Technology, I proposed in passing that Congress might consider prohibiting the sale of organs for transplantation. Congressman Gore interjected: "Dr. Childress, you recommended a legal prohibition on the sale of organs and body parts. You don't think that there is a sufficient record of such practices to warrant legislative action at this time, do you?" I responded: "I have no evidence to suggest that there is. I think in a way this would be a *preemptive action* to avoid some problems down the road, in part because of the increased publicity about the need for organs." While indicating that he and his colleagues would consider such a ban, Congressman Gore stressed that they were "becoming increasingly skeptical of preemptive laws."[4]

Within a matter of months, as reflected in several subsequent hearings as well as in public statements, Congressman Gore had changed his mind, along with several of his colleagues, not because of arguments offered by ethicists and others who testified, but because of a proposal by an ex-physician in Virginia to set up a brokerage firm to purchase organs, particularly in the third world, for transplantation in the United States.[5] Outrage was instantaneous and widespread.[6] As a result, the 1984 National Organ Transplant Act (NOTA) made it illegal "for any person to knowingly acquire, receive, or otherwise transfer any human organ for valuable consideration for use in human transplantation if the transfer affects interstate commerce." (This federal statute limits the prohibition to human organs for use in "human transplantation," whereas the original UAGA covered education and research as well; while the amended UAGA now also prohibits sales or purchases of cadaveric organs, it does not cover sales by living vendors unless the "removal of the part is intended to occur after the death of the decedent.") Several states also prohibited the sale of organs.

What happened between 1968 when the UAGA was promulgated, without a prohibition of the sale of organs, and 1984 when NOTA and several states prohibited the sale of organs? Though kidneys were occasionally offered for sale during that period, sales played virtually no role in obtaining organs (even if now and then financial transactions occurred between family members), and the number of organ transplants increased very slowly—for example, from 1975 to 1980, kidney transplants increased from 3,730 to 4,697, heart transplants from 23 to 36, and liver transplants from 9 to 15. However, when organ transplantation really started to gain momentum in the early 1980s (by 1984, there were 6,968 kidney transplants, 346 heart transplants and 308 liver transplants), particularly as a result of improvements in immunosuppressive medication, the problems of scarcity became more troubling, the growth potential more obvious, and the limitations of the system of voluntary, altruistic donations more evident.[7]

The legal prohibition of sales in 1984 did not silence the discussion, and numerous proposals have emerged since then to institute some sort of market in organs, at least on the side of procurement if not on the side of distribution. These proposals have in part reflected the 1980s' style of competition, markets, and deregulation. They have also reflected the stagnation in organ donation and procurement, which had been expected to increase substantially as a result of laws and policies of required request directed at decedents' families. However, to take just one example, kidney transplantation experienced substantial increases for five consecutive years from 1981 to 1986 (increases of 475, 754, 856, 1027, and 1280), but it declined slightly in 1987. A significant increase in kidney transplants in 1990 resulted mainly from changes in the criteria of eligible "donors," rather than from improved rates of donation from previously eligible "donors." And the small increase in kidney transplants in 1991 resulted from additional donations from living donors. Further modest increases follow similar patterns.[8] Each year there are about 4,300 to 4,800 acts of donation of cadaveric organs, including kidneys, out of approximately 12,000 possible (brain-dead) sources of cadaveric organs. In this context, how should we respond to proposals to buy and sell organs for transplantation?

I will offer a conceptual and normative analysis of modes of transfer of organs for transplantation in terms of property and property rights, examining critically the arguments for some sort of market in organs through expanding the limited property rights in organs now recognized by law to full property rights, including the right to alienate by sales. My intention is to identify several conceptual, normative, and empirical issues that need careful attention as society debates these matters and, along the way, to indicate some deadends as well as some more promising directions.

PROPERTY, PROPERTY RIGHTS, AND MODES OF TRANSFER OF HBPS

My story featured three of the six main possible modes of transfer or acquisition of human body parts (HBPs) identified in the last chapter: (1) *express donation* (by individuals or families); (2) *presumed donation* (by individuals or families); (3) *routine removal;* (4) *abandonment;* (5) *expropriation* or *conscription;* (6) *sale.* The primary debate focuses on the first two and the last (with some uncertainty about whether any policy of "opting out" represents the second or the third). However, the fourth is used on occasion, for example, through the unclaimed body statutes, and the fifth appears in such bodily invasions as coerced autopsies and forced drug testing, but not in transplantation. Only the first two modes of transfer are properly called "donations" or "giving gifts," and only agents who use these two modes can properly be called "donors." (While there is some debate about presumed consent, which I have labeled presumed donation, "passive" giving is still giving.) And, as I argued in the last chapter, the "donor" as the one who makes the donation is not always identical with the "source" of the organs.

Similarly, when someone provides his or her organs for money, that person is not a "donor" but a "seller" or a "vendor." To take one example, among many possible examples, Henry Hansmann, in his important article on markets, notes that in prohibiting sales NOTA and the states focused "particularly on sales by living donors" and that "the principal motivation for adopting a payment scheme . . . would be to increase donation rates."[9] This mixture of categories of transfer—donation and sales—obscures their different personal and sociocultural meanings, which need to be identified clearly for purposes of evaluation.

Each of these six modes of transfer can be interpreted in terms of "property" and "property rights." The concept of property is that of a system of rules to govern control of and access to some resources, whether material or incorporeal.[10] And property rights are basically "the rights of ownership."[11] There are, however, many different conceptions of property and of rights of ownership, even within private property. The fundamental questions about property rights and ownership are: *Who* are the *rights-holders,* and *what* are the *rights held?* For example, rights-holders could be an individual, his or her family, or the state, and they could have, among other rights, the rights to use, to destroy, and to transfer by various means, such as donations or sales. Since some of these rights were analyzed in the previous chapter apart from the language and framework of property, we will consider throughout the advantages and disadvantages of this language and this framework.

According to Judith Jarvis Thomson, "ownership" is a "cluster of claims, privileges, and powers in respect of the thing owned. If we have similar rights in relation to our bodies that we have in relation to things we own, why not call our bodies property and our rights in relation to our own bodies property rights?" "No doubt it sounds odd," Thomson continues, "to say that people own their bodies. How could a person X be thought to *own* something that has as intimate a relation to X as X's body has to X? But ownership really is not more than a cluster of claims, privileges, and power; and if the cluster of rights that person X has in respect of his or her body is sufficiently like the clusters of rights people have in respect of their houses, typewriters, and shoes, then there is *no objection in theory* [emphasis mine] to saying that X does own his or her body—however odd it may sound to say so, however unaccustomed we may be to saying so."[12] Thomson wants to stress both the similarities and the differences between viewing my body as my property and viewing everything else I own as my property.

Two observations: First, while noting the differences between my body as my property and my typewriter as my property, she wonders, "what is there left to me if I sell my body?" In focusing on the intimate relation between person and body, she fails to consider what is left if a person sells only a part, and how selling a part fits with this intimate connection. Perhaps the whole living body is not alienable, by gift or by sales, precisely because of this intimate connection; but HBPs may be alienable in either way without the loss of personal identity. Also, this passage fails to address postmortem transfer of the whole body or its parts. Second, while Thomson notes that there is "no objection *in theory*" to talking about one's ownership of one's own body, this formulation leaves open the possibility that there could be objections in practice. And there are serious objections in practice at least to recognizing the full complement of property rights, including the right to sell, even if there are no conceptual or theoretical barriers to doing so.

Most often, of course, when the language of ownership, property, and property rights is introduced, it is assumed that sales are one acceptable mode of transfer of one's own property. While this is true for most private property, the same features need not characterize all private property. For example, the language of "quasi-property" or "quasi-property rights" emerged in the common law to indicate a form of dispositional authority over the corpse that excluded transfer by sale.[13] The language of quasi-property suggests a metaphorical or analogical approach—human body parts as property *resemble* but are not quite fully property, or put differently, the rights that owners of this sort of property have are not identical to—though they overlap with—the rights that owners of other sorts of property have.

Lawrence C. Becker identifies three levels of justification of property rights.[14] The most *general* justification aims at establishing property rights in general; *specific* justification involves establishing specific sorts of property rights, for example, full property rights in typewriters or partial property rights in human body parts (HBPs); finally, at the *particular* level, justification establishes that a particular person has particular property rights in some thing, such as his or her own HBPs. These justifications proceed within a moral and/or legal system of principles, rules, and values. I will focus on *moral* justifications, that is, arguments within a framework of moral considerations, including both principles and case-judgments, from which we reason analogically, as well as sentiments, symbols, etc.

In Becker's language, I want to concentrate on the second level of justification, with passing attention to the third level, in order to consider whether it is justifiable to treat HBPs as property and, if so, who has what rights over HBPs. Within property rights, I will focus on the right to *transfer* property to others, particularly by donation and sale. To assert that HBPs are "quasi-property," that is, similar to property in some respects but not in others, especially because of what Michael Walzer calls "blocked exchanges,"[15] is to beg the question. For the question is whether and how a "blocked exchange," such as the prohibition of transfer of organs by sale, can be justified. Now, the assertion of a "quasi-property" right is not problematic *if* it simply offers an accurate statement of legal rights. However, as a statement of what the law ought to be, it is inadequate.

In what follows, a conception of "quasi-property" rights or limited property rights, including inalienability by sale, will be labeled *LPRs*, while a conception of full property rights, including the right to alienate by sale, will be labeled *FPRs*. To permit sales of HBPs is to recognize property rights in them, but to recognize property rights in HBPs is not necessarily to authorize sales since those property rights could be limited (LPRs). It is thus possible to have property and even private property where ownership of certain property does not entail the right to alienate that property through sales. The specific justification would include the limitation; or, stated differently, a set of LPRs could be justified on the specific level for some property, such as HBPs. But is it justifiable to limit the modes of transfer of an individual's body parts for transplantation to express or presumed donation (and occasionally abandonment), without the possibility of sales? (For brevity I will not draw a sharp distinction between living and cadaveric transfers, though it is obviously important in many respects for our topic.)

George Mavrodes has helpfully directed us to consider both the sale of various goods and the transfer of organs as contexts for considering the sale of HBPs.[16] In the first context, the question is what is morally problematic about the sale of *organs* among the various goods that we already sell; in the second context, the question is what is morally problematic about the *sale* of human organs among the various ways we already transfer organs. Or, stated differently, if we sell (or permit the sale of) other goods, why not organs; and if we transfer (or permit the transfer of) organs in other ways, why not by sales? If the response is that we do not *own* our bodies and thus cannot sell their parts, it is then incumbent on the respondent to show how people can give or donate what they do not own. Indeed, as I noted, all the modes of transfer of human organs from one person to another presuppose some notion of property and property rights. Or at least they presuppose some of the cluster of rights associated with property. Here again, the appeal to "quasi-property" is not sufficient, for it is necessary to indicate why the sale of organs is morally problematic.

ASSESSMENT OF MORAL ARGUMENTS
FOR FPRS IN HBPS

I want to examine some of the moral arguments for and against treating HBPs as property in the full sense, that is, as involving FPRs rather than LPRs. Three distinguishable but related kinds of judgment are crucial here, just as they were in the previous chapter: Is a system of FPRs in HBPs, including transfer by sale, ethically acceptable? Is it ethically preferable to other ethically acceptable systems, including ones that recognize only LPRs in HBPs? Is a system of FPRs in HBPs *feasible*, that is, can it actually be implemented and realize its aims?

Both of the first two ethical questions are essential, because several modes of transfer may fall within the range of ethical acceptability; but there may be reasons, such as the promotion of altruism and a sense of community, for favoring one over the others.[17] The last question is also important, because if the ethically preferable system cannot be realized, that fact is an important part of the justification for a set of rules for a nonideal social context. My analysis and evaluative comments will proceed by implicit or explicit appeal to various principles, rules, and values embedded in our social practices, policies, and laws, as well as by analogical reasoning about different practices, policies, and laws. The various available arguments do not convince me that the act of selling an organ is intrinsically immoral. However, there may be extrinsic reasons for prohibiting or at least not encouraging as a matter of social practice what is not intrinsically immoral. (In this discussion I will concentrate on acquiring organs; I will separate the distribution of organs from their procurement and assume that the distribution of procured organs occurs in a fair and equitable way—otherwise I believe there would be a decisive objection to the sale of organs on the grounds of an unfair distribution of organs.)

Two main arguments are offered for permitting or encouraging the transfer of organs and tissues by sale. One is based largely on the principle of respect for persons, particularly their autonomy; the other is based largely on the principle of utility. The first argument for a system of FPRs in HBPs holds that autonomous agents should have the liberty to dispose of their body parts as they choose, including in the market place. The second argument holds that society should recognize a system of FPRs for HBPs because it is the most effective and efficient way to increase the supply of organs to save human lives. The second argument has emerged, especially recently, in response to the chronic shortage of organs. In brief, utilitarian organ procurers have joined forces with libertarians; but their union is only one of convenience, and the utilitarian argument for a market in organs provides the occasion for much of the recent debate. An important difference is this: The libertarian argument, at least in its absolutist form, holds whatever the consequences, including a decrease in the number of available organs, while the utilitarian argument holds only if the good consequences can be realized and outweigh the bad consequences.

In fact, libertarian and utilitarian arguments vary within their own structures. For instance, while some libertarian arguments are absolutist, others are more contextual. Among the latter, Lori Andrews in response to an earlier version of this chapter contends that accepting a view of the body as property, with FPRs, gives people clearer "control over what is done to their bodies. In our market-based society, our laws are often more protective of property than persons. By calling body parts property, there is a legal basis for a remedy when actions are taken with respect to our bodily parts that may not exist under themes of privacy, autonomy, assault, or infliction of emotional distress."[18] She is particularly interested in individuals' rights to control "extracorporeal biological materials" that have been removed from their bodies, and contends that their control, as well as their remedy for violations, rests more securely in property rights than in personal rights. Andrews further stresses that recognizing individuals' autonomy to treat their own parts, particularly their regenerative parts, as property provides them with potential psychological, physical, and economic benefits."[19] Apparently these benefits accrue to individuals as a result of their property rights (of control and remedy), but it is not clear how they function in the specific and particular justifications of FPRs in HBPs.

Still another argument proposes libertarian policies toward the human body "by default." H. Tristram Engelhardt, Jr., and Kevin Wm. Wildes contend that all understandings of appropriate uses of the human body and its parts are particular, that is, they operate in the context of particular narratives with their content-full interpretations and rankings of substantive values. As a consequence, in the postmodern world, where moral strangers are unable to convince each other of validity of their respective moral narratives, they "will by default be libertarian" regarding the uses and modes of transfer of the human body and its parts. They cannot convincingly appeal to such categories as "dignity," "respect," "sanctity," and the like, all of which presuppose particular (and, at most, only partially shared) narratives. Hence, any state that recognizes the limits of secular moral reason cannot enforce a particular narrative but must instead allow particular individuals and communities to make their own choices, including, if they choose, to sell HBPs.[20]

Both of these perspectives suggest that libertarian arguments need not be foundational and absolutist, but may in fact be quite contextual regarding the way FPRs in HBPs can express the principle of respect for autonomy under specific historical circumstances. As a result some of the counterarguments may be more persuasive for some libertarian positions than for others. Neither of the libertarian positions just discussed can simply dismiss several of the counterarguments that follow. On the one hand, Andrews's claims about the advantages of FPRs in HBPs are subject to challenge under particular historical circumstances, including threats to providers of HBPs and the possibility or probability of major shifts in attitudes and practices toward living persons. On the other hand, claims by Engelhardt and Wildes about the lack of consensus among strangers in a postmodern age may neglect shared fragments regarding human dignity, and their own commitment, by default, to consent and autonomy may actually require various state interventions to protect both, which may under some circumstances be threatened by a system of FPRs in HBPs. Neither position can blithely dismiss such contingencies the way an absolutist position on autonomy can.

Challenges to libertarian arguments. Several rejoinders have been offered to libertarian arguments that a market in organs (at least for procurement, but not necessarily for distribution) would respect personal autonomy. First, some critics charge that the principle of autonomy is too limited and is better conceived as the principle of respect for persons, including their autonomy as embodied. How the principle of respect for embodied persons is interpreted will depend in part on whether embodiment is viewed as *essential* to or only *incidental* to personhood.[21] The notion of embodiment may set some constraints on respect for autonomous choices, but much will depend on the content given to notions of violation of the body's integrity, disrespect for the remains of an embodied person, degradation, debasement, and the like, notions that are notoriously difficult to specify. And it is difficult to defend this criticism without falling into a form of questionable paternalism.

Second, there are questions about the vendors' lack of voluntariness, especially if they are poor, economically vulnerable, and subject to exploitation. In general, there is no cogent reason to deny that people can make a sufficiently autonomous choice to sell an HBP, if they can make other choices about risky employment and the like. It is not even clear that poverty should be viewed as coercing people to sell HBPs, for "a hard choice is not a non-choice."[22] If we consider the first and second reasons together, we confront what Margaret Jane Radin calls the "dilemma of commodification."[23] Prohibition of the sale of organs is a form of "personal prophylaxis" to protect agents from degrading their personhood when they want to sell body parts because of their desperate circumstances. The dilemma is this: "We cannot honor our intuition of what is required for society to respect personhood, either by permitting sales or by banning them." The problem is the larger social context in which the dilemma occurs. If we ban sales, we keep desperate people from relieving their desperation; if we fail to ban sales, they engage in acts that are variously held to degrade themselves, to degrade the goods so transferred, to degrade the relationships involved, and to degrade the society in which they occur.

Third, there are risks to sellers or to others whose HBPs may be transferred by vendors after they are killed or allowed to die. However, it is possible to reduce the risks for living vendors by regulating sales, for example, by not letting people sell kidneys under some risky conditions. It is also possible to reduce the risks to potential cadaveric sources of HBPs by prohibiting the sale of their HBPs unless they make the sale themselves, perhaps through a futures contract, or by making certain that they are not killed or allowed to die in order to provide a source of revenue.

Fourth, there are concerns about risks to recipients of purchased organs. However, these risks can be reduced and controlled in a system of sales in the same way as in a system of gifts.

Fifth, a major (and frequently overlooked) problem in the ethical analysis of a market in organs is that people do not simply sell a HBP to another person. These are not merely private orderings, with no implication of the society. Indeed, the transfer and use of organs cannot occur without the heavy involvement of third parties. Not only is the provision of health care a social act but organ transplantation in particular involves whole teams of professionals who would have to effectuate the market transfer by removing the sold/purchased organ for transplantation into another person. It is risky to invest the health care system with the power to effectuate these market transfers. And it is not surprising that many professional organizations in transplantation have questioned the moral acceptability of a market in organs.[24]

Finally, a market in HBPs is frequently criticized because of its *commodification* of HBPs. This concept is notoriously unclear in contemporary debates about markets in organs. Defenders of a market in organs tend to agree with Hansmann that "it is difficult . . . to find a clear statement of precisely what is meant by commodification or why it is undesirable."[25] If commodification simply means that the material in question has been made a commodity and exchanged for valuable consideration, it is simply another term to describe the transaction; it fails to add any reason for not having a financial transaction. However, critics frequently use the term to include various negative consequences, especially those that result from objectification (making one's body parts an object), for the sociocultural context, including ways of thinking about bodies and organs, attitudes, sentiments, and practices. Terms of debasement and degradation, which were introduced earlier, also play a role. In the process of objectification, what is treated as a commodity, when it shouldn't be, is altered and becomes somehow inferior. Commodification appears to damage the "humaneness of social life," which, Becker argues, is one of the goals served by property laws and one of the most decisive objections to a market in body parts.[26] (The putative effect of commodification on altruism will be considered in relation to the second argument for FPRs in HBPs.)

Some versions of the commodification argument appear to commit the fallacy of composition, which claims that people will start to view whole persons, themselves and others, as commodities if some HBPs are treated in that way.[27] According to Andrews, the kind of "cultural transformation" feared by critics of a procurement market is unlikely to occur because the putative "perceptual changes" are harder to effect than many acknowledge since we have important "anchoring concepts—in this case, about the treatment of people."[28] There is little evidence, Joel Feinberg argues, to suggest that the alleged "coarsening effects" of treating the primary object (the HBP) as a commodity will actually transfer to the secondary object (the human person).[29] If such a "cultural transformation" were to occur, its probable mechanism would be our mode of discourse, particularly the images and metaphors we use in thinking about and experiencing the human body, our own and others'. It is difficult, however, to determine just how probable such a fundamental change in perception would be under a system of FPRs in HBPs. And yet judgments about particular policies often presuppose such predictions, which are very complicated, in part because HBPs themselves vary so much.

Even if selling some HBPs, such as solid organs, is potentially depersonalizing or degrading to the society, there is legitimate debate about whether these results of commodification would flow from the sale of every HBP (defined as "an organ, tissue, eye, artery, blood, fluid, or other portion of a human body").[30] It may even be plausible to view the sale of renewable tissues as the provision of a service, as some state laws have construed the sale of blood, rather than as the sale of a commodity.[31] Observations about markets in blood and sperm may not be generalizable to other HBPs, because both are replenishable and their transfer does not impose risks on sellers. While bone marrow is replenishable too, the risks of bone marrow removal are more comparable to kidney removal than to blood removal. In addition, some proponents of a market would distinguish living vendors from cadavers and exclude situations in which a potential conflict of interest exists (e.g., the sale of aborted fetuses or fetal tissues, the possibility of which, some have argued, would provide financial incentives for abortions). Certainly, in the case of cadavers, the standard of death required may also be relevant. Finally, it would be possible to distinguish types of valuable consideration, such as direct payments and indirect incentives (a topic I will discuss below).

Such variations in the appropriate treatment of different HBPs may be consistent with and even required by a Kantian position based on human dignity. Stephen Munzer contends that it is not easy to move from Kant's conception of dignity associated with human beings as such, and as distinguished from price on a market, to moral arguments, much less legal rules, against markets in (all) HBPs. According to Munzer, any argument against FPRs in HBPs, even a Kantian argument, must avoid what he calls the "fallacy of division" which involves arguing that "what is true of a whole must also be true of its parts."[32] (The fallacy of division is the reverse of the fallacy of composition.) Hence, it is not possible to move directly from human dignity, as the dignity of the whole person, to the dignity of particular HBPs, or to the claim that selling some particular HBP necessarily infringes the dignity of the whole. Efforts to avoid the fallacy of division, Munzer suggests, take one or both of two related strategies. Focusing on the way the living human being is made up of the "unified organization of the various body parts," the integration strategy can grade or rank HBPs in terms of their seriousness for human integration, considering, for instance, their replenishability or renewability and the risks of their removal. For this strategy, which Kant in part used, any moral assessment of selling HBPs must consider the nature of the part involved as well as the seller's reasons for selling it. The *derived-status* strategy, by contrast, focuses on the status of particular HBPs in terms of their roles in the whole, for example, their connection with personal identity, and also considers their ultimate use following transfer (e.g., the use of purchased skin for upholstery rather than for a graft for a burn victim). Because of its attention to use, this strategy finds some donations to be as problematic as some sales. According to Munzer's "uneasy" dignity-based argument, HBPs as such lack dignity, but treating them as commodities may, under some circumstances, offend personal dignity by directly affronting, insulting, or demeaning dignity or by reducing the sense of dignity through degradation.

Munzer's Kantian argument based on human dignity is instructive in several respects. It opens the door for variable responses to different body parts, depending on such considerations as a part's essential role in the integrated, functioning body, its connection to personal identity, the risks involved in removing it, and the use to which it will be put. Furthermore, his argument indicates that sometimes the opposition to sales of HBPs on grounds of offense to human dignity actually depends less on the mode of transfer and more on other factors, such as the use of the HBP in question. Finally, however, Munzer's argument does not adequately address the cadaver and its relation—

particularly its symbolic relation—to the living whole person, as well as the relation of cadaver body parts to the cadaver as a whole.

Challenges to the utilitarian arguments. What is defensible in the rejoinders to the libertarian argument for allowing a (procurement) market in organs will also apply to the utilitarian arguments. In addition, the latter involves debates about effectiveness, as well as cost-effectiveness and cost-benefit analysis. For many pondering our persistent shortage of organs, the obvious solution is simply to expand people's property rights in their bodies, from LPRs to FPRs. Following is a typical statement, specifically for the procurement of organs: "The market solution is more effective, more efficient, and more robust than the alternatives. . . ."[33] While it is unclear exactly what kind of claim this is, it is certainly not empirically based. We simply do not know whether a market to obtain organs would meet these standards; indeed, there are strong reasons to think it would not. Hence, rather than arguing that the market is intrinsically wrong, because it necessarily violates human nature, dignity, flourishing, and the like, my argument is that it will probably be ineffective, perhaps counter-productive (by reducing donations and perhaps even the overall number of organs available for transplantation), potentially exploitative (the poor as the primary target for purchases), and probably productive of a shift in attitudes and practices more in the direction of the commodification of the human body and its parts. A market is, furthermore, unnecessary because other approaches, discussed in detail in the previous chapter, are feasible as well as ethically preferable.

To begin to determine whether buying and selling organs (with economic agents having FPRs in their own HBPs and perhaps also in cadavers under their dispositional authority) might be effective in increasing the supply of organs, we have to consider the reasons for *failures to donate* in our current context. In predicting human behavior under a market in this sensitive area, we can learn more from what people have done in donating—or not donating—organs than from generalizations about what they have done in other spheres of their lives, not involving intimate HBPs, which, as I have stressed, are associated with complex sentiments, attitudes, symbols, and rituals. In their attempts to be realistic by economizing on morality, economic interpreters sometimes oversimplify the motives of human action. For example, Henry Hansmann says: "Presumably [people] fail to donate only because of inertia, mild doubts about their preferences, a slight distaste for considering the subject, or the inconvenience involved in completing or carrying a donor card."[34] His statement ignores the evidence of various opinions polls and qualitative analyses, which show the importance, in addition to the reasons he notes, of some fundamental fears, distrust, and mistrust. Fears of being declared dead prematurely or having one's death hastened in order to provide organs play a major role.[35] Thus, it does not follow, as Hansmann supposes, that "a relatively modest financial incentive would improve donation rates substantially."

In fairness to Hansmann, he does return later to other fears and says that they "might not be easy to allay." However, it is my contention that the fears and attitudes that reduce the effectiveness of a system based on voluntary donations can reasonably be expected to destroy a system based on sales, such as a futures market, in which individuals contract now for delivery of organs upon their death.[36] (And such a futures market is the most promising arrangement for transfer by sale, both in terms of the protection of autonomy and potential effectiveness.) If someone is reluctant now to sign a donor card for fear that he may not receive proper care in the hospital, it is easy to imagine his fears about accepting money (even in the form of current health insurance reductions, or modest payments, or payments to his estate or other designated beneficiary) for delivery of usable organs upon his death. After all, the contract has been effected, and all that remains for the fulfillment of all of its terms is his death followed by the removal of usable organs. Defenders of a futures market might respond by stressing that sales and purchases would be made through a regulated market, with various controls and safeguards; but after all, the system of donation has been regulated (indeed, some critics contend, overregulated) and yet this is the context in which these fears both emerged and persist.

One major concern about commodification is its putative impact on attitudes and sentiments, particularly altruism, not only in organ transplantation but in other areas of life as well. However, it is unclear why this one area of HBPs, which for solid organs from cadaveric sources involves only about 4,500 to 4,800 *acts of donation* each year (of course, there are many more acts of donation for corneas, skin, bones, etc.), should be deemed crucial for society's quota of altruism, solidarity, and community. Even if altruism in organ donation is as important as the critics of a market suggest, altruistic acts can continue in the market—indeed, so the argument goes, acts of donation could be even more altruistic, because the donors could renounce their right to monetary compensation while transferring their rights to the organs. Nevertheless, I believe the critics are correct when they claim that a market in organs would drive out, or very substantially reduce, organ donations, in part because those acts of donation would have been redefined. No longer would donors provide the "gift of life"; rather, they would provide the gift of the equivalent of the market value for the organ donated.[37] Adequate attention to the complexity of human motivations would disclose the other-directedness of many allegedly self-interested actions in the market, where economic agents often act for family members and loved ones. For example, reports of sales of kidneys frequently stress the agents' dramatic actions on behalf of their needy families. Furthermore, the *act* of donation is still right, good, and even praiseworthy, even if the agents' *motives* are often highly self-interested (e.g., the agent desires to relieve guilt or to gain honor). In simplistically focusing on motives of agents, commentators typically overestimate the altruism in organ donation, in effect sentimentalizing organ donation, and underestimate the altruism in the market.

An example of the overestimation of altruism in organ donation appears in Jeffrey Prottas's fine recent study, *The Most Useful Gift: Altruism and the Public Policy of Organ Transplants:* "The voluntary decision to donate must be based on altruistic motives; otherwise, it is not permitted."[38] That is a serious overstatement. Indeed, for cadaveric donors, there is no inquiry into motives as long as financial compensation does not appear to be involved. Donors may have all sorts of mixed motives for donating—ranging from altruism, which is clearly important, to a sense of obligation, to an effort to find redemptive meaning in a tragic set of circumstances, to a desire for praise, honor, fame, and the like.

Even though the motive of altruism often appears to be present and even dominant in organ donation—in contrast to being necessary for organ donation—it does not provide a good reason for ruling out modes of acquisition of organs other than donation. At most, it may provide a reason for ethically preferring a system of donation, probably express donation, over sale, other things being equal (including their ability to provide the needed organs). It does not, I believe, offer a cogent argument for declaring a market ethically unacceptable. Indeed, it would be odd to design a system to promote altruistic behavior, with the ultimate goal of saving human lives, if another system based on compensation or financial incentives could save more lives without violating anyone's rights. Altruistic but ineffective conduct is not an end in itself. In fact, in contrast to providing opportunities for individual or familial altruism, *societal altruism* may be best expressed in establishing an effective system to procure organs to save human lives. (This is not to say that societal lifesaving efforts are overwhelmingly important and should not be constrained by moral and other sentiments, principles, etc., which would perhaps rule out expropriation or conscription of organs.)

Defenders of the market have not, however, proposed an effective system to obtain HBPs. In short, the main rejoinder to the utilitarian argument for a (procurement) market in order to increase the supply of organs is that it would probably be ineffective and counterproductive and could have other social costs. In addition, its financial costs would not be negligible, particularly if on grounds of justice a market is used to obtain organs but not to distribute them.[39] Furthermore, as the previous chapter argued, our system of express donation (opting in) can be made more effective; and it has features, including its connection with altruism, that make it ethically preferable, other things being equal. While a system of presumed donation is not ethically unacceptable in principle, making it ethically acceptable in practice (i.e., ensuring that silence expresses adequately informed consent) would require a commitment of resources that would probably be sufficient to make a system of express donation quite effective. One common proposal to make the system of express donation more productive is to provide financial incentives.

FINANCIAL INCENTIVES FOR DONATIONS

Short of actually buying and selling organs (even for future delivery), can financial incentives be used legitimately "as tools to improve the decision for donation"?[40] Can financial incentives be used without obliterating the distinction between vendors or sellers and donors of HBPs? Can some financial incentives for organ donation be distinguished from purchases, and can they be implemented without opening the door to purchases?

As already suggested, proponents of a system of organ donation often appear to suppose that pure altruism marks the donation of organs to the community for those in need and suspect that the presence of any other motives vitiates the gift. For example, according to one statement, "avoidance of self-interest on behalf of the donor must be implicit in the process."[41] Gifts of all kinds, including organs, frequently involve mixed motives, and it is difficult, if not impossible, to sort out which motives are necessary and/or sufficient for the act of giving. Gifts in families, for example, may involve a sense of obligation that cannot be reduced to altruism. And our tax laws encourage charitable contributions through the allowance of deductions.

The meaning of the act of transfer—whether sales or donation—is not totally reducible to the motives of agents, however important these are. For example, an individual who signs a "document of gift," such as a donor card, or family members who donate a relative's organs may have a wide variety of motives for participating in a practice that is designed to benefit others. The question is whether the use of financial incentives to supplement altruism and whatever other nonmonetary incentives may be present, such as a sense of obligation, a feeling of guilt, or a desire for honor, will blur the boundary between giving and selling.

It is important to distinguish removing financial disincentives from providing financial incentives. Currently, among the financial disincentives for living related donors are the loss of compensation, child care support, and travel expenses.[42] It is possible and desirable to remove these disincentives in order to enable the agent to make the decision on its own terms, without having to worry about certain financial burdens. Likewise, there should be no financial disincentives to cadaveric donations. However, some propose adding financial incentives to encourage living and cadaveric donation. The question is whether some of these financial incentives amount to indirect sales and purchases.

William F. May argues against "rewarded gifts" on the grounds that "if a donor receives a financial reward only if he gives the gift, the transaction differs very little from an outright sale." And, "whether rewarded gift or sale, the marketplace fails to respect the human meaning of the body and the awesomeness of the demise."[43] However, "very little differences" and differences of degree, as well as of kind, often mark significant boundaries in our interactions. One crucial question is whether the "very little difference" is sufficient to keep the transfer of organs within the framework of donation and out of the framework of sale.

Several recent proposals confuse financial incentives for donation (a donation or gift framework) with payment for organs (a market framework) and in the process appear to eliminate these "very little differences." One example appears in a recent proposal from the Council on Ethical and Judicial

Affairs of the American Medical Association, which proposes a future contract under the guise of offering a "financial incentive to encourage the *donation* of organs." While opposing an open, unregulated market as potentially exploitative and degrading, it proposes a "pilot program" to determine whether a future contract would be effective, all the while supposing that such a future contract remains within a framework of "donation."[44] The mixture of socio-moral practices is evident in the following statement: "Incentives should be limited to future contracts offered to prospective donors. By entering into a future contract, an adult would agree while still competent to donate his or her organs after death. In return, the appropriate state agency would agree to give some financial remuneration to the donor's family or estate after the organs have been retrieved and judged medically suitable for transplantation." According to other conditions proposed by the AMA Council, incentives would not be provided for organs from living donors, they would not be offered to anyone other than the source of the organs, and they would be of "moderate value and . . . the lowest amount that can reasonably be expected to encourage organ donation." In the previous chapter I argued that the AMA Council's proposal of mandated choice reflected an excessively individualistic, rationalistic, and legalistic approach to organ donation. Similar deficiencies mark the current proposal, particularly in its reckless combination of social practices with different moral and psychosocial meanings while supposing that its proposal remains within the framework of making donations or offering gifts. Even worse, the AMA Council holds that combining its two policy proposals, each of which I have argued is misguided, would be the most effective strategy: "Thus, it is likely that mandated choice and future contracts, implemented together, would have the greatest success in increasing the availability of organs." Can other proposals for financial incentives remain within a framework of donation and also possibly be effective?

Thomas G. Peters has proposed a pilot program to test the effects of providing a one-time death benefit of $1,000 paid through organ procurement organizations for recoverable donations (i.e., not for the act of donation).[45] Yet again it is hard to see how this program, which focuses on recoverable donations, can be distinguished from purchasing usable organs. Perhaps the line between giving and selling could be maintained if the death benefit were conceived as a tangible expression of societal gratitude for the *act of donation* rather than for the recoverable organs themselves. The reluctance to provide the benefits where the donated organs are not usable is understandable, but this product-oriented approach may threaten the framework of donation.

A slightly modified approach might provide financial incentives and yet preserve the psychological, social, and cultural meaning of the act of donation: As a regularized expression of gratitude, which, after all, is the appropriate attitudinal and behavioral response following the acceptance of gifts, the society would cover the decedent's funeral expenses up to a certain level, perhaps $1,000 or more. In this way, the community would recognize with gratitude the decedent's and/or the family's act of donation and also pay respect to the donor and/or source of the organs by sharing in the disposition of his/her final remains, after the removal of donated parts.

Any particular proposal for "rewarded gifting" will require the most careful scrutiny, in part because organ donation is a highly sensitive area, marked by complex beliefs, symbols, attitudes, and sentiments. Some of the policy reforms from the mid-1980s to the present have probably been ineffective and even counterproductive at times because of their lack of sensitivity to psychosocial norms and attitudes. At the very least careful surveys of public opinion about "rewarded gifting" would be helpful. According to one recent survey, "responders were asked whether some form of financial or non-financial compensation should be offered in the USA in an effort to increase the number of organs for donation: 52% of responders said it should."[46] While public opinion is not the final arbiter of ethical policies, it is relevant in determining social-moral norms as well as the feasibility of different approaches to organ procurement. Finally, some carefully conceived pilot experiments in obtaining organs, such as the provision of death benefits, may be justifiable, but these will be seriously limited by current laws and, in any event, should be undertaken only with the greatest caution.

COMMUNAL OWNERSHIP OF DONATED ORGANS

In retrospect I feel somewhat embarrassed that it took me so long, as vice-chair of the federal Task Force on Organ Transplantation, to appreciate that the Task Force's debate about the distribution and allocation of donated organs—who should make the decisions and by what criteria?—was in part a debate about ownership: Who owns and thus has dispositional authority over donated organs?[47] Living donors of kidneys and some other tissues usually designate the donee or recipient, and the original UAGA and the 1987 revision both identified several possible donees of anatomical gifts, including "a designated individual for therapy or transplantation needed by him," as well as institutions or health care professionals. Indeed, some well-known cases of cadaveric donation have involved the designation of specific donees in need; for example, in California, parents of a teenager who died suddenly and unexpectedly donated his heart to his sick girlfriend as he had indicated he wanted done if anything happened to him. Acceptance of the gift requires abiding by the donor's terms, within medical, legal, and social limits. In some situations, especially where the donation is marked for a particular group, such as veterans, or a race, there are good reasons for rules to prevent the transplant team from accepting the donation on behalf of that group. Most often, however, the donor of cadaveric organs does not designate a donee: "If a donee is not designated or if the donee is not available or rejects the anatomical gift, the anatomical gift may be accepted by any hospital."[48] Who then *owns* the donated organ?

One of the most important contributions of the federal Task Force on Organ Transplantation was to clarify the social meaning of ownership of donated organs in the absence of designated donees. It held that *donated* organs belong to the "community." From this perspective, organ procurement and transplant teams *accept* donated organs as *trustees* and *stewards* for the community as a whole, and they should determine who will receive available organs according to public criteria that have been developed by a publicly accountable body with public representation and that reflect principles of justice as well as medical standards. Selecting recipients of donated organs involves "decisions about how a scarce public resource should be used."[49]

I cannot here rehearse all the ambiguities associated with communal ownership or dispositional authority, in particular the questions that arise regarding the boundaries of the community—for example, local, regional, national, or international—on behalf of which teams accept organs for transplantation. Some of these ambiguities were addressed in my earlier discussion of fairness in the distribution of organs for transplantation, which also spelled out the implications of communal ownership of donated organs for particular policies of distribution. (See chapter 12 above.) As also noted in that context, the public's perception that these policies are fair, equitable, and just appears to influence quite significantly its willingness to donate organs.

CONCLUSIONS

The theoretical objections to viewing HBPs as property are not cogent, and practically speaking, it may be useful to note (in Thomson's language) the overlaps of clusters of rights and to view HBPs as property where the similarities are sufficiently great between the rights associated with ordinary material property and those associated with HBPs. Even though we can, I believe, assert, recognize, and enforce major rights relating to the body and its parts—including such matters as dispositional authority—without resorting to property and property rights, there are points where viewing HBPs as property may be helpful and even illuminating in trying to sort out *rights-holders* and *rights held.* However—and this is the critical point—a recognition of HBPs as property, and even as private property, does not entail that the rights-holder has (or should have) the usual property right to alienate by sale or that a market may or should be developed in organs in order to increase their supply for transplantation. That is, a recognition of HBPs as property does not entail that their transfer and use should be governed by FPRs rather than LPRs.

I do not hold that the sale of organs is intrinsically wrong—for example, I find it difficult to say that a poor father or mother in Calcutta is acting wrongly in selling one of his or her kidneys to provide funds for their children. Nevertheless, a society may be justified in prohibiting such sales and in avoiding a market as a way to increase the supply of organs. My arguments against a societal policy to use the market in an effort to increase the supply of organs focus on its probable ineffectiveness, counterproductivity, and other costs in view of widespread fears and sentiments. However, in addition to other potentially helpful ways to increase organ donations discussed in the last chapter, there may be ways to augment and enhance the incentives for organ donation by adding financial incentives without crossing into sales. At any rate this possibility merits exploration. Just as the economic model of human motivation is deficient, so is the model that sees only unadulterated altruism in the donation of organs. A more realistic conception of human motivations, sentiments, symbols, and actions, particularly in regard to HBPs, should guide all public policies in this area.

Nevertheless, changes in policy and practice should be undertaken only with extreme caution because of the difficulty of predicting social consequences, both positive and negative. The failure to attend to various motivations, sentiments, and the like, as both morally and practically significant, may have diminished the effectiveness of efforts to increase the supply of organs over the last several years. Efforts to enlarge LPRs to FPRs in HBPs are even blinder to these critical factors and extend an unchastened market ideology into an area where its fit is problematic.[50]

CHAPTER SIXTEEN

□

Ethics, Public Policy, and Human Fetal Tissue

Transplantation Research

Over the last several years controversy has raged in the United States about the use of human fetal tissue, following elective abortions, in transplantation research, particularly for patients suffering from Parkinson's disease. In this controversy various ethical arguments have been hurled back and forth by participants in the debate about public policies, primarily about whether the federal government should fund such research. Here I want to analyze and assess arguments for and against the use of human fetal tissue following elective abortions in transplantation research and for and against the use of federal funds for such research. I will focus on the deliberations of the Human Fetal Tissue Transplantation Research Panel (on which I served as a member from September to December 1988) in its larger political, social, and cultural context.[1] I will first provide a case study, then examine the major arguments for and against human fetal tissue transplantation research, particularly as developed in the Panel's deliberations, and finally reflect on those deliberations through the analytical lens of the concept of consensus before offering a conclusion and a postscript to indicate subsequent developments.

CASE STUDY

It is important to distinguish (1) the use of tissue from dead fetuses in research, from (2) fetal research, that is, research on the living (viable or nonviable) fetus, in utero or ex utero. Fetal research is subject to careful scrutiny and, according to federal regulations, can be justified only under very stringent conditions.[2] However, fetal tissue has been used in research for many years; one major, widely noted example of its use was in the development of the polio vaccine.[3] In 1987 the National Institutes of Health (NIH) provided over $11 million in 116 grants and contracts which directly or indirectly used human fetal tissue.[4] The use of fetal tissue in transplantation is a special instance of research using fetal tissue; transplantation research uses fetal tissue *as* therapy rather than to develop therapy.[5] Transplantation of human fetal thymus is an established therapy for DiGeorge's syndrome, a rare condition of varying degrees of severity.[6] Another experimental procedure is the transplantation of fetal pancreatic tissue into patients with diabetes.[7] And on the basis of animal research in the United States and elsewhere, several researchers in other countries experimentally transplanted human fetal neural tissue into patients with Parkinson's disease.[8]

However, in October 1987, when researchers at the NIH asked for permission to conduct such research, using tissue from electively aborted fetuses, the director of NIH, James Wyngaarden, chose not to exercise his legal authority and instead sought approval from the office of the Secretary of the Department of Health and Human Services (DHHS) because of the politically sensitive nature of the research.[9] Whereas Wyngaarden's request focused on the potential public controversy, the staff of the Assistant Secretary for Health, Robert Windom, stressed the ethical issues. In March 1988, the Assistant Secretary declared a moratorium on the use of federal funds in human fetal tissue transplantation research (hereafter HFTTR) with tissue from induced abortions until NIH could convene "special outside advisory committees" to hear testimony, deliberate, and offer recommendations. He asked the committees to address ten questions, which focused mainly on the connection or linkage between abortion and the use of human fetal tissue in research.

During the summer of 1988 NIH appointed the Human Fetal Tissue Transplantation Panel (hereafter Panel or HFTTR Panel) consisting of twenty-one ethicists, lawyers, biomedical researchers, clinical physicians, public policy experts, and religious leaders who had been recommended by various sources, including members of Congress, members of the executive branch, and organizations interested in this research. While defenders of HFTTR worried about the role on the Panel of strong opponents of abortion, critics of HFTTR objected to what they viewed as the Panel's overall bias in favor of such research.

Before the HFTTR Panel's first meeting, a draft executive order, which was leaked from the White House, proposed banning HFTTR with tissue from electively aborted fetuses. However, Dr. Otis Bowen, Secretary of the

Department of Health and Human Services, decided not to impose new restrictions on HFTTR until the advisory committees reported or until the president issued a direct order. The next several weeks saw vigorous efforts to persuade the president to promulgate the executive order. In separate letters, fifty members of Congress and several hundred physicians and others strongly supported the executive order. However, the president took no action.

Then in three public meetings (September 14–16, October 20–21, and December 5) the HFTTR Panel heard scientific, legal, and ethical testimony from over fifty invited speakers as well as over fifteen representatives of public interest groups, drafted "responses" to the Assistant Secretary's questions, and offered "considerations" for those responses. Although the Panel had originally been expected to complete its deliberations in the first scheduled meeting of three days, that time was inadequate, and a second meeting was arranged to consider a draft report that offered relatively brief "responses" to the Assistant Secretary's ten questions without providing substantial justification for the responses. After some discussion about the possibility of circulating and approving justificatory statements without a third meeting, the Panel decided to submit only what had been developed and accepted by the end of the second meeting. However, a few minutes before the second meeting adjourned, two Panel members, James Bopp and James T. Burtchaell, distributed to the Panel a substantial dissent to the report. Panelists in the majority later expressed their concern that such a long and eloquent dissent would overwhelm the report's brief responses, which lacked adequate justificatory statements; they requested a third meeting, which was scheduled for December 5, when the Panel could consider drafts, prepared by different panelists, of "considerations" for each "response" to the ten questions and could put the report into final form. In addition to the "responses" and "considerations," the final report contains three concurring statements, including a long one signed in whole or in part by eleven panelists, and dissenting statements by four panelists. A second volume of the report contains the written testimony received by the panel.

The Panel concluded that it would be "acceptable public policy" for HFTTR to proceed with funds from the federal government if several "guidelines" or "safeguards" were put into place to separate as much as possible the pregnant woman's decision to abort from her decision to donate fetal tissue following the abortion.

On December 14, 1988, the report was submitted to the NIH Director's Advisory Committee, with oral presentations by nine of the ten Panel members who attended (another absent panelist's statement was entered into the record). The Advisory Committee quickly concluded that "the consensus of the Panel reflected the consensus of the country itself, where widely divergent views are held about the morality of elective abortions and about the use of fetal materials derived from such abortions for the purposes of research." It then unanimously recommended (1) acceptance of the report and its recommendations, (2) lifting of the moratorium on federal funding of HFTTR by the Assistant Secretary for Health, and (3) acceptance of current laws and regulations governing human fetal tissue research along with the development of "additional policy guidance as appropriate, to be prepared by NIH staff, to implement the recommendations of the Human Fetal Tissue Transplantation Research Panel."[10] However, no action was taken in the last days of the Reagan administration, or in the Bush administration until November 1989, when the Secretary of the Department of Health and Human Services, Dr. Louis Sullivan, indefinitely extended the moratorium.[11]

THE MORAL STATUS OF THE FETUS AND THE MORALITY OF ABORTION

Fetal tissue may have some special advantages in treatment of Parkinson's disease. It is immunologically more naive than developed tissue, and it grows and differentiates rapidly. In addition, there are over 1.5 million legal abortions each year in the United States. Some of these abortions would be viewed as ethically justified by most moral and religious traditions. For example, in the United States in 1985 there were approximately 75,000 ectopic pregnancies; about half of such pregnancies end early in spontaneous abortions and the others in induced abortions to save the pregnant woman.[12] Protestants, Catholics (through the rule of double effect), and Jews (through the primacy of maternal health) generally accept abortions in such cases as ethically justified. However, these and other abortions that are generally viewed as ethically justifiable would not suffice to provide fetal tissue. And spontaneous abortions are often marked by fetal conditions that preclude the use of the tissue. Thus, usable fetal tissue would often come from abortions that many would not consider ethically justifiable. Moral opposition to many abortions is the main source of the controversy about HFTTR.

There are three major views about the status of fetal life—the fetus may be viewed as (1) mere tissue, (2) as potential human life, or (3) as full human life. For those who defend the first view of the fetus, abortions pose no serious moral problems, and the moral framework for HFTTR is that of tissue donation by a living donor. The fetal tissue belongs to the pregnant woman, who may dispose of it as she chooses, just as she may dispose of other products of conception, such as the placenta. For those who accept either the second or the third view of fetal life, the fetus has some independent moral standing. For the second view, abortions are not absolutely wrong, but they are prima facie or presumptively wrong and thus stand in need of moral justification. Proponents of this view disagree about which reasons are adequate to justify abortions. And for the third view, abortions may be generally wrong, except perhaps to protect maternal life.

Those who defend the second or third views of the fetus may either oppose or accept HFTTR, depending on whether they believe it is possible to *separate*, morally and practically, the use of fetal tissue from the (immoral) abortions that produce the tissue. Proponents of the third view of fetal life tend to oppose

HFTTR more strongly than those who defend the second view, but even some proponents of third view can accept HFTTR because it uses tissue from *dead* fetuses. For example, Judge Arlin Adams, who chaired the HFTTR Panel, strongly opposes abortion except in very limited circumstances; but he believes that it is possible to separate HFTTR from abortion decisions and, furthermore, that such separation can be more readily secured through regulations associated with federal funding of the research (pp. 25–26).

For those who defend the third view of fetal life, the fundamental issue is what *equal respect* entails. The HFTTR Panel insisted on procedures that would accord dead human fetuses "the same respect accorded other cadaveric human tissue entitled to respect" (p. 1). Although the Panel did not develop its implications, this requirement of equal respect entails that the dead human fetus not be subjected to procedures that would be considered inappropriate because undignified or disrespectful toward "cadaveric human tissue." However, acceptance of the requirement of equal respect does not necessarily presuppose a conviction that the fetus is a full human being. It may rest simply on the conviction that (1) the fetus is a potential human being and that the dead fetus has symbolic significance, or that (2) respect for human fetal tissue is appropriate to avoid offense to those who view the fetus as full (or potential) human life. At any rate, against opponents of HFTTR, the principle of equal respect implies that if it is justifiable to use any "cadaveric human tissue" in transplantation research—for example, after accidents or homicides—then it is justifiable to use cadaveric fetal tissue after abortions. However, it is also necessary to address other arguments offered by opponents of HFTTR to distinguish the use of fetal tissue following elective abortions from the use of other "cadaveric human tissue," and I will undertake that task in the remainder of the essay.

Just as it is possible to defend different views of the status of the fetus and of the morality of abortion and still accept HFTTR, so it is possible to defend guidelines or safeguards to separate decisions to abort from decisions to donate fetal tissue without necessarily accepting the immorality of (most) abortions. These guidelines or safeguards—to be discussed below—are designed to reduce the chance that the possibility of providing fetal tissue for transplantation would influence the pregnant woman's decision to abort. Some panelists invoked reasons other than the immorality of abortion to support these guidelines and safeguards. These reasons include an interest in (1) reducing the chance of exploitation and coercion of pregnant women as sources of fetal tissue, and (2) avoiding fanning the flames of moral controversy about abortion. The *Report* notes that "a decisive majority of the panel found that it was acceptable public policy to support transplant research with fetal tissue either because the source of the tissue posed no moral problem or because the immorality of its source could be ethically isolated from the morality of its use in research" (p. 2). In short, the majority rejected the view that HFTTR should not receive federal funds because it is, morally speaking, inextricably linked to or would lead to immoral abortions.

COMPLICITY, COLLABORATION, AND
ASSOCIATION IN MORAL EVIL

In Roman Catholic moral theology, there is a long tradition of reflection on degrees of complicity in moral evil, that is, complicity in the moral wrongdoing of others. A complicated set of moral distinctions has evolved in the face of the fact that "in a complex world of sinful persons, total detachment from evil is impossible."[13] These distinctions focus on various forms of cooperation with evil, which is viewed as "concurrence with another in some immoral action." In formal cooperation the agent both consents to and actively participates in the wrongdoing of others, while in material cooperation the agent does not consent but nonetheless concurs through an action that, while not in itself evil, somehow contributes to another's evil action. Material cooperation can be immediate or mediate, and it can be proximate or remote. An example of major disputes in Roman Catholic moral theology concerns the extent to which an assistant may cooperate in what is viewed as an intrinsically "evil" operation, such as an abortion or a sterilization. Even though actions of association, collaboration, or complicity with wrongdoing are not always as problematic as the wrongdoing itself, there are good reasons to break some connections through disassociation from the wrongdoing.

In the Panel's deliberations, in a dissent (with James Bopp), and in separate articles, James T. Burtchaell, a theologian at Notre Dame University, invoked this traditional discussion to stress the impossibility of a separation, at least in practice, of the use of fetal tissue from the (immoral) abortions that produce the tissue. (The language of complicity also appears in Rabbi J. David Bleich's dissent.) Many of Burtchaell's statements are heavily metaphorical and require careful scrutiny in order to determine exactly how agents who are involved in the use of fetal tissue following elective abortions may be complicitous in the abortions themselves. For example, he uses such phrases as "supportive alliance," "acquiescent partnership," "silent but unmistakable alliance," "accomplices," "confederates," and "ally."[14] From this standpoint, there may be nothing intrinsically wrong about using fetal tissue, just as there may be nothing intrinsically wrong about using the tissue of other dead human beings. It is the connection with acts and practices of abortion that is morally problematic.

In a very careful analysis of the issues of complicity in the context of HFTTR, a report from the University of Minnesota's Center for Biomedical Ethics identifies four major types of involvement that may be alleged to render an agent complicit in the wrongdoing of others.[15] The first two involve causal responsibility—for particular immoral abortions, or for the practice of abortion. In order to identify what is distinctive about the complicity arguments offered by Burtchaell, among others, I will reserve arguments about causal responsibility for separate consideration, analyzing them as arguments (1) that HFTTR would lead some women to abort when they would not otherwise have aborted, and (2) that HFTTR would provide societal legitimation of abortion practices, thus rendering the society less willing to restrict abortion practices.

The Minnesota report identifies two other versions of complicity: activities that imply approval of immoral abortions, and acceptance of benefits from immoral abortions. Both versions are particularly important for identifying distinctive features of the complicity argument, because the agents are not causally and morally responsible for the evil actions of others (i.e., immoral abortions), but through their own actions they are complicitous with those evil actions, even after those actions have occurred. Complicity need not always be causative. Thus, it is important to distinguish various forms of cooperation that involve causal actions, links in a chain of actions, or parts of a composite action—for example, driving the getaway car after a robbery—from actions that only symbolize, convey, or express approval of the wrong actions of others without materially contributing to the wrongdoing. Particularly important for Burtchaell is a form of indirect association that implies moral approval: "It is the sort of association which implies and engenders approval that creates moral complicity."[16]

Burtchaell invokes various analogies to establish moral complicity. In one analogy, "the Florida banker who accepts for deposit large sums of money from drug transactions that are already completed becomes complicit in the drug trade: after the sales, after the fact," even though it was complete before his subsequent involvement.[17] And the banker differs little from the researcher who visits the abortionist each week to obtain fetal tissue and who each time expresses his disapproval of abortion while noting that it will continue whatever he does and making plans to return the following week. Burtchaell claims that even though the wrongdoing occurs apart from the complicitous agent's actions, his or her actions involve an association that both yields benefits and conveys approval. Even if the agents expressly condemn wrongful actions, their association symbolically eviscerates those condemnations. The agents' actions involve complicity in the moral wrongdoing of others, whether drug trafficking or abortion, whatever the agents say. The agents become accessories after the fact.

Sometimes, as his examples suggest, Burtchaell stresses the *institutionalization* of the relationship with the wrong-doer and the wrongdoing. For example, he and Bopp write in their dissent from the Panel's report: "Our argument, then, is that whatever the researcher's intentions may be, by entering into an institutionalized partnership with the abortion industry as a supplier of preference, he or she becomes complicit, though after the fact, with the abortions that have expropriated the tissue for his or her purposes" (p. 70). Similarly, in written testimony to the Panel, the Bishops' Committee for Pro-Life Activities of the National Conference of Catholic Bishops observed: "It may not be wrong *in principle* for someone unconnected with an abortion to make use of a fetal organ from an unborn child who died as a result of an abortion; but it is difficult to see how this *practice can be institutionalized* [including arrangements to ensure informed consent] without threatening a morally unacceptable collaboration with the abortion industry."[18] Which agents are complicitous with the wrongdoing of abortion in HFTTR? Presumably all agents who have this symbolic association—patients receiving transplants, physicians, nurses and others performing the transplants, and those involved in tissue procurement. In addition, the use of federal funds to support HFTTR would involve the federal government (and ultimately each taxpayer) in complicity as a sponsor.

There are at least two possible rebuttals to the charge of moral complicity in the wrongdoing of others. One is to deny that the primary action, in this case abortion, is morally wrong; another is to deny that the use of fetal tissue implies approval or acquiescence. Although the issue of the morality of abortion hovered over its proceedings, the Panel did not try to resolve it and certainly would have encountered major disagreements about which abortions are immoral. Instead the majority took the second approach and insisted that it is at least possible to draw a moral line between the use of fetal tissue and the abortions that make the tissue available, such that unacceptable moral cooperation does not occur (p. 2). For example, it is possible to use organs and tissues from homicide and accident victims—and thus to accept benefits from those tragedies and moral evils—without implying approval of homicides and accidents and without diminishing efforts to reduce their occurrence. Even if abortion is viewed as immoral, it does not follow that using fetal remains involves moral complicity in the "abortions that occur prior to or independent of later uses of fetal remains."[19] Agents involved in the use of fetal tissue may deny that they have "dirty hands," and insist that, even in HFTTR, they are sufficiently disassociated from acts of abortion.[20] Numerous abortions are already being performed, and fetal tissue that could benefit others is now being destroyed rather than used.

Opponents of HFTTR often invoke the analogy with Nazi research in order to show complicity in the wrongdoing of others. This analogy, which is presented in many different ways, focuses on Nazi research on living subjects who were destined for death, on researchers' use of dead bodies and parts, and on subsequent uses of research data.[21] Critics of the analogy stress several morally relevant differences between the use of tissue from dead fetuses following debatably immoral (but legal) abortions and the clearly immoral actions of the Nazi investigators in experimenting on living subjects against their will.[22] Still there are questions about whether individuals and the society should renounce the benefits (if any) that might be gained from the Nazi experiments in order to avoid the moral taint of those experiments.[23] Regarding HFTTR a majority of the panelists noted in a concurring statement that the complicity claim "is considerably weakened when the act making the benefit possible is legal and its immorality is vigorously debated, as is the case with abortion" (p. 33). Thus, there are reasons to be cautious about basing public policy concerning HFTTR on perceptions of complicity, apart from establishing that using human fetal tissue in transplantation research would actually lead to (immoral) abortions that would not otherwise have occurred. It is "a human estimate," as Burtchaell concedes, "how close and operative complicity actually is."[24] The Panel recognized that individuals, whether patients or professionals, might feel that their participation in HFTTR would involve moral complicity in abortion; it thus recommended disclosure of information that the tissue came from aborted fetuses (p. 1).

The charge of complicity in the moral wrongdoing of others could actually be directed against those who invoke it to oppose HFTTR. For example, failure to provide sex education, contraceptives, and social support for pregnant women could be construed as modes of complicity in the actions of abortion.[25] Here the alleged complicity also involves causal responsibility through acts of omission.

I have analyzed complicity mainly in terms of symbolic connections and associations that morally taint the agent. For many critics of HFTTR the charge of complicity is distinctive in that it focuses on the corruption of the agent apart from any further consequences.[26] By contrast, the remaining two major arguments against HFTTR focus on other negative consequences of the research (or developed therapy): societal legitimation of abortion practices and provision of additional incentives for individual abortions.

SOCIETAL LEGITIMATION OF ABORTION PRACTICES

To some extent, the analytical categories used in this discussion tend to collapse into one another; in particular, societal legitimation of abortion practices is at least partially reducible to issues of complicity in and encouragement of abortions. However, it is useful to consider societal legitimation separately, apart from the charges of moral complicity and actual causal responsibility for individual abortion decisions. To "legitimate an act or practice," Dorothy E. Vawter et al. note, "is to justify or promote it in such a manner that others will become more inclined to regard it as acceptable and to engage in it."[27] Here I want to focus on societal legitimation (rather than individual approval) of abortion practices (rather than individual decisions).

In one version of the societal legitimation arguments, critics contend that federally funded HFTTR following elective abortions would tend to legitimate abortion practices because of the difficulty—or even the impossibility— of separating within the expenditure of federal funds (1) approval of the use of fetal tissue from elective abortions, and (2) approval of the elective abortions that produced the fetal tissue. Rabbi J. David Bleich argues that "federal funding conveys an unintended message of moral approval for every aspect of the research program" (p. 40, n. 2). It creates an "aura of moral acceptability." In response, defenders of the research could argue that the provision of federal funds in the treatment of end-stage kidney failure through renal transplantation does not constitute approval of the homicides, suicides, and accidents that provide the occasions for organ donations (p. 35, n. 23). Furthermore, there is simply no evidence of an abatement of efforts to reduce such tragedies in order to maintain or increase the supply of needed organs in the face of scarcity. Thus, it is important to consider all of the society's actions that bear upon such matters as the use of fetal tissue following elective abortions. In addition, symbolic societal legitimation can be reduced, at least in part, by the kinds of separation measures discussed elsewhere in this essay.

A second version of the societal legitimation argument focuses on *societal acceptance* of the benefits of human fetal tissue provided through elective abortions rather than on *governmental funding*. It is difficult, perhaps even impossible, critics argue, for the society to accept the benefits of HFTTR without becoming increasingly inclined to accept as legitimate the abortions that make the benefits possible. Particularly if HFTTR is successful, even through protocols supported only by private funds, the society would become less likely to de-legitimate abortion and restrict its performance, even if future Supreme Court decisions make such restrictions more possible. In this version of the argument, societal legitimation cannot be avoided by the proposed separation measures since the major impact occurs through acceptance of the positive results of HFTTR following elective abortions; hence it can be avoided only by renouncing the benefits.

It is important to note that the distinction between *private* and *public* funding of HFTTR is not adequate to avoid the second version of societal legitimation, however much it may help avoid the first version. Dr. Louis Sullivan, Secretary of the Department of Health and Human Services, indefinitely extended the moratorium on the use of federal funds for HFTTR on the grounds that it would encourage or promote abortion; but he also noted that the private sector can continue HFTTR to generate "whatever biomedical knowledge" might emerge.[28] However, as John Fletcher suggests, it is odd for Health and Human Services officials to hold that it would be wrong for the federal government to fund HFTTR without condemning (and even apparently accepting) privately funded HFTTR and its benefits.[29] For societal acceptance of the benefits—if they can be produced by HFTTR—would provide societal (though not necessarily governmental) legitimation of abortion practices. And there are potential moral costs in the separation of private and public sectors when the federal government sits on the sidelines. It is more likely that the proposed measures to separate decisions to abort and decisions to provide fetal tissue can be implemented under the requirements and sanctions attached to federal funding of HFTTR.[30]

A final criticism of societal legitimation of abortion through HFTTR appears in a dissenting letter by Daniel N. Robinson, who argues that "induced abortion is a moral wrong and that it cannot be redeemed by any actual or potential 'good' secured by it. Thus, the possible medical benefits held out by research on tissues obtained by such measures cannot be exculpatory" (p. 73).[31] Defenders of HFTTR can respond that it in no way "redeems" or "exculpates" the abortions themselves, however much it may reduce feelings of guilt and tragedy. Bringing some good out of evil or tragedy does not cancel the evil or tragedy. HFTTR only involves the use of tissue that would otherwise be discarded or incinerated, without implying approval (or disapproval) of abortion practices; and whatever their rationale, the separation measures set a symbolic gulf between abortion practices and the use of fetal tissue.

INCENTIVES FOR ABORTIONS

Despite the importance of the complicity and legitimation arguments just addressed, the main criticism of federal funding of HFTTR is that it would lead to abortions that would not otherwise have occurred. (This argument may also extend to private funding, especially if HFTTR becomes routine.) Proponents of this argument of causal responsibility need not contend that there would be a net increase in the number of abortions; they need only contend that (some) women who would not otherwise have had (some) abortions will choose to do so in part because of the possibility of providing tissue to benefit others through transplantation. Both this argument and the societal legitimation argument probably carry greater weight in the context of therapy than in the context of research, but they are also offered for HFTTR.

The argument that HFTTR would lead to additional abortions hinges in part on matters that should be resolvable by empirical data—the reasons women choose to have abortions. The Panel's report noted "that the reasons for terminating a pregnancy are complex, varied, and deeply personal," and regarded it as "highly unlikely that a woman would be encouraged to make this decision [to abort] because of the knowledge that the fetal remains might be used in research" (p. 3). In addition, the Panel underlined the lack of any evidence that over the last thirty years the possibility of donating fetal tissue for research purposes has resulted in an increase in the number of abortions (p. 3). Still, as noted earlier, opponents of HFTTR might contend that there is an important difference between using fetal tissue to develop treatments and using fetal tissue as treatments,[32] and that so far the latter has been rare.

Defenders of the causal responsibility argument identify three potential incentives for abortion provided by HFTTR. At least two of these incentives could largely be avoided by the guidelines and safeguards proposed by supporters of the research, while the third remains largely speculative.

Financial incentives. HFTTR could provide a motive for women to have an abortion through financial incentives for the provision of fetal tissue. During the deliberations of the HFTTR Panel, the National Organ Transplant Act was amended to include, under the "human organs" that may not be bought or sold, "any other human organ (or any subpart thereof, including that derived from a fetus) specified by the Secretary of Health and Human Services by regulation."[33] There is some debate about whether this prohibition covers fetal *neural tissue,* which may not be an organ or a subpart of an organ. There is also some uncertainty about whether an abortion clinic that receives reimbursement for its expenses in procuring fetal tissue, as is permitted by the National Organ Transplant Act, can then reduce the price of the abortion, thereby passing the earnings on to the pregnant women.[34] The HFTTR Panel held that "it is essential . . . that no fees be paid to the woman to donate, or to the clinic for its efforts in procuring fetal tissue (other than expenses incurred in retrieving fetal tissue)" (p. 9). There is general agreement on the prohibition of the sale or purchase of fetal tissue, and it can be justified on several grounds—to protect fetuses from abortion, to protect women from exploitation and coercion, to reduce moral controversy about abortion, and also to avoid societal commodification of "human" body parts. (For a fuller discussion of sales and financial incentives, see the previous chapter.)

Specific benevolence or altruism. Already some women have volunteered to become pregnant in order to abort and donate fetal tissue to help a beloved family member.[35] Their motivation can be called "specific benevolence" or "specific altruism," as distinguished from "general benevolence" or "general altruism," which will be considered below. Specific benevolence is beneficence directed toward specific, known individuals, usually (but not necessarily) in a loving relationship.

The Panel held that it is both desirable and feasible to prevent such motives from leading to abortions that would not otherwise have occurred. In contrast to the directed donations now permitted by the Uniform Anatomical Gift Act, the woman donating the fetal tissue for federally funded HFTTR should not be permitted to designate the beneficiary/recipient. This requirement of anonymity between the donor (i.e., the woman donating fetal tissue) and the recipient should protect both parties—the donor should not know who will receive the tissue and the donor's identity should be concealed from the recipient and from the transplant team (p. 4). Here as elsewhere the rationale may involve maternal welfare and freedom from coercion and exploitation, as well as concerns about the morality of abortion. Furthermore, questions can be raised about the unfairness of the distributional pattern that results from directed donations.

The Panel also noted that there is no evidence at this time that the requirement of anonymity between donor and recipient would have a negative impact on transplantation, which often involves matching donors and recipients. However, because of the uncertainty about the possible future value of matching donor and recipient, the Panel proposed a procedural solution and urged the Secretary for Health and Human Services to "review these recommendations at regular intervals" (p. 8). In a concurring statement, several panelists even noted that the ban on donor designation of recipients—and even "aborting for transplant purposes"—should be reexamined if the supply of fetal tissue from elective abortions proves to be inadequate: "The ethical and legal arguments in favor of and against such a policy [i.e., the ban] would then need careful scrutiny to determine whether such a policy remains justified" (p. 38).

General benevolence or altruism. This third possible motivation for pregnant women to terminate their pregnancies because of the possibility of HFTTR is the most controversial. It is both the most speculative and the least subject to control. The question is whether the possibility of donating fetal tissue to benefit unrelated and unknown patients would play a role in women's decisions to abort. However, neither the defenders nor the critics of HFTTR can find strong evidence to support their claims about the potential impact on individual abortion decisions of the knowledge of the possibility of donating fetal tissue (p. 3). Thus the debate hinges on speculation, often shaped by sexist perspectives, about women's abortion decisions and on answers to the moral question about which way society should err in such a situation of doubt.

Critics of HFTTR stress that women are often ambivalent and vacillate about abortion and that this additional motive of general benevolence may lead to some abortions that would not otherwise have occurred. For example, Secretary Sullivan of Health and Human Services insisted that providing pregnant women, who "arrive at the abortion decision after much soul searching and uncertainty," with "the additional rationalization of directly advancing the cause of human therapeutics cannot help but tilt some already vulnerable women toward a decision to have an abortion."[36] By contrast, I tend to downplay the potential role of general benevolence in pregnant women's decision to abort; it appears that pregnant women usually choose to abort because of their perception of the burdens of the pregnancy or of the future child and that, while often present, their benevolence toward others is mainly directed toward identifiable persons in a relationship rather than toward anonymous beneficiaries.[37]

I was brought up short, however, by a telephone call from a professor in another state who had heard that I had defended HFTTR on a national television program. He indicated that after his wife had delivered their second child, he had a vasectomy because they didn't want to have any more children. However, the vasectomy was not successful, and his wife was now pregnant for the third time. They wanted to have an abortion, but they also wanted to donate fetal tissue in order "to reduce their troublesome thoughts about abortion." Having referred him (and his wife) to others who could help, I do not know whether his wife chose to have an abortion and, if she did, whether the possibility of donating fetal tissue (to help others) was really significant in her decision. Even if it helped to relieve her (and his) guilt, it may not have been a necessary or sufficient motive in her decision to abort. That is, she may have had sufficient motivation to abort without this consideration. As C. D. Broad has reminded us, it is very difficult to determine whether any motive is necessary and/or sufficient in our (or others') decisions.[38] It is important not to confuse primary considerations and secondary considerations, where they can be distinguished, and evidence presented to the Panel indicated that the primary reasons for termination of pregnancy would leave little room for a significant role for general benevolence.[39]

In contrast to efforts to eliminate motives of financial gain and specific benevolence, it is not possible to develop guidelines and safeguards to completely eliminate the possibility that the opportunity to donate fetal tissue, based on general benevolence, will play a role in some abortion decisions.

However, with the majority of the Panel, I believe that this risk is worth taking in view of the possible benefits of HFTTR. I disagree with the Assistant Secretary for Health that the risk of even one additional abortion would be too high a price to pay for these benefits: "If just one additional fetus were lost because of the allure of directly benefiting another life by the donation of fetal tissue, our department would still be against federal funding. . . ."[40] It is difficult to attribute full moral seriousness to such claims when the agents fail to perceive that other policies contribute more predictably to abortions that would not otherwise have occurred—for example, the failure to provide sex education, contraceptives, and psychosocial support for pregnant women.

It is also important to distinguish means, ends, and consequences. Additional elective abortions do not constitute a *means* to the *end* of HFTTR. At most they are a possible *consequence* of the use of HFTTR. Hence, risk-benefit analysis and assessment are required. Risk is a probabilistic notion, and it focuses on the probability of a negative outcome of some magnitude. It is thus necessary to judge how probable a negative outcome is as well as its magnitude. The risk of additional fetal deaths—that is, ones that would not otherwise have occurred—is comparable to other losses of life in the pursuit of important societal goals, such as automobile design, highway engineering, and bridge building. As noted in John Robertson's concurring statement, joined by ten other panelists, the "risk that *some* lives will be lost, however, is not sufficient to stop those projects when the number of deaths is not substantial, when the activity serves worthy goals and when reasonable steps to minimize the loss have been taken" (p. 34). And it is not justifiable to adopt a more stringent policy for HFTTR "just because the risk is to prenatal life from *some* increase in the number of legal abortions" (pp. 34–35).

The concurring statement notes that the risk of additional fetal deaths is speculative and stresses that it is possible to identify similar speculative risks that the society might encourage or legitimate deaths from homicide, suicide, and accidents in order to gain organs for transplantation. However, these speculative risks are not considered a sufficient reason to stop using organs from such sources (p. 35, n. 23). In a later response, James Mason, then the Assistant Secretary for Health, contended that the concurring statement simply threw out moral and ethical considerations altogether in the argument just presented.[41] However, signers of the concurring statement, including this writer, view it as offering a different balance of moral and ethical considerations rather than a denial of those considerations. When there is no evidence to support the claim of a high probability of a large number of additional fetal deaths, and when those deaths, if they occur, are the indirect (and unintended) result of legitimate and promising medical research, the rule of multiple effects would appear to justify proceeding with HFTTR. In addition, the Panel held that its recommended guidelines and safeguards would reduce the probability of any increase in the number of abortions.

A different risk-benefit calculation appears in the dissent by Rabbi J. David Bleich, who held that the safeguards proposed by the Panel will not prevent "an increment in the total number of abortions performed" (p. 39). By contrast, the duty to rescue human life through fetal tissue transplants is weakened because (1) at this point these are *research* protocols with undetermined and remote benefits for future patients, rather than therapeutic protocols with high probabilities of immediate benefits for current patients, and (2) because the "moral harm" of the increase in the number of abortions is virtually certain and immediate. Hence, "on balance, the duty to refrain from a course of action that will have the effect of increasing instances of feticide must be regarded as the more compelling moral imperative" (pp. 42–43). Rabbi Bleich's formulation, which closely follows the Jewish tradition's reasoning about the duty to rescue, appears to leave open the possibility of a different balance if the transplantation procedure reaches the point, without federally funded research, of providing an immediate and certain benefit to current patients. By contrast, the majority of the panel held that the increase in the number of abortions is not certain and immediate and that the probability and the number of additional abortions can be greatly reduced by the proposed guidelines or safeguards.

DONORS AND DONATIONS: DISPOSITIONAL AUTHORITY OVER FETAL REMAINS

As noted above, one of the difficult questions surrounding HFTTR is the mode of transfer or acquisition of human fetal tissue following induced abortions. A society could adopt one of several possible modes of transfer or acquisition of fetal tissue—express donation; presumed donation; abandonment; expropriation; and sale, all of which are discussed in the two preceding chapters.[42] Donation is the dominant mode of transfer or acquisition of cadaveric organs and tissue in the United States. *Express donation* by the decedent or by the decedent's next of kin is the main version of donation, but *presumed donation* also exists—several states authorize procurement teams to remove corneas without express consent unless the decedent or the next of kin has dissented or opted out. (I use the term "donor" throughout this volume to refer to the decision maker, that is, the one who makes the gift, rather than to the source of the organs if that source could not or did not make a decision to give.)

There is some evidence that fetal tissue has sometimes been viewed as abandoned or unclaimed and used without explicit maternal consent (p. 11); the ethical justification for such a practice depends in part on expectations about disposition following removal. While such uses of fetal tissue could have been based on presumed consent or donation, it is difficult to justify presumed donation ethically unless there is good reason to believe that silence actually reflects understanding and choice. Perhaps the moral basis is routine salvaging, with the possibility of opting out. The Panel heard no evidence that fetal tissue has been expropriated (i.e., taken against the wishes of the next of kin). The reasons for opposing transfer by sales were discussed above, and the Panel heard no evidence to indicate that it is being practiced. By and large, then, express donation of fetal tissue prevails. And studies indicate that when asked over 90 percent of pregnant women agree to donate fetal tissue for research.[43]

The HFTTR Panel held that, in general, the framework of express donation provided by the Uniform Anatomical Gift Act (UAGA) should also apply to the transfer/acquisition of fetal tissue: "Express donation by the pregnant woman after the abortion decision is the most appropriate mode of transfer of fetal tissues because it is the most congruent with our society's traditions, laws, policies, and practices, including the Uniform Anatomical Gift Act and current Federal research regulations" (p. 6).[44] This response in favor of the sufficiency (within certain limits) of maternal consent/donation attracted the smallest majority of any answer to any of the Assistant Secretary's ten questions: 17 yes, 3 no, 1 abstention.

Much of the debate hinges on what donation of tissues presupposes about the relationship between the donor and the cadaveric source of the tissue. Critics of HFTTR contend that when the pregnant woman "resolves to destroy her offspring, she has abdicated her office and duty as the guardian of her offspring, and thereby forfeits her tutelary powers" (p. 47). From this perspective the abortion decision deprives the pregnant woman of any subsequent authority over the disposition of the fetal remains. This perspective appears to require an absolute separation between the decision to abort and the decision to use fetal tissue by disqualifying the woman who decides to abort from making a decision about use. However, it would also disqualify others on the grounds of their failure to exercise their protective powers and their complicity in the moral wrongdoing of others.[45]

By contrast, the Panel argued that a woman's choice of a legal abortion does not disqualify her legally and should not disqualify her morally from serving "as the primary decisionmaker about the disposition of fetal remains, including the donation of fetal tissue for research" (p. 6). The Panel rejected arguments that the decision to abort leaves only biological kinship, with no moral authority:

> Disputes about the morality of her decision to have an abortion should not deprive the woman of the legal authority to dispose of fetal remains. She still has a special connection with the fetus and she has a legitimate interest in its disposition and use. Furthermore, the dead fetus has no interests that the pregnant woman's donation would violate. In the final analysis, any mode of transfer other than maternal donation appears to raise more serious ethical problems. (P. 6)

A concurring statement (written by John Robertson and signed by a majority of the panelists) argues that the guardianship model "mistakenly assumes that a person who disposes of cadaveric remains acts as a guardian proxy for the deceased, who has no interests, rather than as a protector of their own interests in what happens to those remains" (p. 36). Of course, where the decedent has expressed his or her wishes, the situation is different.

It may also be instructive to consider societal practices where there is suspected or admitted parental abuse resulting in a child's death. According to one study, "there is no automatic legal disqualification of parents as proxy agents for organ donation when they are suspected of or have admitted playing a role in the death of their child."[46] Even in such cases, whatever moral queasiness may exist, organ procurement teams still ask the parents for organ donation. Of course, decisions about the cadaver should be distinguished from decisions about withholding or withdrawing life-support systems from a living person, prior to the determination of death, for it is important to avoid conflicts of interest in life-death decisions. However, after the determination of death, there are no cadaveric interests to be protected. Respectful and dignified treatment and disposal of cadaveric remains are required, in part to prevent offense to the living. Nevertheless, this requirement is consistent with the use of cadavers in education, transplantation, and research, including, for example, crash tests. And there is no reason on grounds of *equal respect* to disqualify the woman having a legal abortion from making a decision about donation of cadaveric fetal tissue.

While accepting the structure of the UAGA (revised 1987) as generally adequate, the Panel did recommend a modification in policy for the donation of fetal tissue in federally funded research. As applied to fetal tissue, the UAGA allows either parent to donate unless the other parent is known to object. However, the Panel concluded that "the pregnant woman's consent should be *necessary* for donation—that is, the father should not be able to authorize the donation by himself, and the mother should always be asked before fetal tissue is used. In addition, her consent or donation should be *sufficient*, except where the procurement team knows of the father's objection to such donation" (p. 7). Noting that there is no legal or ethical obligation to seek the father's permission, the Panel held that there is "a legal and ethical obligation not to use the tissue if it is known that he objects (unless the pregnancy resulted from rape or incest)" (p. 7).

In its recommendations of guidelines for federal funding of HFTTR, the Panel also stressed compliance with state laws and noted that at least eight states have statutes that prohibit the experimental use of cadaveric fetal tissue from induced abortion (p. 13).[47] Another important and unresolved issue is whether state required request laws apply to fetal remains; over forty states now have laws that require institutions to set up mechanisms and procedures to ensure that each family of a decedent who is a possible source of organs and tissues for transplantation is asked about donation.

The debates surrounding this question also focused on ways to *separate* the abortion decision, made by the pregnant woman, from the decision about the use of fetal tissue. If the woman undergoing the abortion may donate fetal tissue following the abortion, further difficult questions arise about limits on her decisional authority and about limits on the disclosure of information to her. For various reasons already identified—the desire not to provide incentives for abortions, the desire to protect pregnant women from exploitation and coercion, and the desire not to fan the flames of the abortion controversy—the

Panel supported some limits on the pregnant woman's autonomy, without restricting the abortion decision itself. The Panel's projected "wall of separation" between the abortion decision and the donation and use of cadaveric fetal tissue was "built of procedural bricks."[48]

Under the UAGA there is no obligation to accept donated tissue or organs; hence the woman's right to donate fetal tissue does not create an obligation on anyone else to accept the fight. Furthermore, the Panel recommended that directed donations not be accepted and supported the prohibition on the sale of fetal tissue. Stressing the express donation model embodied in the UAGA, the U.S. Panel's recommendations would allow the pregnant woman to choose whether to donate fetal tissue for research or some other purpose and to receive as much information as she requests regarding donation, without allowing her to know or to designate the recipient of the fetal tissue. By contrast, the Polkinghorne report in the United Kingdom recommended indeterminate donation to the extent of providing the pregnant woman "no knowledge of what will actually happen to the fetus or fetal tissue"—in order to make it even less likely that the possibility of beneficial use of tissue will influence her decision to have an abortion—and not allowing her "to make any direction regarding the use of her fetus or fetal tissue."[49]

Regarding potential pressures to modify the timing and method of abortion to secure more viable fetal tissue, the U.S. Panel's report underlined the current absence of pressure for later abortions but insisted that "to the extent that Federal sponsorship or funding is involved, no abortion should be put off to a later date nor should any abortion be performed by an alternate method entailing greater risk to the pregnant woman in order to supply more useful fetal materials for research" (p. 14). However, the Panel did not address what should be permitted and done if the pregnant woman requests information about the advantages of changes in timing and methods of abortion for the later use of donated fetal tissue and then requests those changes.

I cannot go into detail regarding additional information that should be provided to the woman considering donation. The policies and practices surrounding donation do not, in general, require the level of disclosure of information that would be expected in informed consent for therapy or research, but the pregnant woman may, of course, request any information she wants prior to making a decision to donate. Among the information that should be disclosed to the pregnant woman who is contemplating donation is information about any tests that will be conducted on her in order to determine the quality of the fetal tissue—for example, HIV antibody tests—and any modifications of the abortion procedure being considered because of the anticipated donation. Both of these matters may have an impact on her well-being and should be disclosed.[50] Since she is donating for transplantation research, she may also have an interest in knowing any plans for commercialization.

When should there be an inquiry about donation? The HFTTR Panel held that "ideally, permission to use tissues from the aborted fetus would not even be sought until the abortion itself had been performed" (p. 10). However, because postmortem tissue deteriorates quickly and cryogenic storage is not fully developed—and, in any event, as a departure from the normal expectations regarding disposition, may require permission—"the pregnant woman must be consulted before the abortion is actually performed" (p. 10).[51] However, the Panel held that no information about the donation and use of fetal tissue in research should be provided prior to the pregnant woman's decision to abort, *unless* she specifically requests that information (pp. 3–4). There are also good reasons for separate consent forms for the abortion and for the donation in order to reduce the chance that the latter will influence the former and will be simply one part of a single decision. Similarly, in determining *who* should make the inquiry about donation, there should be a separation of parties to make sure that there is no conflict of interest and that the woman experiences no pressure to donate.

Regarding disclosure of information to other parties, the U.S. Panel recommends the disclosure of the source of the tissue, that is, that it came from a fetus or fetuses, so that the potential recipient of the transplant could choose not to participate, while the British report recommends against such disclosure. However, both reports recommend disclosure of information about the source to health care professionals.

THE PANEL'S CONSENSUS AND SOCIETAL CONSENSUS

The nature and goal of consensus. A science writer who observed the HFTTR Panel's meetings described the Panel's process as one of seeking consensus.[52] Indeed, much of the Panel's discussion was devoted to debating and modifying the formulation of particular responses in an effort to gain as much consensus as possible. However, the members then voted on responses to each question, listing the totals for, against, and abstaining. In most cases a small minority of one to three voted against the carefully constructed compromises. Even though claims about the Panel's consensus became important in subsequent policy debates, so did the numbers actually voting for and against particular responses.[53] And the dissents published along with the majority report and concurring statements played an important role in the reception and subsequent use of the report.

Consensus is a slippery concept, and its contemporary use often reflects unclarity about the extent of agreement, the nature of the agreement (e.g., what or why), and the process that generates the agreement. Within the boundaries of any group, there is a question about how many of its members have to agree before their agreement is properly described as a "consensus." Clearly if there is unanimity, there is consensus; but in the absence of unanimity, how many must affirm a position before it represents a consensus? The term "consensus" was used very loosely by the NIH Advisory Committee, which held that the Panel's consensus reflected "the consensus of the country."

It is unclear how there can be a "consensus of the country" that HFTTR is "acceptable public policy" within certain "safeguards"—which is what the Panel held—if the country is also marked, as the Advisory Committee notes, by "widely divergent views . . . about the use of fetal materials derived from such abortions for the purposes of research."[54] And because of the dissents from the Panel's report, some commentators have referred to its "substantial consensus" or "near consensus."[55]

However defined, consensus is not unquestionably valuable, either as an end or as a means, but requires a more nuanced evaluation. While consensus can increase our confidence in particular and general judgments, any consensus may be mistaken.[56] Furthermore, the significance of the consensus reached in any group such as the HFTTR Panel depends in part on the fairness of the process, the range of information presented, the open-mindedness of the panelists, etc. It is also important not to overlook the psychosocial dynamics of group formation and interaction.[57] In the final analysis, selection of the participants will often determine what kind of agreement will emerge. Critics of the Panel charge that its consensus was foreordained because of the selection process, which resulted in the appointment of panelists who had previously supported HFTTR, who accepted abortion, or who had been beneficiaries of NIH. Thus, according to participant and dissenter Burtchaell, "there was no surprise whatsoever in the final vote of the panel."[58] By contrast, defenders of the Panel stress its fair and open-minded discussion.[59] Nevertheless, important questions arise about how representative any such public body must be of the diverse views held in the society at large.

Consensus: ethical or political? Was the "consensus" or "near consensus" reached by the HFTTR Panel an ethical consensus or a political consensus? There was general consensus about what should or could be done as a matter of "acceptable public policy," but it is not clear whether this was ethical consensus in contrast to a normative consensus about public policy. What is a consensus about *what* ought to be done, *which* policies ought to be adopted, without a consensus about *why?* Certainly, as I noted earlier, the various reasons panelists gave, for example, for the recommended "safeguards" and "guidelines" indicate great diversity about *why* they ought to be adopted. However, the inclusion of ethical premises and analyses in reports of governmentally appointed bodies such as the HFTTR Panel has perhaps fostered, in the words of panelist Patricia King, "the illusion that these bodies have achieved consensus at the level of ethical principle or even ethical analysis. In fact, the consensus usually comes at the level of practice and policy. Moreover, it is not clear whether an effort to reach consensus at the level of principle is either possible or desirable."[60]

While agreeing that federally funded HFTTR, within certain safeguards, is "acceptable public policy," the Panel vigorously debated whether to add the modifier "ethically," as in "ethically acceptable," but finally settled for "acceptable public policy," in part because the chairman, a strong opponent of abortion in most cases, vigorously opposed such a modifier.[61] Without this concession, he probably would have dissented and thus reduced the political significance of the report.

However, in retrospect, the Panel's concession regarding "ethically acceptable" played into the hands of its critics, who charged that the Panel simply abandoned ethical considerations altogether. It could, for instance, lend credence to the claim, offered by Assistant Secretary for Health James Mason, that the majority of panelists indicated that "moral and ethical considerations were not central to their view of the issue."[62] Since, far from neglecting "moral and ethical considerations," the majority simply offered a different view of their implications for HFTTR, such gross (mis)interpretations have led some panelists to wonder whether they should have pressed for stronger language in the report—for example, "ethically acceptable" rather than merely "acceptable public policy"—because the efforts to find compromise language to gain the support of more panelists left the report vulnerable at points and subject to neglect, misuse, and misquotation.[63]

And efforts to achieve consensus are often costly. One cost for the HFTTR Panel was the excessive amount of time spent trying to find formulations and words that would gain the agreement of more and more panelists. The carefully crafted responses, worked out through the group process that consumed energy as well as large amounts of the limited available time, usually resulted in divided votes (but with only small minorities). Seeking this consensus also may have diminished the intellectual and rhetorical quality of the final product. However, as I have suggested, any effort to short-circuit the process of consensus formation—perhaps by pressing for earlier votes on more sharply worded formulations—could have produced a less effective product from a political standpoint, for panelists in the minority might have withdrawn in protest and thereby de-legitimated the Panel's report.

Clearly, then, here was a significant trade-off, for the Panel's report with its brief "responses," followed by its brief "considerations," is not intellectually or rhetorically satisfactory. The report may fail to persuade because it does not clearly and fully explain the reasons that in collective deliberation could lead people with different views about the status of the fetus and about the morality of abortion to support federal funding for HFTTR using tissue from deliberate abortions within certain "guidelines." "It was probably necessary," as King notes, "to describe the process that resulted in acceptance of this point rather than merely stating it." Perhaps a deeper problem, King continues, is that in its "drive to achieve consensus, the panel gave insufficient attention to diverse views, to raising new questions, to stimulating debate, and to furthering societal discussion of controversial matters. Perhaps consensus was achieved at the expense of other functions that these national bodies ought to perform."[64]

Presupposed and predicted societal consensus. I have already noted the ambiguity of the NIH Advisory Committee's claim that the Panel's consensus reflected the consensus of the country. Certainly the panelists often presupposed and sometimes even appealed to a preexisting societal consensus, on at least some matters and at some level of generality. For instance, panelists accepted the negative paradigm case of Nazi experimentation and the principles embedded in the Nuremberg Code, but for the majority these did not dictate a particular conclusion on HFTTR with cadaveric fetal tissue from deliberate abortions. Furthermore, in its argument for recognizing the woman who chooses an abortion as the appropriate decision maker about the donation of fetal tissue, the Panel stressed the congruity between her role in this context and in other societal practices of organ and tissue donation, which apparently reflect a societal consensus. And the Panel implicitly appealed to other general principles and values, such as respect for persons and justice. The Panel did not, however, focus specifically on public sentiment, except briefly in a discussion of what the public believes about abortion.[65]

While the Panel thus did not explicitly appeal to an existing societal consensus about funding HFTTR, its debate often focused on a *possible new* societal consensus regarding *abortion*. Opponents of HFTTR worried, as I have stressed, that the provision of government funds and acceptance of the benefits of successful HFTTR would lead to a new societal consensus that abortion is acceptable, while the majority insisted that it is possible for the society to fund and to accept benefits from HFTTR without approving of the abortions that produce the tissue, just as it now funds and accepts organ transplantation without approving of the homicides or accidents that make the organs available.

Regulatory and prophetic roles. Sometimes bioethicists who participate in such governmentally appointed bodies or who focus their writings on public policy appear to be playing only a *regulatory* role, not a *prophetic* role.[66] Can such a body, seeking consensus, function only in a regulatory mode? Can such a body ever be prophetic? Prophetic judgments may (1) appeal to transcendent norms by which to criticize and evaluate current social norms and practices, (2) offer a different interpretation of the meaning and ranking of current social norms, or (3) call for conformity to accepted social norms. The first form of prophecy is not a likely prospect for a governmentally appointed panel, but the other two are possibilities. Reflecting the last two approaches, Michael Walzer describes much Hebrew prophecy as "social criticism because it challenges the leaders, the conventions, the ritual practices of a particular society and because it does so in the name of values recognized and shared in that same society."[67] Prophecy in the second and third senses presupposes some consensus, rather than only opposing an established consensus.

The prophetic approach may also sow the seeds of a different future consensus. Walter Harrelson has observed that DHHS received "what it desired" but not "what it most needed" in the HFTTR Panel's report. What it most needed was "a rhetorically and aesthetically attractive report . . . [with] a language and a set of images that will help a polarized community begin to build elements of consensus."[68] Such a report could have contributed more significantly to the societal conversation and formation of a consensus.

The extensive concurring statement, prepared by John Robertson and signed in whole or in part by ten other panelists, including myself, was intended to counterbalance the dissenting statement by Bopp and Burtchaell. However, it may not have fully achieved its goal, in part because some of its language and conclusions provided materials that critics could use, even if inappropriately, to further substantiate their charge that the majority had sacrificed ethical considerations for "significant medical goals."[69]

Framing and reframing problems. The Panel sought consensus in its responses to the Assistant Secretary's definition of the problem. His definition of the problem for the Panel appeared in his ten questions, which focused on the linkage or connection of HFTTR with abortion.[70] This linkage is understandable, because without the connection between HFTTR and abortion, the director of NIH would have approved the research without involving the Assistant Secretary for Health. However, in defining the problem so exclusively in relation to abortion, the ten questions severely constrained the Panel's deliberations.

Certainly a more neutral set of questions—for example, "Under what set of circumstances, if any, should the federal government support human fetal tissue transplantation research?"—would have allowed the exploration of a broader range of perspectives and frameworks.[71] Or the Panel could have offered another and perhaps more promising reformulation, one that is already implicit in its report. It could have started from the well-established social practice of using cadaveric human tissue in research, transplantation, and education—as previously noted, cadaveric fetal tissue has already been used in all these ways, too (though not widely for transplantation). The Panel could then have extended the principles embedded in these practices and moved by analogical reasoning to the use of cadaveric fetal tissue in transplantation research, probing for consistency or inconsistency between this proposed use in transplantation research and the established principles and paradigm cases.[72] It could then have addressed the Assistant Secretary's ten questions in a new context. After laying out the paradigm of the use of cadaveric tissue from various sources, the Panel could have asked whether the fact that the tissue came from an aborted fetus, or aborted fetuses, makes any relevant moral difference. While reframing the problem this way would certainly not eliminate the abortion issue, it would put the abortion issue in a different context. The linkage to abortion could be addressed secondarily in trying to determine whether the cause of death is not only "morally relevant" but also "morally decisive" in the use of cadaveric human tissue. Reframing the problem in this way would also make it easier to explain why the Bopp-Burtchaell effort to invoke the Nazi analogy and the principles undergirding research with human subjects fails. By directly (rather than merely indirectly) reframing the problem, the Panel could probably have produced a more significant report, even if it would probably have been no more effective, at least immediately.

It may not be fair to judge the Panel's report on the basis of its immediate fate. The Department of Health and Human Services in effect adopted the minority dissent rather than the majority report. (In retrospect, the critics of HFTTR on the Panel may have used their platform, particularly in their

dissents, in more politically astute ways than those who supported HFTTR.)
At congressional hearings in April 1990, Representative Henry Waxman (D-CA) challenged Assistant Secretary Mason about precedents for the DHHS
rejection of a *unanimous* advisory committee report—the Advisory Commit-
tee to the Director of NIH had unanimously endorsed the HFTTR Panel's
majority report and recommendations.[73] The Assistant Secretary conceded
that such a rejection is rare. However, would a stronger report by a majority of
panelists, with a less general consensus, actually have changed public policy?
Almost certainly not; but the societal conversation, with the possibility of a
future new consensus, would probably have been better served by an intellectually and rhetorically richer report that directly reframed the problem it
addressed.

The significance of international consensus. So far my discussion has focused
on the United States. However, LeRoy Walters, who chaired the ethics
discussion of the HFTTR Panel, notes that the Panel's position is in accord
with the international ethical consensus on this research, as reflected in the
recommendations of various committees or deliberative bodies around the
world.[74] (At least nine such reports had appeared by December 1988 when the
HFTTR Panel made its report, and several more appeared subsequently.)
Despite important cultural differences in the countries involved, the "remarkable similarities" of these reports represent "an impressive international consensus on the ethical standards that should govern the use of fetal tissue for
research." (Of course, questions may arise about whether the other reports
reflect a societal consensus in their countries of origin.) According to Walters,
the "remarkable ethical consensus" in the fourteen reports available by April
1990 is that HFTTR is "ethically acceptable in principle." Then they add
conditions to ensure the ethical conduct of such research, in practice, by
limiting use to cadaver tissue and by attempting "to insulate the humanitarian
use of fetal tissue from the abortion decision on the one hand and the
commercial sphere on the other." While a consensus, whether on a panel or
within a country or around the world, does not guarantee that a position is
"ethically correct," Walters contends that "we are less likely to make a serious
moral mistake when numerous groups of conscientious men and women from
around the world have sought to study the issue with great care and have
reached virtually identical conclusions about appropriate public policy."

Walters further argues that the "burden of proof on the ethics of fetal tissue
research rests on DHHS officials who would have U.S. public policy differ
from this impressive [international] ethical consensus."[75] The political strength
of the right-to-life movement in the United States is certainly one factor, but
critics of HFTTR can also point to special conditions in the United States that
might even justify departing from the international consensus.

First, U.S. abortion laws are less restrictive than those in much of
Europe.[76] Thus, there may be more reason to fear the impact of HFTTR on
abortion decisions and on the societal acceptance of abortion in the United
States than in many other countries.

Second, the United States has allowed more commercialization of and imposed less regulation on both abortion clinics and tissue procurement than some other countries. Furthermore, friendly critics have suggested that the HFTTR Panel did not adequately appreciate the institutional pressures that exist in the United States.[77] By contrast, the U.K. Polkinghorne report called for a national "intermediary" organization, with governmental funding, to separate the *practice* of abortion and the use of fetal tissue—a recommendation that goes beyond the U.S. panel's efforts to separate the *decisions* about abortion and donation. In view of these societal and institutional factors, one question in the United States is whether the limited consensus on what should be done if HFTTR goes forward (whether with private or public funds) can actually be implemented. In view of the Panel's consensus, based on a variety of reasons, about separating abortion decisions and practices from the donation and use of fetal tissue in transplantation research, questions of feasibility become important for an ethical assessment of public policy. The chair of the HFTTR Panel, Judge Adams, a strong opponent of abortion, finally supported federal funding of HFTTR, largely to have a way to implement the "safeguards" endorsed by the Panel (pp. 25–26).

CONCLUSION

In conclusion, despite the moral and political perplexities associated with HFTTR following induced abortions, I believe that such research should go forward, with federal funding, as long as the guidelines and safeguards identified above are in place. These guidelines and safeguards should prevent unacceptable complicity in abortion (or at least allow those who perceive such complicity to opt out, except for the payment of taxes), and, in conjunction with other policies and practices, they should reduce the likelihood of societal legitimation of abortion practices and of providing incentives to pregnant women to have abortions they would not otherwise have had. To be sure, there are no guarantees that these consequences can be prevented, but a reasonable risk analysis/assessment indicates these risks are largely speculative. On the benefit side, the animal studies are promising and may justify human clinical trials, at least for Parkinson's disease and juvenile diabetes (pp. 14 and 19–22). However, my justification is couched in general terms, and it addresses the risks that critics identify as moral obstacles to proceeding with HFTTR, whatever the prospective benefits. The justification of any particular human clinical trials must look more carefully at the scientific evidence for probable benefits and other risks.

It is not implausible to suggest that U.S. public policy regarding HFTTR has often been held hostage to society's uneasiness about abortion. The continuing societal debate about HFTTR concerns in part how to proceed in a situation of doubt and uncertainty and thus which side has the moral and political burden of proof when there is a lack of irrefutable evidence that it is possible rigidly to separate abortion decisions and practices from decisions and practices about the use of fetal tissue following abortions. In such a context the debate is to a great extent about which analogies from current, justifiable practices should govern decision making. The U.S. panel chose the analogy with organ and tissue donation as the most defensible one for approaching human fetal tissue following elective abortions but then argued for some special guidelines and safeguards because of societal concerns about abortion.

While the Panel, sometimes only implicitly, put HFTTR in the context of the transfer and use of any cadaveric human tissue, some other ethicists and policy analysts put the controversy about HFTTR in the context of the broader public controversy about "fetal research."[78] That move is finally a huge conceptual and normative mistake, and it plays into the hands of opponents of HFTTR. For both moral and political reasons, it is important to underline the differences between the use of cadaveric human tissue, whether from a fetus or not, and "fetal research." It is possible to support the use of federal funds for transplantation (and other) research using cadaveric fetal tissue, after deliberate abortions, without necessarily supporting federally funded research that experiments on living fetuses or that creates embryos for research purposes.

POSTSCRIPT (1995)

The Panel's report gained renewed life when President Clinton on his very first day in office in 1993 lifted the ban on the use of federal funds in HFTTR, when Congress in June 1993 passed the NIH Reauthorization Act (P.L. 103–43), and when government bureaucrats drafted regulations for the use of cadaveric human fetal tissue in transplantation research. However, these shifts probably had less to do with the report itself than with other important changes in perceptions of HFTTR among the public in general and policy makers in particular.

Policy analyst Steven Maynard-Moody recently located the "turning point" in the debate in the powerful testimony of antiabortionists Guy and Terri Walden before a House Committee in April 1991, and a Senate committee in November 1991.[79] Having also testified at the November hearings—though much less effectively, I might add—I can attest directly to the powerful impact of the Waldens' moving personal testimony about the loss of two of their children to Hurler's disease, a genetic disease which prevents the production of an adequate supply of an enzyme that helps cells break down some metabolic byproducts and thus causes severe retardation, developmental, and motor problems over time, with regression setting in after about five years and death occurring, in most cases, before the age of ten from pneumonia or heart failure. Then when their son, Nathan, was diagnosed in utero with the same genetic disease, they decided to try an experimental procedure that transplanted fetal tissue (fetal liver stem cells) from an aborted fetus (a case of life-threatening tubal pregnancy) into Nathan while still in utero. This fetus-to-fetus transplant was noteworthy in part because the Waldens staunchly oppose abortion on religious grounds and would not consider abortion as an option even when the fetus has a fatal genetic condition. As "Bible-believing Christians"—Guy Walden is a pastor—they concluded that God actually favors such transplants since Genesis reports that God performed the first transplant in creating Eve from Adam's rib. Their testimony was personal and powerful, as they first circulated photographs of their three children with Hurler's because, as they put it, "a picture is worth a thousand words, and this puts faces to the situation," and then movingly described their children (and their own) plight as well as their decision to accept the first (reported) fetus-to-fetus transplant. And, again in personal rather than ideological terms, they supported action to lift the ban on federally funded HFTTR.[80] (Some research was obviously continuing, at least in minimal ways, in the private sector.)

Other personal testimony that day was also crucially important. Former Senator Morris ("Mo") Udall's daughter, Anne Udall, reported on her father's deterioration from Parkinson's disease and his desire for an experimental fetal tissue transplant, which he was unable to receive because his deterioration was too severe. And Carol Lurie, the founder of the Juvenile Diabetes Foundation, reported her struggle after learning, years before, that her ten-year-old son had diabetes. These testimonies put vital personal faces on the statistics about patients who might benefit from HFTTR. These personal narratives demonstrated exactly what HFFTR might mean in human, rather than ideological, terms. And no one could challenge the Walden's "pro-life" stance.

Other spokespersons, who were clearly opposed to abortion, contributed to the shift in the debate. Among them was Senator Strom Thurmond (R-SC), a staunch antibortionist, who told about his own daughter's struggle with diabetes and who denied that lifting the ban on the use of federal funds in HFTTR would "encourage abortion." And Senator Mark Hatfield (R-OR), who found Anne Udall's testimony moving, indicated that "allowing fetal tissue research is a pro-life position." This redefinition of HFTTR, made possible by the personal narratives, accomplished what the Panel—and others—could not; and both the House and the Senate voted by wide margins to lift the ban in 1992. President Bush vetoed the NIH Reauthorization Bill in order to "prevent taxpayer funds from being used for research that many Americans find morally repugnant and because of its potential for promoting and legitimizing abortion."[81] Supporters of HFTTR did not have enough votes in the House to override the veto.

The Bush administration, nevertheless, also attempted to find compromises that might break the connection between *abortion,* or at least *immoral abortions,* and HFTTR. Perhaps in part in response to the Walden case, which had involved fetal tissue from an ectopic pregnancy, the Bush administration indicated that termination of ectopic pregnancies—what Assistant Secretary for Health James Mason called "treatment for an ectopic pregnancy"—did not count as abortions, and hence that fetal tissue made available as a result of such medical procedures, as well as from spontaneous abortions, could be used in federally funded transplantation research.[82]

Perhaps the administration drew, knowingly or unknowingly, on developments in Catholic moral theology, which has accommodated medical procedures to end tubal pregnancies on the grounds that they do not count as abortions since the fetal deaths in those cases are the indirect and unintended results of a legitimate medical procedure to deal with the maternal health problem. The Panel could perhaps have explored the limited "middle" ground regarding what some would call justifiable abortions and others would call medical treatment of ectopic pregnancies (with the unintended effect of fetal death). Virtually all religious and humanistic traditions, as noted earlier, would concur on the legitimacy of such actions. Hence, there would be no moral or political problem in using the resulting fetal tissue. However, questions arise about how much usable tissue could be procured in this way.

In an effort to head off the congressional move to authorize federal funds for HFTTR, using tissue from deliberately aborted fetuses, President Bush signed, in May 1992, an executive order to require the NIH to establish several "fetal tissue banks," which would collect, freeze, and store fetal tissue "obtained exclusively from ectopic pregnancies and spontaneous abortions" with the expectation—which most scientists viewed as unrealistic—that usable tissue would be obtained from 1,500 to 2,000 fetuses each year.

Then came the election of President Clinton and the lifting of the ban, followed in 1994 by NIH grants for HFTTR. It is too early to determine whether such research will ultimately be beneficial, whether some of the opponents' fears will be realized, and whether the winds of moral and political change may yet restore the ban. To reiterate, the change in the larger context that made it easier for President Clinton to lift the ban resulted, to a great extent, from the very personal portraits of patients and the families of patients in need of such therapy. As the real "turning point" in the public and policy debate, these personal narratives helped to create the climate for a different risk-benefit analysis, one that now stressed the need for, and possible fruits of, HFTTR, and at the same time minimized the potential impact on abortion attitudes and practices. HFTTR became "pro-life," rather than "pro-abortion." It is not clear that the Panel's arguments and recommendations played much of a role, if any, in this shift, but then it is hard to determine which factors do play such roles, other than what is explicitly invoked (as in the case of the personal narratives). However, many of the Panel's proposed "guidelines" and "safeguards" made their way, at least in modified form, into the regulations that currently govern federally funded HFTTR.

NOTES

1. METAPHOR AND ANALOGY IN BIOETHICS

1. My efforts to combine these directions over a number of years were perhaps best expressed in my Summer Seminars for College Teachers, under the auspices of the National Endowment for the Humanities, on "Principles and Metaphors in Biomedical Ethics," and in my work on paternalism in health care. See especially James F. Childress, *Who Should Decide? Paternalism in Health Care* (New York: Oxford University Press, 1982).

2. George Lakoff and Mark Johnson, *Metaphors We Live By* (Chicago: University of Chicago Press, 1980), p. 5.

3. Janet Martin Soskice, *Metaphor and Religious Language* (Oxford: Clarendon Press, 1985).

4. See, for instance, Max Black, "Metaphor," in *Models and Metaphors: Studies in Language and Philosophy* (Ithaca: Cornell University Press, 1962), pp. 25–47; Paul Ricoeur, *The Rule of Metaphor,* trans. Robert Czerny (Toronto: University of Toronto Press, 1977); and Soskice, *Metaphor and Religious Language.*

5. Black, "Metaphor."

6. George Miller, "Images and Models, Similes and Metaphors," in *Metaphor and Thought,* ed. Andrew Ortony (Cambridge: Cambridge University Press, 1979), pp. 202–50.

7. Black, "Metaphor," and Black, "More about Metaphor," in *Metaphor and Thought,* ed. Ortony, pp. 19–43.

8. Soskice, *Metaphor and Religious Language.*

9. Black, "Metaphor," and Lakoff and Johnson, *Metaphors We Live By.*

10. Black, "More about Metaphor," p. 31.

11. Childress, *Who Should Decide?* chap. 1. See also Childress, *War as Reality and War as Metaphor: Some Moral Reflections,* The Joseph A. Reich, Sr., Distinguished Lecture, November 18, 1992 (Colorado Springs: U.S. Air Force Academy, 1993). For other discussions of military metaphors in health care, in addition to the ones that appear in these notes, see Virginia Warren, "A Powerful Metaphor: Medicine as War" (unpublished paper), and Samuel Vaisrub, *Medicine's Metaphors: Messages and Menaces* (Oradell, NJ: Medical Economics Company, 1977).

12. Susan Sontag, *Illness as Metaphor and AIDS and Its Metaphors* (New York: Doubleday Anchor Books, 1990), p. 72. Her essay "Illness as Metaphor" was first published separately in the *New York Review of Books* in 1978 and later that year by Farrar, Straus and Giroux, which also published the essay *AIDS and Its Metaphors* separately in 1989 before the combined edition appeared.

13. Ibid., p. 68.

14. Lawrence Pray, *Journey of a Diabetic* (New York: Simon and Schuster, 1983), and "How Diabetes Became My Teacher," *Washington Post*, July 31, 1983.

15. Anne Hunsaker Hawkins, *Reconstructing Illness: Studies in Pathography* (West Lafayette: Purdue University Press, 1993).

16. William F. May, *The Physician's Covenant* (Philadelphia: Westminster Press, 1983), p. 66.

17. For an explication of the just-war tradition, see James F. Childress, *Moral Responsibility in Conflicts* (Baton Rouge: Louisiana State University Press, 1982), esp. chap. 3 on "Just-War Criteria."

18. Richard Rettig and Kathleen Rohr, "Ethical Dimensions of Allocating Scarce Resources in Medicine: A Cross-National Study of End-Stage Renal Disease," unpublished manuscript (1981).

19. Franz Ingelfinger, Editorial, *New England Journal of Medicine* 287 (December 7, 1982): 1198–99.

20. Anne R. Somers, "The 'Geriatric Imperative' and Growing Economic Constraints," *Journal of Medical Education* 55 (February 1980): 89–90, which is also the source of the figures in the remainder of this paragraph.

21. For these points, see James F. Childress, "Ensuring Care, Respect, and Fairness for the Elderly," *Hastings Center Report* 14 (October 1984): 27–31.

22. Howard F. Stein, *American Medicine as Culture* (Boulder: Westview Press, 1990).

23. George J. Annas, "Reframing the Debate on Health Care Reform by Replacing Our Metaphors," *New England Journal of Medicine* 332 (March 16, 1995): 744–47.

24. See Judith Ross, "Ethics and the Language of AIDS," in *The Meaning of AIDS: Implications for Medical Science, Clinical Practice, and Public Health Policy*, ed. Eric Juengst and Barbara A. Koenig (New York: Praeger, 1989), pp. 30–41; and Sontag, *Illness as Metaphor and AIDS and Its Metaphors*. See also Judith Ross, "The Militarization of Disease: Do We Really Want a War on AIDS?" *Soundings* 72 (1989): 39–50.

25. James Hunter, *Before the Shooting Begins: Searching for Democracy in America's Culture Wars* (New York: Free Press, 1994), and *Culture Wars: The Struggle to Define America* (New York: Basic Books, 1991).

26. James Hunter, "America at War with Itself: Those Debates over Sitcoms and Motherhood Aren't Froth—This Is a Cultural Showdown," *Washington Post*, September 11, 1992, C1 and C4.

27. Peter Steinfels, "Beliefs: Metaphors Are Flying . . . ," *New York Times*, December 12, 1991, p. 10.

28. James F. Childress and Mark Siegler, "Metaphors and Models of Doctor-Patient Relationships: Their Implications for Autonomy," *Theoretical Medicine* 5 (1984): 17–30, which is reprinted below as chapter 3. See also Mark Siegler, "The Physician-Patient Accommodation: A Central Event in Clinical Medicine," *Archives of Internal Medicine* 142 (1982): 1899–1902.

29. May, *Physician's Covenant*, and Robert M. Veatch, "Models for Ethics in Medicine in a Revolutionary Age," *Hastings Center Report* 2 (June 1972): 5–7. See also May, "Code, Covenant, Contract, or Philanthropy," *Hastings Center Report* 5 (December 1975): 29–38.

30. Nancy M. P. King, Larry R. Churchill, and Alan W. Cross, eds., *The Physician as Captain of the Ship* (Dordrecht: D. Reidel, 1988).

31. K. Danner Clouser, "Veatch, May, and Models: A Critical Review and a New View," in *The Clinical Encounter*, ed. Earl E. Shelp (Dordrecht: D. Reidel, 1983), pp. 89–103.

32. David Eerdman, "Coleridge as Editorial Writer," in *Power and Consciousness*, ed. Conor Cruise O'Brien and William Dean Vanech (New York: New York University Press, 1969), p. 197.

33. Gerald Winslow, "From Loyalty to Advocacy: A New Metaphor for Nursing," *Hastings Center Report* 14 (June 1984): 32–40; Ellen W. Bernal, "The Nurse as Patient Advocate," *Hastings Center Report* 22 (July/August 1992): 18–23.

34. Winslow, "From Loyalty to Advocacy," and Bernal, "Nurse as Patient Advocate."

35. Bernal, "Nurse as Patient Advocate."

36. Winslow, "From Loyalty to Advocacy."

37. David Heyd, *Genethics: Moral Issues in the Creation of People* (Berkeley: University of California Press, 1992), p. 4 (emphasis in original).

38. Paul Ramsey, *The Patient as Person* (New Haven: Yale University Press, 1970), chap. 7, esp. pp. 256 and 259. Ramsey waffles, sometimes holding that human beings should not play God, at other times (as in this context) suggesting that there are proper ways to play (and not to play) God.

39. See the discussion in the President's Commission for the Study of Ethical Problems in Medicine and Biomedical and Behavioral Research, *Splicing Life* (Washington, D.C.: U.S. Government Printing Office, 1983).

40. Paul Ramsey, *Fabricated Man: The Ethics of Genetic Control* (New Haven: Yale University Press, 1970), p. 138.

41. Edmund L. Erde, "Studies in the Explanation of Issues in Biomedical Ethics: (II) On 'On Playing God', Etc.," *Journal of Medicine and Philosophy* 14 (1989): 593–615.

42. Paul Ramsey, *Ethics at the Edges of Life* (New Haven: Yale University Press, 1978), pp. 306–307.

43. For fuller analyses and assessments of wedge and slippery slope arguments, from which some of these points have been drawn, see Tom L. Beauchamp and James F. Childress, *Principles of Biomedical Ethics*, 4th ed. (New York: Oxford University Press, 1994), pp. 228–31.

44. Arthur L. Caplan, "Ethical Engineers Need Not Apply: The State of Applied Ethics Today," *Science, Technology and Human Values* 6 (1980): 30.

45. Larry Churchill, "The Ethicist in Professional Education," *Hastings Center Report* 8 (December 1978): 13–15.

46. Sontag, *Illness as Metaphor and AIDS and Its Metaphors.*

47. Donald Schoen, "Generative Metaphor: A Perspective on Problem-Setting in Social Policy," in *Metaphor and Thought,* ed. Ortony.

48. Black, "Metaphor."

49. Ricoeur, *The Rule of Metaphor.*

50. Black, "More about Metaphor."

51. Soskice, *Metaphor and Religious Language.*

52. Dorothy Emmet, *The Nature of Metaphysical Thinking* (New York: St. Martin's Press, 1945), p. 6.

53. See David H. Helman, ed., *Analogical Reasoning: Perspectives of Artificial Intelligence, Cognitive Science, and Philosophy* (Dordrecht: Kluwer, 1988); and Mark T. Keane, *Analogical Problem Solving* (Chichester: Ellis Norwood, 1988).

54. Arthur L. Caplan, "The Doctors' Trial and Analogies to the Holocaust in Contemporary Bioethical Debates," in *The Nazi Doctors and the Nuremberg Code: Human Rights in Human Experimentation,* ed. George J. Annas and Michael A. Grodin (New York: Oxford University Press, 1992), chap. 14. The remainder of this paragraph draws on Caplan.

55. Albert R. Jonsen and Stephen Toulmin, *The Abuse of Casuistry: A History of Moral Reasoning* (Berkeley: University of California Press, 1988), p. 252.

56. Ibid., p. 316.

57. Ramsey, *Patient as Person,* chap. 7.

58. Lisa Sowle Cahill, "Abortion and Argument by Analogy," *Horizons* 9 (1982): 271–87.

59. Ibid.

60. Judith Jarvis Thomson, "A Defense of Abortion," *Philosophy and Public Affairs* 1 (1972): 47–66.

61. John T. Noonan, "How to Argue about Abortion," (New York: The Ad Hoc Committee in Defense of Life, 1974).

62. See Susan Mattingly, "Viewing Abortion from the Perspective of Transplantation: The Ethics of the Gift of Life," *Soundings* 67 (1984): 399–410; and Patricia Beattie Jung, "Abortion and Organ Donation: Christian Reflections on Bodily Life Support," *Journal of Religious Ethics* 16 (1988): 273–305.

63. Cahill, "Abortion and Argument by Analogy."

64. Ibid., p. 283.

65. See Max Weber, *Max Weber on Law in Economy and Society,* ed. and ann. Max Rheinstein, trans. Edward Shils and Max Rheinstein (New York: Simon & Schuster, 1967), p. 1. The translators use the term "purpose-rational" for *zweckrational.*

66. Richard Zeckhauser, "Procedures for Valuing Lives," *Public Policy* (Fall 1975): 447–48. Contrast Richard A. Rettig, "Valuing Lives: The Policy Debate on Patient Care Financing for Victims of End-Stage Renal Disease," *The Rand Paper Series* (Santa Monica: Rand Corporation, 1976).

67. Charles Fried, *An Anatomy of Values: Problems of Personal and Social Choice* (Cambridge: Harvard University Press, 1970), p. 224f. Much of the discussion in these paragraphs on symbolic rationality has been drawn from my *Priorities in Biomedical Ethics* (Philadelphia: Westminster Press, 1981), chap. 4.

68. See Guido Calabresi and Philip Bobbitt, *Tragic Choices* (New York: Norton, 1978).

2. ETHICAL THEORIES, PRINCIPLES, AND CASUISTRY IN BIOETHICS

1. K. Danner Clouser and Bernard Gert, "A Critique of Principlism," *Journal of Medicine and Philosophy* 15 (1990): 219–36.

2. Tom L. Beauchamp and James F. Childress, *Principles of Biomedical Ethics* (New York: Oxford University Press, 1st ed., 1979; 2nd ed., 1983; 3rd ed., 1989; 4th ed., 1994). In offering this reinterpretation and defense of "principlism," I do not want to claim that Tom Beauchamp, who coauthored *Principles of Biomedical Ethics* with me, would agree with all that I argue here; for we do have some important differences that we either submerge or compromise when we cannot convince the other author in the course of preparing various editions of *Principles of Biomedical Ethics.* Our fourth edition attempts to address criticisms of previous editions. All subsequent references will be given in the text as *PBE,* followed by the edition and page numbers as needed.

3. Human Fetal Tissue Transplantation Research Panel, *Report of the Human Fetal Tissue Transplantation Research Panel,* 2 vols. (Bethesda, MD: National Institutes of Health, 1988).

4. See James F. Childress, "Ethics, Public Policy, and Human Fetal Tissue Transplantation Research," *Kennedy Institute of Ethics Journal* 1 (June 1991): 93–121 (reprinted, with additions, as chap. 16 below); and "Deliberations of the Human Fetal Tissue Transplantation Research Panel," in *Biomedical Politics,* ed. Kathi E. Hanna (Washington, D.C.: National Academy Press, 1991), pp. 215–48.

5. Albert R. Jonsen, "Transplantation of Fetal Tissue: An Ethicist's Viewpoint," *Clinical Research* 36 (1988): 215–19.

6. Ibid.

7. K. Danner Clouser, "Statement to the Advisory Committee to the Director, NIH," in Report of the Advisory Committee to the Director, National Institutes of Health, *Human Fetal Tissue Transplantation Research* (Bethesda, MD: National Institutes of Health, 1988), pp. C17–18.

8. Bernard Gert, *Morality: A New Justification of the Moral Rules* (New York: Oxford University Press, paperback edition, 1989).

9. Clouser, "Statement to the Advisory Committee," p. C17.

10. Ibid., p. C18.

11. Childress, "Statement to the Advisory Committee to the Director, NIH," in Report of the Advisory Committee to the Director, National Institutes of Health, *Human Fetal Tissue Transplantation Research* (Bethesda, MD: National Institutes of Health, 1988), pp. C7–8.

12. HFTTR Panel, *Report,* vol. 1, p. 6.

13. Childress, "Statement to the Advisory Committee," pp. C7–8.

14. In this instance, the theorist and the principlist were directly involved in the formulation of a policy that was accepted by the Panel and the NIH Advisory Committee but ultimately rejected by DHHS, while the casuist was offering a proposal for consideration.

15. Clouser and Gert, "A Critique of Principlism," pp. 19–236. For the rest of this section, references to the Clouser and Gert essay will be given in the text by page numbers.

16. For a related theory and critique of principlism, see Ronald M. Green, "Method in Bioethics: A Troubled Assessment," *Journal of Medicine and Philosophy* 15 (1990): 179–97.

17. William Frankena, *Ethics,* 2nd ed. (Englewood Cliffs: Prentice-Hall, 1973).

18. Even though Clouser and Gert use the second edition of *PBE* (1983) rather than the third edition (1989), this probably makes little difference in their critique because, if anything, the third edition is more subject to some of their charges as we move farther away from a strong conception of ethical theory, downplay differences at that level, and affirm a more historicist perspective (on the last point, see *PBE*, 3rd ed., p. 24, n. 20; 4th ed., chap. 1). Clouser and Gert justify their use of the second edition even though the third edition was available on the grounds that the second was better known and that they were criticizing a trend rather than particular works (p. 236. n. 1). I have indicated a few places where the different editions may make a difference in interpretation.

19. At this point, Clouser and Gert would probably find more acceptable the theories offered by H. Tristram Engelhardt, Jr., *Foundations of Bioethics* (New York: Oxford University Press, 1986) and Robert M. Veatch, *A Theory of Medical Ethics* (New York: Basic Books, 1981), because both establish some priorities or weights among these principles. However, Clouser and Gert do not address these positions.

20. Stanley G. Clarke and Evan Simpson, eds., *Anti-Theory in Ethics and Moral Conservatism* (Albany: State University of New York, 1989), p. 4.

21. Tom L. Beauchamp, "On Eliminating the Distinction between Applied Ethics and Ethical Theory," *Monist* 67 (1984): 515–31.

22. Gert, *Morality,* p. 284.

23. Ibid.

24. Ibid., p. 285.

25. Childress, *Who Should Decide?* pp. 237–41.

26. Bernard Gert and Charles Culver, "The Justification of Paternalism," *Ethics* 89 (1979): 199–210.

27. Jonsen, "Transplantation of Fetal Tissue," p. 216.

28. Albert R. Jonsen, Mark Siegler, and William J. Winslade, *Clinical Ethics,* 2nd ed. (New York: Macmillan, 1986).

29. Albert R. Jonsen and Stephen Toulmin, *The Abuse of Casuistry* (Berkeley: University of California Press, 1988). For the remainder of this section, references to this book will be given in the text by page numbers.

30. Stephen Toulmin, "The Tyranny of Principles," *Hastings Center Report* 11 (December 1981): 31–39.

31. National Commission for the Protection of Human Subjects of Biomedical

and Behavioral Research, *The Belmont Report: Ethical Guidelines for the Protection of Human Subjects of Research,* DHEW Publication No. 78–00 (OS) (Washington, D.C.: Department of Health, Education and Welfare, 1978).

32. LeRoy Walters, "Some Ethical Issues in Research Involving Human Subjects," *Perspectives in Biology and Medicine* (Winter 1977): 193–211; and James F. Childress, *Priorities in Biomedical Ethics* (Philadelphia: Westminster Press, 1981), chap. 3.

33. Toulmin, "Tyranny of Principles."

34. R. M. Hare, "Principles," *Essays in Ethical Theory* (Oxford: Clarendon Press, 1989), pp. 49–65.

35. 1st ed., p. 13; 2nd ed., p. 13; 3rd ed., p. 16; 4th ed., p. 23.

36. Henry S. Richardson, "Specifying Norms as a Way to Resolve Concrete Ethical Problems," *Philosophy and Public Affairs* 19 (1990): 279–320.

37. Hare, "Principles," p. 54.

38. David DeGrazia, "Moving Forward in Bioethical Theory: Theories, Cases, Rules, and Specified Principlism," *Journal of Medicine and Philosophy* 17 (1992): 511–39.

39. R. A. Carson, "Interpretive Bioethics: The Way of Discernment," *Theoretical Medicine* 11 (1990): 51–59.

40. John D. Arras, "Getting Down to Cases: The Revival of Casuistry in Bioethics," *Journal of Medicine and Philosophy* 16 (1991): 29–51.

41. Hare, "Principles," p. 65.

42. Judith Jarvis Thomson, "A Defense of Abortion," *Philosophy and Public Affairs* 1 (1971): 47–66.

43. Arras, "Getting Down to Cases," pp. 48–49.

44. Ibid., p. 47.

45. Clouser and Gert, "A Critique of Principlism," p. 219. See also Daniel Wikler, "What Has Bioethics to Offer Health Policy?" *Milbank Quarterly* 69 (1991): 233–51.

46. See Baruch Brody's effort to distinguish good and bad scholarship in bioethics, "Quality of Scholarship in Bioethics," *Journal of Medicine and Philosophy* 15 (1990): 161–78.

47. Carol Gilligan, *In a Different Voice: Psychological Theory and Women's Development* (Cambridge: Harvard University Press, 1982).

48. See, among others, Susan Sherwin, *No Longer Patient: Feminist Ethics & Health Care* (Philadelphia: Temple University Press, 1992).

49. See James F. Childress, "Love and Justice in Christian Biomedical Ethics," in *Theology and Bioethics: Exploring the Foundations and Frontiers,* ed. Earl E. Shelp (Dordrecht: D. Reidel, 1985).

50. Alisa L. Carse, "The 'Voice of Care': Implications for Bioethical Education," *Journal of Medicine and Philosophy* 16 (1991): 5–28.

51. Ibid., p. 11.

52. For a discussion of some of these convictions in my work, see Courtney S. Campbell, "On James F. Childress: Answering Every Person," *Second Opinion* 11 (1989): 118–24.

53. For an older but very pertinent and very helpful theological discussion of norm and context, see James M. Gustafson, "Context versus Principles: A Misplaced Debate in Christian Ethics," *Harvard Theological Review* 58 (1965): 171–202.

3. METAPHORS AND MODELS OF DOCTOR-PATIENT RELATIONSHIPS

1. On metaphor, see George Lakoff and Mark Johnson, *Metaphors We Live By* (Chicago: University of Chicago Press, 1980).

2. See Thomas S. Szasz and Marc H. Hollender, "A Contribution to the

Philosophy of Medicine: The Basic Models of the Doctor-Patient Relationship," *Archives of Internal Medicine* 97 (1956): 585–92; see also, Thomas S. Szasz, William F. Knoff, and Marc H. Hollender, "The Doctor-Patient Relationship and Its Historical Context," *American Journal of Psychiatry* 115 (1958): 522–28.

3. For a fuller analysis of paternalism and its justification, see James F. Childress, *Who Should Decide? Paternalism in Health Care* (New York: Oxford University Press, 1982).

4. Eric J. Cassell, "Autonomy and Ethics in Action," *New England Journal of Medicine* 297 (1977): 333–34 (italics added). Partnership is only one of several images and metaphors Cassell uses, and it may not be the best one to express his position, in part because he tends to view autonomy as a goal rather than as a constraint.

5. According to Robert M. Veatch, the main focus of this model is "an equality of dignity and respect, an equality of value contributions" (Veatch, "Models for Ethical Medicine in a Revolutionary Age," *Hastings Center Report* 2 [June 1972]: 7). See also Veatch, *A Theory of Medical Ethics* (New York: Basic Books, 1981). Contrast Eric J. Cassell, who disputes the relevance of notions of "equality" and "inequality" in *The Healer's Art: A New Approach to the Doctor-Patient Relationship* (Philadelphia: J. B. Lippincott, 1976), pp. 193–94.

6. Szasz and Hollender, "A Contribution to the Philosophy of Medicine," pp. 586–87.

7. See, for example, Paul Ramsey, "The Ethics of a Cottage Industry in an Age of Community and Research Medicine," *New England Journal of Medicine* 284 (1971): 700–706; Ramsey, *The Patient as Person: Explorations in Medical Ethics* (New Haven: Yale University Press, 1970), esp. chap. 1; and Hans Jonas, "Philosophical Reflections on Experimenting with Human Subjects," in *Ethical Aspects of Experimentation with Human Subjects, Daedalus* 98 (1969): 219–47.

8. Veatch, "Models for Ethical Medicine in a Revolutionary Age," p. 7.

9. See Roger Masters, "Is Contract an Adequate Basis for Medical Ethics?" *Hastings Center Report* 5 (December 1975): 24–28. See also William F. May, "Code and Covenant or Philanthropy and Contract?" in *Ethics in Medicine: Historical Perspectives and Contemporary Concerns*, ed. Stanley Joel Reiser, Arthur J. Dyck, and William J. Curran (Cambridge: MIT Press, 1977), pp. 65–76.

10. P. Lain Entralgo, *Doctor and Patient,* trans. from the Spanish by Frances Partridge (New York: McGraw-Hill Book Co., World University Library, 1969), p. 242.

11. Ibid., p. 197.

12. See Charles Fried, *Medical Experimentation: Personal Integrity and Social Policy* (New York: American Elsevier, 1974), p. 76. Our discussion of Fried's position is drawn from that work, from *Right and Wrong* (Cambridge: Harvard University Press, 1978), chap. 7, and from "The Lawyer as Friend: The Moral Foundations of the Lawyer-Client Relation," *Yale Law Journal* 85 (1976): 1060–89.

13. Immanuel Kant, *The Doctrine of Virtue,* part II of the *Metaphysic of Morals,* trans. Mary J. Gregor (New York: Harper and Row, Harper Torchbook, 1964), p. 140.

14. Szasz, Knoff, and Hollender, "The Doctor-Patient Relationship and Its Historical Context," p. 525. See also Veatch, "Models for Ethical Medicine in a Revolutionary Age," p. 5; and Leon Kass, "Ethical Dilemmas in the Care of the Ill: I. What Is the Physician's Service?" *Journal of the American Medical Association* 244 (1980): 1815, for criticisms of the technical model (from very different normative positions).

15. Veatch, "Models for Ethical Medicine in a Revolutionary Age," p. 7.

16. See Stephen Toulmin, "The Tyranny of Principles," *Hastings Center Report* 11 (December 1981): 31–39.

17. On trust and control, see James F. Childress, "Nonviolent Resistance: Trust and Risk-taking," *Journal of Religious Ethics* 1 (1973): 87–112.

18. Talcott Parsons, *The Structure of Social Action* (New York: Free Press, 1949), p. 311.

19. On the factors in the decline of trust, see Michael Jellinek, "Erosion of Patient Trust in Large Medical Centers," *Hastings Center Report* 6 (June 1976): 16–19.

20. Alasdair MacIntyre, "Patients as Agents," in *Philosophical Medical Ethics: Its Nature and Significance,* ed. Stuart F. Spicker and H. Tristram Engelhardt, Jr. (Boston: D. Reidel, 1977).

21. John Ladd, "Legalism and Medical Ethics," *Journal of Medicine and Philosophy* 4 (March 1979): 73.

22. This case has been presented in Mark Siegler, "Searching for Moral Certainty in Medicine: A Proposal for a New Model of the Doctor-Patient Encounter," *Bulletin of the New York Academy of Medicine* 57 (1981): 56–69.

23. See ibid. for a discussion of negotiation. Other proponents of a model of negotiation include Robert A. Burt, *Taking Care of Strangers: The Rule of Law in Doctor-Patient Relations* (New York: Free Press, 1979), and Robert J. Levine, *Ethics and Regulation of Clinical Research* (Baltimore: Urban and Schwarzenberg, 1981).

24. See the discussion in Childress, *Who Should Decide?* chap. 3.

25. Alan W. Cross and Larry R. Churchill, "Ethical and Cultural Dimensions of Informed Consent," *Annals of Internal Medicine* 96 (1982): 110–13.

26. See Oscar Thorup, Mark Siegler, James F. Childress, and Ruth Roettinger, "Voluntary Exit: Is There a Case for Rational Suicide?" *The Pharos* 45 (Fall 1982): 25–31.

4. IF YOU LET THEM, THEY'D STAY IN BED ALL MORNING

1. James F. Childress, "Ensuring Care, Respect, and Fairness for the Elderly," *Hastings Center Report* 14 (October 1984).

2. In the third edition of *Principles of Biomedical Ethics* (New York: Oxford University Press, 1989)—continued in the fourth edition (1994)—Tom L. Beauchamp and I reformulated what we had earlier called "the principle of autonomy" as "the principle of respect for autonomy."

3. See, for example, Gerald Dworkin, "Autonomy and Behavior Control," *Hastings Center Report* 6 (February 1976).

4. This formulation is influenced by Ruth R. Faden and Tom L. Beauchamp, *A History and Theory of Informed Consent* (New York: Oxford University Press, 1986).

5. See, for example, Eric J. Cassell, *Talking with Patients,* 2 vols. (Cambridge: MIT Press, 1985).

6. For a fuller discussion, see James F. Childress, *Who Should Decide? Paternalism in Health Care* (New York: Oxford University Press, 1982).

7. Ibid., pp. 224–25. This case was prepared by Gail Povar, M.D.

8. Bruce Miller, "Autonomy and the Refusal of Life-Saving Treatment," *Hastings Center Report* 11 (August 1981): 22–28. See also Dworkin, "Autonomy and Behavior Control."

9. Childress, *Who Should Decide?*

10. Ibid.

11. See, for example, Daniel Callahan, "Minimalist Ethics," *Hastings Center Report* 11 (October 1981): 19–25.

12. Although specification has been part of the approach taken in *Principles of Biomedical Ethics* in its various editions—e.g., the "derived" rules for relationships between health care professionals and patients that appear in chapter 7 in each edition—in the most recent edition, Beauchamp and I presented it in greater detail in addition to balancing, which has always been featured (Beauchamp and Childress, *Principles of Biomedical Ethics,* 4th ed. [New York: Oxford University Press, 1994]).

13. The argument for this position appears in ibid.

14. For a somewhat different formulation, see ibid., p. 34f., where these justificatory conditions are presented as constraints on the process of balancing.

15. See the discussion in James F. Childress, *Priorities in Biomedical Ethics* (Philadelphia: Westminster Press, 1981).

16. This case was prepared for *Everyday Ethics: Resolving Dilemmas in Nursing Home Life*, ed. Rosalie Kane and Arthur L. Caplan (New York: Springer, 1990) by the volume's editors.

17. On negotiated consent, see Harry R. Moody, "From Informed Consent to Negotiated Consent," *The Gerontologist* 28, Suppl. (1988): 64–70. On compromise, see Martin Benjamin, *Splitting the Difference: Compromise and Integrity in Ethics and Politics* (Lawrence: University Press of Kansas, 1990).

18. H. Richard Niebuhr, *The Responsible Self* (New York: Harper and Row, 1963).

19. See Andrew Jameton, "In the Borderlands of Autonomy: Responsibility in Long Term Care Facilities," *The Gerontologist* 28, Suppl. (1988): 18–23.

20. D. M. Ambrogi and F. Leonard, "The Impact of Nursing Home Admission Agreements on Resident Autonomy," *The Gerontologist* 28 Suppl. (1988): 82–89.

21. Childress, *Who Should Decide?*

22. Ibid.

23. Robert Dahl, *After the Revolution? Authority in a Good Society* (New Haven: Yale University Press, 1970), p. 137.

24. Joel Feinberg, "Voluntary Euthanasia and the Inalienable Right to Life," *Philosophy and Public Affairs* 7 (1978): 92–123.

25. Beauchamp and Childress, *Principles of Biomedical Ethics*.

5. HOW MUCH SHOULD THE CANCER PATIENT KNOW AND DECIDE?

1. James F. Childress, *Who Should Decide? Paternalism in Health Care* (New York: Oxford University Press, 1982).

2. Jay Katz, *The Silent World of Doctor and Patient* (New York: Free Press, 1984).

3. Donald Oken, "What to Tell Cancer Patients: A Study of Medical Attitudes," *Journal of the American Medical Association* 175 (April 1, 1961): 1120–28.

4. Dennis H. Novack et al., "Changes in Physicians' Attitudes toward Telling the Cancer Patient," *Journal of the American Medical Association* 241 (March 2, 1979): 897–900. See also Robert M. Veatch and Ernest Tai, "Talking about Death: Patterns of Lay and Professional Change," *Annals of the American Academy of Political and Social Science* 447 (January 1980): 29–45.

5. Novack et al., "Changes in Physicians' Attitudes toward Telling the Cancer Patient."

6. M. Priscilla Rea, Shirley Greenspoon, and Bernard Spilke, "Physicians and the Terminal Patient: Some Selected Attitudes and Behavior," *Omega* 6 (1975): 291–302.

7. President's Commission for the Study of Ethical Problems in Medicine and Biomedical and Behavioral Research, *Making Health Care Decisions*, vol. 1: *Report* (Washington, D.C.: U.S. Government Printing Office, October 1982); Louis Harris and Associates, "Views of Informed Consent and Decision-Making," in President's Commission for the Study of Ethical Problems in Medicine and Biomedical and Behavioral Research, *Making Health Care Decisions*, vol. 2: *Appendices, Empirical Studies of Informed Consent* (Washington, D.C.: U.S. Government Printing Office, October 1982).

8. Ibid.

9. Veatch and Tai, "Talking about Death," and Michael Blumenfeld, Norman B. Levey, and Diane Kaufman, "The Wish to Be Informed of a Fatal Illness," *Omega* 9 (1978–79): 323–26.

338   NOTES FOR PAGES 80-89

10. John A. Robertson, *Rights of the Critically Ill* (New York: Ballinger, 1983).

11. Marcia Angell, "Respecting the Autonomy of Competent Patients," *New England Journal of Medicine* 310 (April 26, 1984): 1115–16.

12. Jon R. Waltz and Thomas W. Scheuneman, "Informed Consent to Therapy," *Northwestern University Law Review* 64 (1970): 640.

13. George J. Annas, "Breast Cancer: The Treatment of Choice," *Hastings Center Report* 10 (April 1980): 27–29.

14. Barbara J. McNeil, Ralph Weichselbaum, and Stephen G. Pauker, "Speech and Survival: Tradeoffs between Quality and Quantity of Life in Laryngeal Cancer," *New England Journal of Medicine* 305 (October 22, 1981): 982–87.

15. Barbara J. McNeil et al., "On the Elicitation of Preferences for Alternative Therapies," *New England Journal of Medicine* 306 (May 27, 1982): 1259–62.

16. Sidney H. Wanzer et al., "The Physician's Responsibility toward Hopelessly Ill Patients," *New England Journal of Medicine* 310 (April 12, 1984): 955–59.

17. Robertson, *Rights of the Critically Ill.*

18. Marcia Angell, "The Quality of Mercy," *New England Journal of Medicine* 306 (January 14, 1982): 98–99; Eric J. Cassell, "The Nature of Suffering and the Goals of Medicine," *New England Journal of Medicine* 306 (March 18, 1982): 639–45.

19. "House Refuses to Let the Dying Use Heroin," *Washington Post,* September 20, 1984, p. A3.

20. President's Commission for the Study of Ethical Problems in Medicine and Biomedical and Behavioral Research, *Deciding to Forego Life-Sustaining Treatment* (Washington, D.C.: U.S. Government Printing Office, March 1983), p. 234.

21. Susanna E. Bedell et al., "Survival after Cardiopulmonary Resuscitation in the Hospital," *New England Journal of Medicine* 309 (September 8, 1983): 569–76.

22. Robertson, *Rights of the Critically Ill.*

23. Steven S. Spencer, "'Code' or 'No Code': A Non-legal Opinion," *New England Journal of Medicine* 300 (January 18, 1979): 138–40.

24. Susanna E. Bedell and Thomas L. Delbanco, "Choices about Cardiopulmonary Resuscitation in the Hospital," *New England Journal of Medicine* 310 (April 26, 1984): 1089–93.

25. Angell, "Respecting the Autonomy of Competent Patients."

26. Bernard Lo and Robert L. Steinbrook, "Deciding whether to Resuscitate," *Archives of Internal Medicine* 143 (1983): 1561–63.

27. President's Commission, *Making Health Care Decisions,* vol. 1.

28. Stuart J. Eisendrath and Albert R. Jonsen, "The Living Will: Help or Hindrance?" *Journal of the American Medical Association* (April 15, 1983): 2054–58.

29. President's Commission, *Making Health Care Decisions,* vol. 1.

30. Robertson, *Rights of the Critically Ill.*

31. Bernard Lo and Albert R. Jonsen, "Ethical Decisions in the Care of a Patient Terminally Ill with Metastatic Cancer," *Annals of Internal Medicine* 92 (1980): 107–11.

32. Tom L. Beauchamp and James F. Childress, *Principles of Biomedical Ethics,* 2nd ed. (New York: Oxford University Press, 1983), chap. 4 (also other editions).

33. My reflections on narrative have grown largely out of a seminar, "Narrative in Theology and Ethics," which I cotaught with Larry Bouchard in the spring of 1995 at the University of Virginia. I am most grateful to him (and to the other participants) for what I learned in this seminar.

34. Tod S. Chambers, "The Bioethicist as Author: The Medical Ethics Case as Rhetorical Device," *Literature and Medicine* 13, no. 1 (Spring 1994): 60–78. All references to Chambers's work are to this essay.

35. In addition to this essay, he refers to Beauchamp and Childress, *Principles of Biomedical Ethics,* 3rd ed. (New York: Oxford University Press, 1989) and our use of the *Tarasoff* case.

36. In ibid.

37. See Dena Davis, "The Ethics of Thick Description," *Hastings Center Report* 21 (July/August 1991): 12–17.

38. Leslie J. Blackhall, Sheila T. Murphy, Gelya Frank et al., "Ethnicity and Attitudes toward Patient Autonomy," *Journal of the American Medical Association* 274 (September 13, 1995), pp. 820–25.

39. Joseph A. Carrese and Lorna A. Rhodes, "Western Bioethics on the Navajo Reservation: Benefit or Harm?" *Journal of the American Medical Association* 274 (September 13, 1995), pp. 826–29.

40. See Blackhall et al., "Ethnicity and Attitudes toward Patient Autonomy," p. 822.

41. Joel Feinberg, "Voluntary Euthanasia and the Inalienable Right to Life," *Philosophy and Public Affairs* 7 (Winter 1978): 93–123.

6. MANDATORY HIV SCREENING AND TESTING

1. I will define a liberal society as one that recognizes various individual rights and liberties, including free speech, freedom of association, liberty of action, freedom of conscience, etc. These are identified, with varying emphases, by many interpreters, including John Rawls, *A Theory of Justice* (Cambridge: Harvard University Press, 1971), and *Political Liberalism* (New York: Columbia University Press, 1993).

2. For the history of developments in the United States, see Randy Shilts, *And the Band Played On: Politics, People, and the AIDS Epidemic* (New York: Penguin Books, 1987), and Ronald Bayer, *Private Acts, Social Consequences: AIDS and the Politics of Public Health* (New York: Free Press, 1989).

3. For these principles, see Tom L. Beauchamp and James F. Childress, *Principles of Biomedical Ethics*, 4th ed. (New York: Oxford University Press, 1994).

4. Committee for the Study of the Future of Public Health, Division of Health Care Services, Institute of Medicine, *The Future of Public Health* (Washington, D.C.: National Academy Press, 1988), p. 1.

5. For a fuller discussion of the impact on justifiable policies of the relevant differences between HIV infection and tuberculosis, see Beauchamp and Childress, *Principles of Biomedical Ethics*, chap. 7.

6. See James F. Childress, *Who Should Decide? Paternalism in Health Care* (New York: Oxford University Press, 1982); Childress, "An Ethical Framework for Assessing Policies to Screen for Antibodies to HIV," *AIDS & Public Policy Journal* 2 (Winter 1987): 28–31; and Childress, "The Place of Autonomy in Bioethics," *Hastings Center Report* 20 (January/February 1990): 12–17.

7. Childress, *Who Should Decide?*

8. See Beauchamp and Childress, *Principles of Biomedical Ethics*, esp. chaps. 1 and 2.

9. See Childress, "An Ethical Framework for Assessing Policies to Screen for Antibodies to HIV," from which parts of this essay have been drawn with permission. For other justificatory frameworks, with important similarities to the current one, see Ronald Bayer, Carol Levine, and Susan M. Wolf, "HIV Antibody Screening: An Ethical Framework for Evaluating Proposed Programs," *Journal of the American Medical Association* 256 (October 3, 1986): 1768–74; Carol Levine and Ronald Bayer, "The Ethics of Screening for Early Intervention in HIV Disease," *American Journal of Public Health* 79 (December 1989): 1661–67; Maura O'Brien, "Mandatory HIV Antibody Testing Policies: An Ethical Analysis," *Bioethics* 3 (October 1989): 273–300; and Martha A. Field, "Testing for AIDS: Uses and Abuses," *American Journal of Law and Medicine* 16 (1990): 33–106.

10. Field, "Testing for AIDS," p. 54.

11. Sissela Bok, *Lying: Moral Choice in Public and Private Life* (New York: Pantheon Books, 1978).

12. Robert Nozick, "Moral Complications and Moral Structures," *Natural Law Forum* 13 (1968): 1–50.

13. Margaret Sommerville, "The Case Against HIV Antibody Testing of Refugees and Immigrants," *Canadian Medical Association Journal* 141 (November 1, 1989): 869–94.

14. Frank S. Rhame and Dennis G. Maki, "The Case for Wider Use of Testing for HIV Infection," *New England Journal of Medicine* 320 (May 11, 1989): 1253.

15. Willard Cates, Jr., and H. Hunter Handsfield, "HIV Counseling and Testing: Does It Work?" *American Journal of Public Health* 78 (December 1988): 1533–34; Jane McCusker et al., "Effects of HIV Antibody Test Knowledge on Subsequent Sexual Behaviors in a Cohort of Homosexually Active Men," *American Journal of Public Health* 78 (April 1988): 462–67.

16. Larry O. Gostin, "A Decade of a Maturing Epidemic: An Assessment and Directions for Future Public Policy," *American Journal of Law & Medicine* 16 (1990): 9–10; Field, "Testing for AIDS," pp. 58–59.

17. For a survey of published reports, see Marshall H. Becker and Jill G. Joseph, "AIDS and Behavioral Change to Reduce Risk: A Review," *American Journal of Public Health* 78 (April 1988): 394–410.

18. Bernard Lo, Robert L. Steinbrook, Molly Cooke et al., "Voluntary Screening for Human Immunodeficiency Virus (HIV) Infection: Weighing the Benefits and Harms," *Annals of Internal Medicine* 110 (May 1989): 730.

19. For these risks and possible benefits, see ibid., pp. 727–33.

20. Peter Marzuk et al., "Increased Risk of Suicide in Persons with AIDS," *Journal of the American Medical Association* 259 (1988): 1333–37.

21. Bayer, *Private Acts, Social Consequences*, pp. 158–61.

22. For an important discussion of coercion, see Alan Wertheimer, *Coercion* (Princeton: Princeton University Press, 1987).

23. See John C. Fletcher, "AIDS Screening: A Response to Gary Bauer," *AIDS & Public Policy Journal* 2 (Fall/Winter 1987): 5–7.

24. See LeRoy Walters, "Ethical Issues in HIV Testing during Pregnancy" (publication information in n. 41, below).

25. See A. John Simmons, "Tacit Consent and Political Obligation," *Philosophy and Public Affairs* 5 (Spring 1976): 274–91. For an analysis of varieties of consent, with special reference to medicine and health care, see Childress, *Who Should Decide?* chap. 4.

26. Martha S. Swartz, "AIDS Testing and Informed Consent," *Journal of Health Politics, Policy and Law* 13 (Winter 1988): 607–721.

27. See Keith Henry and Kent Crossley, "Analysis of the Use of HIV Antibody Testing in a Minnesota Hospital," *Journal of the American Medical Association* 259 (January 8, 1988): 264–65.

28. Julie Louise Gerberding, Gary Littell, Ada Tarkington et al., "Risk of Exposure of Surgical Personnel to Patient's Blood during Surgery at San Francisco General Hospital," *New England Journal of Medicine* 322 (June 21, 1990): 1788.

29. Field, "Testing for AIDS," p. 104.

30. Gary Bauer, "AIDS Testing," *AIDS & Public Policy Journal* 2 (Fall/Winter 1987): 1.

31. Larry O. Gostin, "Screening for AIDS: Efficacy, Cost, and Consequences," *AIDS & Public Health Journal* 2 (Fall/Winter 1987): 16.

32. Ibid., p. 17.

33. Bernard J. Turnock and Chester J. Kelly, "Mandatory Premarital Testing for Human Immunodeficiency Virus: The Illinois Experience," *Journal of the American Medical Association* 261 (June 16, 1989): 3415–18.

34. Field, "Testing for AIDS."

35. Turnock and Kelly, "Mandatory Premarital Testing." See also the earlier

predictions by Paul D. Cleary et al., "Compulsory Premarital Screening for the Human Immunodeficiency Virus: Technical and Public Health Considerations," *Journal of the American Medical Association* 258 (October 1987): 1757–62.

36. Edward A. Belongia, James M. Vergeront, and Jeffrey P. Davis, "Premarital HIV Screening" (Letter to the Editor), *Journal of the American Medical Association* 261 (April 21, 1989): 2198.

37. Stephen C. Joseph, "Premarital AIDS Testing: Public Policy Abandoned at the Altar," *Journal of the American Medical Association* 261 (June 16, 1989): 3456.

38. See, for example, Pamela Boyer et al., "Factors Predictive of Maternal-Fetal Transmission of HIV-1," *Journal of the American Medical Association* 271 (June 22/29, 1994): 1925–30; Srisakul C. Kliks et al., "Features of HIV-1 That Could Influence Maternal-Child Transmission," *Journal of the American Medical Association* 272 (August 10, 1994): 467–73; Louise Kuhn et al., "Maternal-Infant HIV Transmission and Circumstances of Delivery," *American Journal of Public Health* 84 (July 1994): 1110–15.

39. Kathleen Nolan, "Ethical Issues in Caring for Pregnant Women and Newborns at Risk for Human Immunodeficiency Virus Infection," *Seminars in Perinatology* 13 (February 1989): 55–65.

40. Ibid., p. 59. See also Levine and Bayer, "Ethics of Screening for Early Intervention in HIV Disease."

41. See LeRoy Walters, "Ethical Issues in HIV Testing during Pregnancy," in *AIDS, Women and the Next Generation,* ed. Ruth Faden, Gail Geller, and Madison Powers (New York: Oxford University Press, 1991), chap. 11. Various essays in this important volume, which appeared after I had prepared earlier versions of this chapter, sketch legal, ethical, social, psychological, and medical factors in prenatal and neonatal HIV screening and testing. A consensus statement at the end holds that there is no clinical or public health justification for mandatory prenatal or neonatal screening (p. 341).

42. See "Zidovudine for the Prevention of HIV Transmission from Mother to Infant," *Journal of the American Medical Association* 271 (May 25, 1994): 1567–68; "Birth Outcomes following Zidovudine Therapy in Pregnant Women," *Journal of the American Medical Association* 272 (July 6, 1994): 17.

43. See Nat Hentoff, "Another 'Tuskegee,'" *Washington Post,* May 20, 1995, p. A23. Hentoff also accepts this analogy.

44. Ron Bayer, "It's Not 'Tuskegee' Revisited," *Washington Post,* May 26,1995, p. A27.

45. These and other concerns appear in James J. Goedert and Timothy R. Cote, "Public Health Interventions to Reduce Pediatric AIDS" (Editorial), *American Journal of Public Health* 84 (July 1994): 1065–66.

46. See Ruth R. Faden et al., "Prenatal HIV-Antibody Testing and the Meaning of Consent," *AIDS and Public Policy Journal* 9 (Fall 1994): 151–59.

47. Randal C. Mutter, Richard M. Grimes, and Darwin Labarthe, "Evidence of Intraprison Spread of HIV Infection," *Archives of Internal Medicine* 154 (April 11, 1994): 793–95.

48. Ruth Macklin, "Predicting Dangerousness and the Public Health Response to AIDS," *Hastings Center Report* 16 (December 1986): 22.

49. Gostin, "Screening for AIDS," p. 21; Field, "Testing for AIDS," pp. 81–91.

50. Jon K. Andrus et al., "HIV Testing in Prisoners: Is Mandatory Testing Mandatory?" *American Journal of Public Health* 79 (July 1989): 842.

51. Macklin, "Predicting Dangerousness and the Public Health Response to AIDS," p. 22.

52. Field, "Testing for AIDS," p. 94.

53. Carol Leslie Wolchok, "AIDS at the Frontier: United States Immigration Policy," *Journal of Legal Medicine* 10 (1989): 128.

54. Somerville, "The Case Against HIV Antibody Testing of Refugees and Immigrants," p. 891.

55. Larry O. Gostin, Paul D. Cleary, Kenneth H. Mayer et al., "Screening Immigrants and International Travelers for the Human Immunodeficiency Virus," *New England Journal of Medicine* 322 (June 4, 1990): 1745–46.

56. Wolchok, "AIDS at the Frontier," p. 142.

57. Gostin, Cleary, Mayer et al., "Screening Immigrants and International Travelers for the Human Immunodeficiency Virus."

58. Ibid., p. 1746.

59. "Possible Transmission of Human Immunodeficiency Virus to a Patient during an Invasive Dental Procedure," *Morbidity and Mortality Weekly Report* 39 (July 27, 1990): 489.

60. Ban Mishu et al., "A Surgeon with AIDS: Lack of Evidence of Transmission to Patients," *Journal of the American Medical Association* 264 (July 25, 1990): 467–70.

61. G. M. Dickinson et al., "Absence of HIV Transmission from an Infected Dentist to His Patients: An Epidemiologic and DNA Sequence Analysis," *Journal of the American Medical Association* 269 (April 14, 1993): 1802–06. See also A. S. Rogers et al., "Investigations of Potential HIV Transmission to the Patients of an HIV-Infected Surgeon," *Journal of the American Medical Association* 269 (April 14, 1993): 1795–1801; C. F. von Reyn et al., "Absence of HIV Transmission from an Infected Orthopedic Surgeon: A 13-year Look-Back Study," *Journal of the American Medical Association* 269 (April 14, 1993): 1807–11; Ban Mishu and W. Schaffner, "HIV-infected Surgeons and Dentists: Looking Back and Looking Forward," *Journal of the American Medical Association* 269 (April 14, 1993): 1843–44.

62. Frank S. Rhame, "The HIV-Infected Surgeon," *Journal of the American Medical Association* 264 (July 25, 1990): 507.

63. Ibid., pp. 507–508.

64. Larry O. Gostin, Letter to the Editor, *Journal of the American Medical Association* 264 (July 25, 1990): 452–53.

65. Rhame, "The HIV-Infected Surgeon."

66. Gary James Wood, "The Politics of AIDS Testing," *AIDS & Public Policy Journal* 2 (Fall/Winter 1987): 35.

67. Gostin, "Screening for AIDS," p. 21.

68. Monroe E. Price, *Shattered Mirrors: Our Search for Identity and Community in the AIDS Era* (Cambridge: Harvard University Press, 1989), pp. 81, 84.

69. Larry Kramer, "A 'Manhattan Project' for AIDS," *New York Times,* Monday, July 16, 1990.

70. On the war metaphor, see Childress, *Who Should Decide?* p. 7; Childress, "Ensuring Care, Respect, and Fairness for the Elderly," *Hastings Center Report* 14 (October 1984): 27–31; Judith Wilson Ross, "Ethics and the Language of AIDS," in *The Meaning of AIDS: Implications for Medical Science, Clinical Practice, and Public Health Policy,* ed. Eric Juengst and Barbara A. Koenig (New York: Praeger, 1989), pp. 30–41; and Ross, "The Militarization of Disease: Do We Really Want a War on AIDS?" *Soundings* 72 (1989): 39–58. See also William F. May, *The Physician's Covenant* (Philadelphia: Westminster Press, 1983).

71. Bayer, *Private Acts, Social Consequences.*

72. See Price, *Shattered Mirrors,* p. 82; Ross, "Ethics and the Language of AIDS."

73. Price, *Shattered Mirrors,* p. 82.

74. Justice Douglas, quoted in ibid., p. 84.

75. Childress, *Moral Responsibility in Conflicts* (Baton Rouge: Louisiana State University Press, 1982).

76. Susan Sontag, *Illness as Metaphor and AIDS and Its Metaphors* (New York: Doubleday, Anchor Books, 1990).

77. Childress, "Ensuring Care, Respect, and Fairness for the Elderly." See also the emphasis on care and caring in feminist moral and social thought.

78. Price, *Shattered Mirrors.*

79. Christopher Collins, "The Case against AIDS Testing," *AIDS & Public Policy Journal* 2 (Fall/Winter 1987): 10.

80. Price, *Shattered Mirrors,* p. 5.

7. "WHO IS A DOCTOR TO DECIDE WHETHER A PERSON LIVES OR DIES?"

1. Most of the quotations from persons involved in the case of Dax Cowart are often taken from the film/videotape, *Dax's Case,* and from the videotape, *Please Let Me Die,* unless otherwise indicated. For information about these sources, see n. 32 below.

2. For a fuller discussion of paternalism, see James F. Childress, *Who Should Decide? Paternalism in Health Care* (New York: Oxford University Press, 1982).

3. "A Happy Life Afterward Doesn't Make Up for Torture," *Washington Post,* June 26, 1983.

4. Bruce Miller, "Autonomy and the Refusal of Life-Saving Treatment," *Hastings Center Report* 11 (August 1981): 22–28.

5. Sharon Imbus and Bruce Zawacki, "Autonomy for Burn Patients when Survival Is Unprecendented," *New England Journal of Medicine* 300 (August 11, 1977): 308–11.

6. See *Medical World News,* January 22, 1979, p. 37. For the proceedings of the NIH (National Institutes of Health) Consensus Development Conference in Supportive Therapy in Burn Care, see Seymour I. Schwartz, "Consensus Summary on Fluid Resuscitation," *Journal of Trauma* 19 (November 1979), Suppl., pp. 876–77.

7. Robert Nozick, *Anarchy, State, and Utopia* (New York: Basic Books, 1974), pp. 28–35.

8. John Stuart Mill, *On Liberty,* ed. Gertrude Himmelfarb (Harmondsworth: Penguin Books, 1976).

9. See Gerald Dworkin, "Autonomy and Behavior Control," *Hastings Center Report* 6 (February 1976), and Miller, "Autonomy and the Refusal of Life-Saving Treatment."

10. *Ten Years After a Severely Burned Patient's Request to Die: A Discussion with Dax Cowart,* a videotape of "Medical Center Hour" at the University of Virginia, Charlottesville, March 28, 1984.

11. Ibid.

12. "A Happy Life Afterward."

13. *Ten Years After.*

14. "A Happy Life Afterward."

15. Burt, *Taking Care of Strangers,* chap. 1 (publication information in n. 32).

16. *Ten Years After.*

17. Lionel Trilling, "Manners, Morals, and the Novel," *The Liberal Imagination: Essays on Literature and Society* (New York: Viking Press, 1950), p. 221.

18. See William F. May, "Dealing with Catastrophe," which appeared in the same volume as the Childress and Campbell essay, *Dax's Case: Essays in Medical Ethics and Human Meaning,* ed. Lonnie D. Kliever (Dallas: Southern Methodist University Press, 1989), pp. 131–50, and the revised version in William F. May, *The Patient's Ordeal* (Bloomington: Indiana University Press, 1991), chap. 1: "The Burned." All references in the text will be to the latter.

19. Sumner B. Twiss, "Alternative Approaches to Patient and Family Medical Ethics: Review and Assessment," *Religious Studies Review* 21 (October 1995): 263–76; Paul Lauritzen, "Ethics and Experience: The Case of the Curious Response," *Hastings Center Report* 26 (January/February 1996): 6–15.

20. Childress, *Who Should Decide?* p. ix.

21. Lauritzen, "Ethics and Experience."

22. Twiss makes a similar argument in "Alternative Approaches," but he then uses categories that may be excessively rationalistic, such as "life plan."

23. Dennis J. Stouffer, *Journeys through Hell: Stories of Burn Survivors' Reconstruction of Self and Identity* (Lanham, MD: Rowman & Littlefield, 1995).

24. Twiss, "Alternative Approaches," p. 274.

25. Lonnie D. Kliever, "Rage and Grief: Another Look at Dax's Case," in *Chronic Illness: From Experience to Policy*, ed. S. Kay Toombs, David Barnard, and Ronald A. Carson (Bloomington: Indiana University Press, 1995), p. 63.

26. Ibid., p. 73.

27. I borrow this question from Sue E. Estroff, who uses it for a different purpose in "Whose Story Is It Anyway? Authority, Voice, and Responsibility in Narratives of Chronic Illness," in *Chronic Illness*, ed. Toombs et al., chap. 5.

28. See my "Narrative(s) versus Norm(s): A Misplaced Debate in Ethics," which I prepared for a conference on narrative ethics sponsored by the Society for Health and Human Values, April 1996 (forthcoming).

29. Adam Zachary Newton, *Narrative Ethics* (Cambridge: Harvard University Press, 1995), p. 58.

30. J. Hillis Miller, "Narrative," in *Critical Terms for Literary Study*, ed. Frank Lentricchia and Thomas McLaughlin (Chicago: University of Chicago Press, 1990), p. 74.

31. The phrase "wounded storyteller" is drawn from the title of Arthur W. Frank's superb book, *The Wounded Storyteller: Body, Illness, and Ethics* (Chicago: University of Chicago Press, 1995).

32. This case was prepared by James Tubbs for use in Childress, *Who Should Decide? Paternalism in Health Care*. It is based on the videotape *Please Let Me Die* and the transcript of that videotape in Robert A. Burt, *Taking Care of Strangers: The Rule of Law in Doctor-Patient Relations* (New York: Free Press, 1979), pp. 174–80. Another version of the case also appears with discussion in Robert B. White and H. Tristram Engelhardt, Jr., "A Demand to Die," *Hastings Center Report* 5 (June 1975): 9–10, 47. The videotape *Please Let Me Die* (1974) is available for rental or purchase from Robert B. White, M.D., Department of Psychiatry, University of Texas Medical Branch, Galveston, Texas. The follow-up videotape, *Dax's Case* (1985), is available for purchase or rental from Concern for Dying, 250 West 57th St., New York, NY 10016, or Filmmakers Library, 124 East 40th St., New York, NY 10107. The latter served as the basis for a series of essays, *Dax's Case*, ed. Kliever, in which this essay first appeared.

8. MUST PATIENTS ALWAYS BE GIVEN FOOD AND WATER?

1. John A. Robertson, "Dilemma in Danville," *Hastings Center Report* 11 (October 1981): 5–8.

2. T. Rohrlich, "2 Doctors Face Murder Charges in Patient's Death," *Los Angeles Times*, August 19, 1982, p. A1; Jonathan Kirsch, "A Death at Kaiser Hospital," *California* 7 (1982): 79ff; Magistrates' findings, *California v. Barber and Nejdl*, No. A 925586, Los Angeles Mun. Ct. Cal., (March 9, 1983); Superior Court of California, County of Los Angeles, *California v. Barber and Nejdl*, No. AO 25586, tentative decision May 5, 1983.

3. *In re* Infant Doe, No. GU 8204–00 (Cir. Ct. Monroe County, Ind., April 2, 1982), writ of mandamus dismissed *sub nom.* State ex rel. Infant Doe v. Baker, No. 482 S140 (Indiana Supreme Court, May 27, 1982).

4. Office of the Secretary, Department of Health and Human Services, "Nondiscrimination on the Basis of Handicap," *Federal Register* 48 (1983), 9630–32. [Interim

final rule modifying 45 C.F.R. #84.61.] See Judge Gerhard Gesell's decision, *American Academy of Pediatrics v. Heckler,* No. 83–0774. U.S. District Court, D.C., April 24, 1983; and also George J. Annas, "Disconnecting the Baby Doe Hotline," *Hastings Center Report* 13 (June 1983): 14–16.

5. *In re* Claire C. Conroy, N.J. Sup. Ct. (Chancery Div-Essex Co. No. P-19083E) February 2, 1983; *In re* Claire C. Conroy, N.J. Sup. Ct. (Appellate Div. No. 4-2483-82T1) July 8, 1983.

6. The President's Commission for the Study of Ethical Problems in Medicine and Biomedical and Behavioral Research, *Deciding to Forego Life-Sustaining Treatment* (Washington, D.C.: U.S. Government Printing Office, 1982).

7. G. E. M. Anscombe, "Ethical Problems in the Management of Some Severely Handicapped Children: Commentary 2," *Journal of Medical Ethics* 7 (1981): 117–24, at 122.

8. See, e.g., the President's Commission for the Study of Ethical problems in Medicine and Biomedical and Behavioral Research, *Making Health Care Decisions* (Washington, D.C.: U.S. Government Printing Office, 1982).

9. President's Commission, *Deciding to Forego Life-Sustaining Treatment,* pp. 171–96.

10. Joyce V. Zerwekh, "The Dehydration Question," *Nursing* 83 (January 1983): 47–51, with comments by Judith R. Brown and Marion B. Dolan.

11. James F. McCartney, "The Development of the Doctrine of Ordinary and Extraordinary Means of Preserving Life in Catholic Moral Theology before the Karen Quinlan Case," *Linacre Quarterly* 47 (1980): 215ff.

12. President's Commission, *Deciding to Forego Life-Sustaining Treatment,* pp. 82–90. For an argument that fluids and electrolytes can be "extraordinary," see Carson Strong, "Can Fluids and Electrolytes be 'Extraordinary' Treatment?" *Journal of Medical Ethics* 7 (1981): 83–85.

13. The Sacred Congregation for the Doctrine of the Faith, *Declaration of Euthanasia,* Vatican City, May 5, 1980.

14. Paul Ramsey contends that "when a man is irreversibly in the process of dying, to feed him and to give him drink, to ease him and keep him comfortable—these are no longer given as means of preserving life. The use of a glucose drip should often be understood in this way. This keeps a patient who cannot swallow from feeling dehydrated and is often the only remaining 'means' by which we can express our present faithfulness to him during this dying" *(The Patient as Person* [New Haven: Yale University Press, 1970], pp. 128–29). But Ramsey's suggestion would not apply to a patient in a deep irreversible coma, and he would be willing to disconnect the IV in the Quinlan case; see Ramsey, *Ethics at the Edges of Life: Medical and Legal Intersections* (New Haven: Yale University Press, 1978), p. 275. Bernard Towers describes an appropriate approach to comfort and dignity: "When a patient is conscious to even the smallest degree, and if he appears to be thirsty and to have a swallowing reflex, and if there is no contraindication to oral fluids, his comfort and dignity would surely demand that he be given nourishing liquids, or at least water. If he lapses into coma, good nursing practice has traditionally required sponging the mouth and moistening the lips. Now, if he lapses into deep coma and is on a dying trajectory, would we try to 'push' fluids by mouth or nasogastric tube? If we did, dignity would surely suffer. The 'comfort' of the patient would, of course, be unaffected if the coma were deep enough and irreversible" ("Irreversible Coma and Withdrawal of Life Support: Is It Murder if the IV Line Is Disconnected?" *Journal of Medical Ethics* 8 [1982]: 205).

15. See Kenneth C. Micetich, Patricia H. Steinecker, and David C. Thomasma, "Are Intravenous Fluids Morally Required for a Dying Patient?" *Archives of Internal Medicine* 143 (May 1983): 975–78.

16. Robert and Peggy Stinson, *The Long Dying of Baby Andrew* (Boston: Little, Brown and Company, 1983), p. 355.

17. One article has discussed a hypothetical case of maintaining a dying, comatose patient on a respirator while withdrawing IV fluids. The authors contend that this approach is not ironic because withdrawal of the respirator "creates the immediate consequence of death for which we must take responsibility. It represents an extreme form of abandonment." Nevertheless, they were willing to stop IV fluids, knowing that death would occur before long. As the article's survey reported, other physicians would have provided nutrition and fluids. See Micetich, Steinecker, and Thomasma, "Are Intravenous Fluids Morally Required for a Dying Patient?"

18. We are grateful to Haavi Morreim and Steven DalleMura for their helpful comments on an earlier version of this paper. We are also grateful for the instruction provided Dr. Lynn by the staff and patients of the Washington Home and its Hospice.

9. WHEN IS IT MORALLY JUSTIFIABLE TO DISCONTINUE MEDICAL NUTRITION AND HYDRATION?

1. This case has been adapted with permission from one presented by Dr. Martin P. Albert, Charlottesville, Virginia.

2. David Hilfiker, "Allowing the Debilitated to Die: Facing our Ethical Choices," *New England Journal of Medicine* 308 (1983): 716–19.

3. Ibid., p. 717.

4. See Paul Ramsey, *Ethics at the Edges of Life* (New Haven: Yale University Press, 1978), p. 153, and Robert M. Veatch, *Death, Dying and the Biological Revolution* (New Haven: Yale University Press, 1976), chap. 3.

5. For a discussion of these principles, see James F. Childress, *Who Should Decide? Paternalism in Health Care* (New York: Oxford University Press, 1982). See also the chapters on "Respect for Autonomy" and "Beneficence" in Tom L. Beauchamp and James F. Childress, *Principles of Biomedical Ethics,* 4th ed. (New York: Oxford University Press, 1994).

6. For a defense of this position, see Childress, *Who Should Decide?*

7. Diane L. Redleaf, Suzanne B. Schmitt, and William C. Thompson, "The California Natural Death Act: An Empirical Study of Physicians' Practices," *Stanford Law Review* 31 (May 1979): 913–45.

8. Sandra Bardinella, quoted by John Paris, S.J., "Kaiser, Conroy, and the Withdrawal of IV Feeding: Killing or Letting Die," unpublished paper, p. 1.

9. Department of Health and Human Services, "Proposed Rules 45 CFR Part 84, Nondiscrimination on the Basis of Handicap Relating to Health Care for Handicapped Infants," *Federal Register* 48: (July 5, 1983): 30846–52.

10. See, for example, the ruling by an intermediate appeals court in Massachusetts, *In re* Mary Hier, 18 Mass. App. 200 (June 4, 1984), which held that a ninety-two-year-old incompetent patient, who had pulled out her gastrostomy tube several times, did not have to undergo a surgical procedure to reinsert the tube. The Supreme Judicial Court of Massachusetts declined to review this case. See also *Barber v. Superior Court* 147 Cal. App. 3d 1006, 195 Cal. Rptr. 484 (1983).

11. Joyce V. Zerwekh, "The Dehydration Question," *Nursing* 83 (January 1983): 47–51.

12. Daniel Callahan, "On Feeding the Dying," *Hastings Center Report* 13 (October 1983): 22. All subsequent references to Callahan's position are to this article.

13. Ibid.

14. Joanne Lynn and James F. Childress, "Must Patients Always Be Given Food and Water?" *Hastings Center Report* 13 (October 1983): 17–21 (chapter 8 in this volume).

15. See Beauchamp and Childress, *Principles of Biomedical Ethics,* chap. 4.

16. Joel Feinberg, "Sentiment and Sentimentality in Practical Ethics," *Proceedings and Addresses of the American Philosophical Association* 56, no. 1 (September 1982): 41–42.

17. Jonathan Glover, *Causing Death and Saving Lives* (New York: Penguin Books, 1977), p. 296.

18. Gerald Kelly, S.J., "The Duty to Preserve Life," *Theological Studies* 12 (December 1951): 550–56.

19. Lynn and Childress, "Must Patients Always Be Given Food and Water?"

20. *Bouvia v. County of Riverside*, No. 159780 (Super. Ct., December 16, 1983), and *In re* Plaza Health and Rehabilitation Center (Super. Ct., Onandage County, February 1, 1984) (elderly man allowed to starve himself to death in a nursing home). In mentioning these two cases together I do not imply that there are no important differences between them. For example, in some of her claims, Elizabeth Bouvia appeared to assert a positive right to the assistance of others in her death rather than a negative right to noninterference.

21. See the survey by Kenneth C. Micetich, Patricia H. Steinecker, and David C. Thomasma, "Are Intravenous Fluids Morally Required for a Dying Patient?" *Archives of Internal Medicine* 143 (May 1983): 975–80.

22. This position is taken by Micetich, Steinecker, and Thomasma, ibid.

23. Micetich, Steinecker, and Thomasma report that some physicians in an informal poll took this position. Ibid., p. 975.

24. My ideas in this essay obviously reiterate and build on my article with Joanne Lynn (see n. 14 above), and I am indebted to her for fruitful discussions. In addition, some of the ideas and formulations in this essay have been drawn from "Caring for Symbols and Caring for Patients: Reflections on 'On Feeding the Dying,'" which I coauthored with Steven L. DalleMura in *BioLaw* 1 (August 1986): 501–507. I am also grateful to Mr. DalleMura for helpful discussions. Others are not, of course, responsible for any of the flaws in this essay.

25. For a fuller summary of this case along with references, see Beauchamp and Childress, *Principles of Biomedical Ethics*, 4th ed. (1994), appendix, Case 6, and Lawrence J. Schneiderman and Nancy S. Jecker, *Wrong Medicine: Doctors, Patients, and Futile Treatment* (Baltimore: Johns Hopkins University Press, 1995).

26. John D. Lantos, Peter A. Singer, Robert M. Walker et al., "The Illusion of Futility in Clinical Practice," *The American Journal of Medicine* 87 (July 1989): 82. My analysis is also much indebted to Robert D. Truog, Allan S. Brett, and Joel Frader, "The Problem with Futility," *New England Journal of Medicine* 326 (June 4, 1992): 1560–64.

27. Lantos, Singer, Walker et al., "The Illusion of Futility in Clinical Practice," p. 83.

28. See Beauchamp and Childress, *Principles of Biomedical Ethics*, 3rd and 4th eds., chap. 4.

10. WHO SHALL LIVE WHEN NOT ALL CAN LIVE?

1. George Bernard Shaw, *The Doctor's Dilemma,* from *Complete Plays with Prefaces* (New York: Dodd, Mead & Company, 1963), vol. 1, pp. 132–33.

2. Henry K. Beecher, "Scarce Resources and Medical Advancement," *Daedalus* (Spring, 1969): 279–80.

3. Leo Shatin, "Medical Care and the Social Worth of a Man," *American Journal of Orthopsychiatry* 36 (1967): 97.

4. Harry S. Abram and Walter Wadlington, "Selection of Patients for Artificial and Transplanted Organs," *Annals of Internal Medicine* 69 (September 1969): 615–20.

5. *United States v. Holmes*, Fed. Cas. 360 (C.C.E.D. Pa 1842). All references are to the text of the trial as reprinted in Philip E. Davis, ed., *Moral Duty and Legal Responsibility: A Philosophical-Legal Casebook* (New York: Appleton-Century-Crofts, 1966), pp. 102–18.

6. Edmond Cahn, *The Moral Decision* (Bloomington: Indiana University Press, 1955), p. 71.

7. I am excluding from consideration the question of the ability to pay because most of the people involved have to secure funds from other sources, public or private, in any case. [I discuss ability to pay in other essays in this volume.]

8. Joseph Fletcher, "Donor Nephrectomies and Moral Responsibility," *Journal of the American Medical Women's Association* 23 (December 1968): 1090.

9. Leo Shatin, "Medical Care and the Social Worth of a Man," pp. 96–101.

10. For a discussion of the Seattle selection committee, see Shana Alexander, "They Decide Who Lives, Who Dies," *Life* 53 (November 9, 1962): 102. For an examination of general selection practices in dialysis, see "Scarce Medical Resources," *Columbia Law Review* 69 (1969), and Abram and Wadlington, "Selection of Patients for Artificial and Transplanted Organs."

11. David Sanders and Jesse Dukeminier, Jr., "Medical Advance and Legal Lag: Hemodialysis and Kidney Transplantation," *UCLA Law Review* 15 (1968): 378.

12. "Letters and Comments," *Annals of Internal Medicine* 61 (August 1964): 360. Dr. G. E. Schreiner contends that "if you really believe in the right of society to make decisions on medical availability on these criteria you should be logical and say that when a man stops going to church or is divorced or loses his job, he ought to be removed from the program and somebody else who fulfills these criteria substituted. Obviously no one faces up to this logical consequence." G. E. W. Wolstenholme and Maeve O'Connor, eds., *Ethics in Medical Progress: With Special Reference to Transplantation*, A Ciba Foundation Symposium (Boston: Little, Brown and Company, 1964), p. 127.

13. Helmut Thielicke, "The Doctor as Judge of Who Shall Live and Who Shall Die," in *Who Shall Live?* ed. Kenneth Vaux (Philadelphia: Fortress Press, 1970), p. 172.

14. Ibid., pp. 173–74.

15. Ibid., p. 173.

16. Helmut Thielicke, *Theological Ethics*, vol. 1, *Foundations* (Philadelphia: Fortress Press, 1966), p. 602.

17. Cahn, *The Moral Decision*, p. 71.

18. Paul Freund, Introduction, *Daedalus* (Spring 1969), xiii.

19. Paul Ramsey, *Nine Modern Moralists* (Englewood Cliffs: Prentice-Hall, 1961), p. 245.

20. Charles Fried, "Privacy," in *Law, Reason, and Justice*, ed. Graham Hughes (New York: New York University Press, 1969), p. 52.

21. My argument is greatly dependent on John Rawls's version of justice as fairness, which is a reinterpretation of social contract theory. Rawls, however, would probably not apply his ideas to "borderline situations." See "Distributive Justice: Some Addenda," *Natural Law Forum* 13 (1968): 53. For Rawls's general theory, see "Justice as Fairness," *Philosophy, Politics and Society* (2nd ser.), ed. Peter Laslett and W. G. Runciman (Oxford: Basil Blackwell, 1962), pp. 132–57, and his other essays on aspects of this topic. (Of course, Rawls's *A Theory of Justice*, which appeared after my essay was written, is the major statement prior to his later revisions. See *A Theory of Justice* [Cambridge: Harvard University Press, 1971.])

22. Occasionally someone contends that random selection may reward vice. In "Medical Care and the Social Worth of a Man" (p. 100), Shatin insists that random selection "would reward socially disvalued qualities by giving their bearers the same special medical care opportunities as those received by the bearers of socially valued qualities. Personally I do not favor such a method." Obviously society must engender

certain qualities in its members, but not all of its institutions must be devoted to that purpose. Furthermore, there are strong reasons, I have contended, for exempting SLMR from that sort of function.

23. Nicholas Rescher, "The Allocation of Exotic Medical Lifesaving Therapy," *Ethics* 79 (April 1969): 184. He defends random selection's use only after utilitarian and other judgments have been made. If there are no "major disparities" in terms of utility, etc., in the second stage of selection, then final selection could be made randomly. He fails to give attention to the moral values that random selection might preserve.

24. Harry S. Abram, M.D., "The Psychiatrist, the Treatment of Chronic Renal Failure, and the Prolongation of Life: II," *American Journal of Psychiatry* 126 (1969): 158.

25. I read a draft of this paper in spring 1970 in a seminar titled "Social Implications of Advances in Biomedical Science and Technology: Artificial and Transplanted Internal Organs," sponsored by the Center for the Study of Science, Technology, and Public Policy of the University of Virginia. I am indebted to the participants in that seminar, and especially to its leaders, Mason Willrich, Professor of Law, and Dr. Harry S. Abram, Associate Professor of Psychiatry, for criticisms which helped me to sharpen these ideas. Good discussions of the legal questions raised by selection (e.g., equal protection of the law and due process), which I have not considered, can be found in "Scarce Medical Resources," *Columbia Law Review* 69 (1969); "Patient Selection for Artificial and Transplanted Organs," *Harvard Law Review* 83 (1969); and Sanders and Dukeminier, "Medical Advance and Legal Lag." (For purposes of publication in this volume, this article has been changed only slightly from the way it appeared in 1970.)

26. I prepared an earlier draft of these subsequent reflections for a conference on the origins of bioethics at the University of Washington in 1992. The conference organizer, Albert R. Jonsen, viewed the 1962 controversy about selecting patients for dialysis as the beginning of modern bioethics or biomedical ethics.

27. See n. 19 above.

28. Childress, "Who Shall Live When Not All Shall Live?" *Soundings* 53 (Winter 1970): 339–55; and Frederic B. Westervelt, "A Reply to Childress: The Selection Process as Viewed from Within," *Soundings* 53 (Winter 1970): 356–62.

29. See Ramsey, *The Patient as Person* (New Haven: Yale University Press, 1970).

30. Childress, "Rationing of Medical Treatment," *Encyclopedia of Bioethics,* ed. Warren T. Reich (New York: Macmillan/Free Press, 1978), vol. 4, p. 1414.

31. These quotations from Hecaton's "Moral Duties" appear in Ciecero, *De Officiis,* trans. Walter Miller (Cambridge: Harvard University Press, 1947), p. 365.

32. Ambrose, *Duties of the Clergy,* III.4, *Nicene and Post-Nicene Fathers of the Christian Church,* 2nd ser., vol. 10 (New York: Christian Literature Company, 1896), p. 71. See also Lactantius's discussion of the so-called "board of Carneades" in *The Divine Institutes (The Fathers of the Church, A New Translation,* vol. 49 [Washington, D.C.: Catholic University of America Press, 1964], V17).

33. See Saul Lieberman, "How Much Greek in Jewish Palestine?" in *Biblical and Other Studies,* ed. Alexander Altmann (Cambridge: Harvard University Press, 1963). For other discussions, see Chaim W. Reines, "The Self and the Other in Rabbinic Ethics"; Louis Jacobs, "Greater Love Hath No Man . . . The Jewish Point of View of Self-Sacrifice," in *Contemporary Jewish Ethics,* ed. Menachem Marc Kellner (New York: Sanhedrin Press, Hebrew Publishing Co., 1978); and Jakob J. Petuchowski, "The Limits of Self-Sacrifice," in *Modern Jewish Ethics,* ed. Marvin Fox (Columbus: Ohio State University Press, 1975). The paragraph in the text is drawn largely from James F. Childress, *Priorities in Biomedical Ethics* (Philadelphia: Westminster Press, 1981).

34. A. W. Brian Simpson, *Cannibalism and the Common Law* (Chicago: Univer-

sity of Chicago Press, 1984), p. 164. The discussion in the remainder of this paragraph is drawn from Simpson's careful and thoughtful study.

35. Ibid., p. 176.

36. Albert H. Katz and Donald M. Procter, "Social-Psychological Characteristics of Patients Receiving Hemodialysis Treatment for Chronic Renal Failure," contract no. PH-108-66-95 (Rockville, MD: U.S. Department of Health, Education and Welfare, Kidney Disease Control Program, July 1969).

37. Joseph Fletcher, *The Greatest Good of the Greatest Number: A New Frontier in the Morality of Medical Care,* Sanger Lecture no. 7 (Richmond: Medical College of Virginia, Virginia Commonwealth University, n.d.).

38. Ramsey, *Patient as Person,* p. 259.

39. Gerald R. Winslow, *Triage and Justice: The Ethics of Rationing Life-Saving Medical Resources* (Berkeley: University of California Press, 1982); John F. Kilner, *Who Lives? Who Dies? Ethical Criteria in Patient Selection* (New Haven: Yale University Press, 1990). George J. Annas presented various arguments in a creative fictional legal decision in "Allocation of Artificial Hearts in the Year 2002: Minerva v. National Health Agency," *American Journal of Law and Medicine* 3 (Spring 1977): 59–76.

40. Ramsey, *Patient as Person,* p. 240 and passim.

41. Along with many others, I attempted to address some questions of macro-allocation in "Priorities in the Allocation of Health Care," *Soundings* 62 (Fall 1979): 256–74.

42. Shirley Jackson, "The Lottery," in *The Lottery* (New York: Popular Library, 1949).

43. Ralph P. Forsberg, "Rationality and Allocating Scarce Medical Resources," *Journal of Medicine and Philosophy* 20 (February 1995): 25–42. References to this essay are given in the text in parentheses.

44. Barbara Goodwin, *Justice by Lottery* (Chicago: University of Chicago Press, 1992), p. 178.

45. Ibid.

46. This quotation comes from Evan DeRenzo, a bioethicist at the National Institutes of Health Clinical Center, and appeared in Diane Naughton, "Drug Lotteries Raise Questions: Some Experts Say System of Distribution May Be Unfair," *Washington Post Health,* September 26, 1995, pp. 14–15. See also "AIDS Patients to Enter Lottery for New Drug," *Charlottesville Daily Progress,* June 22, 1995, p. A2.

47. Tamar Lewin, "Prize in an Unusual Lottery: A Scarce Experimental Drug," *New York Times,* January 7, 1994, pp. A1 and A17.

11. TRIAGE IN INTENSIVE CARE

1. For a fuller discussion of "care," see James F. Childress, *Who Should Decide? Paternalism in Health Care* (New York: Oxford University Press, 1982), chap. 2, esp. pp. 34–39; Stanley Hauerwas, "Care, " *The Encyclopedia of Bioethics,* ed. Warren T. Reich (New York: Free Press, 1978); and Milton Mayeroff, *On Caring* (New York: Harper and Row, 1971).

2. Robert A. Burt notes the double-edged meaning of "taking care of" someone. It may not mean to render care, but to hurt and even to kill. For example, a criminal may threaten to "take care of someone." See Burt, *Taking Care of Strangers: The Rule of Law in Doctor-Patient Relations* (New York: Free Press,1979).

3. Peter Budetti et al., *Case Study #10: The Costs and Effectiveness of Neonatal Intensive Care,* Background Paper no. 2, Case Studies of Medical Technologies (Washington, D.C.: Office of Technology Assessment, 1981), p. 7 (hereafter referred to as *Costs and Effectiveness of NIC).*

4. Ibid., p. 9.

5. Clement Smith, "Neonatal Medicine and Quality of Life: An Historical Perspective," in *Ethics of Newborn Intensive Care*, ed. Albert R. Jonsen and Michael J. Garland (Berkeley: University of California, Institute of Governmental Studies, 1976), p. 31.

6. *Costs and Effectiveness of NIC*, pp. 11 and 13.

7. The metaphor of the "commons" is used in Garrett Hardin, "The Tragedy of the Commons," *Science* 162 (1968): 1243–48; Howard H. Hiatt, "Protecting the Medical Commons: Who Is Responsible?" *New England Journal of Medicine* 292 (July 31, 1975): 235–40; and Raymond S. Duff, "On Deciding the Use of the Family Commons," in *Developmental Disabilities: Psychologic and Social Implications*, ed. Daniel Bergsma and Anne E. Pulver, Birth Defects: Original Article Series, vol. 12, no. 4 (New York: Alan R. Liss, 1976), pp. 73–84.

8. The metaphor of the "lifeboat" appears in Garrett Hardin, "Living on a Lifeboat," *Bioscience* 24 (October 1974): 561–68; Paul Ramsey, *The Patient as Person* (New Haven: Yale University Press, 1970), chap. 7; and James F. Childress, "Who Shall Live When Not All Can Live?" *Soundings* 53 (Winter 1970), reprinted in the previous chapter.

9. Robert J. T. Joy, "Triage—Who Is Sorted, and Why?" in *Hard Choices* (Boston: WGBH Educational Foundation, Office of Radio and TV for Learning, 1980), p. 28.

10. Discussions of the evolution of the concept of "triage" appear in Joy, "Triage—Who is Sorted, and Why?" pp. 26–28; Douglas A. Rund and Tondra S. Rausch, *Triage* (St. Louis: C. V. Mosby, 1980), pp. 3–10; Stuart W. Hinds, "On the Relations of Medical Triage to World Famine: An Historical Survey," in *Lifeboat Ethics: The Moral Dilemmas of World Hunger*, ed. George R. Lucas, Jr., and Thomas Ogletree (New York: Harper and Row, 1976), pp. 29–51; and Gerald R. Winslow, *Triage and Justice: The Ethics of Rationing Life-Saving Medical Resources* (Berkeley: University of California Press, 1982), chap. 1. Also, see "triage" in *The Oxford English Dictionary.*

11. Quoted in Rund and Rausch, *Triage*, p. 4 (italics added). Baron Domineque Jean Larrey (1766–1842), who was Surgeon-in-Chief of the imperial armies of Emperor Napoleon and Inspector General of the medical staff of the French armies, first devised a system of sorting patients according to their medical and surgical needs and assigning priorities accordingly. Larrey held that "those who are dangerously wounded must be tended first, entirely without regard to rank or distinction. Those less severely injured must wait until the gravely wounded have been operated upon and dressed. The slightly wounded may go to the hospital in the first or second line; especially the officers, since they have horses and therefore have transport—and regardless, most of these have but trivial wounds." Quoted in Hinds, "On the Relations of Medical Triage to World Famine," p. 32. According to the laws of war, injured and wounded enemy soldiers also have claims to medical resources, but it is unclear how much weight these claims have to the claims of one's fellow soldiers. See Morris Greenspan, *Soldier's Guide to the Laws of War* (Washington, D.C.: Public Affairs, 1969), chap. 3.

12. Rund and Rausch, *Triage*, p. 9.

13. For recent discussions, see Joseph F. Waeckerle, "Disaster Planning and Response," *New England Journal of Medicine* 324 (March 21, 1991): 815–21; and *Health and Medical Aspects of Disaster Preparedness*, ed. John C. Duffy (New York: Plenum Press, 1990).

14. For example, Winslow argues that "an important continuity is the dominant type of justification that generally has been given for triage. When a reason has been stated, it has been usually in terms of doing the greatest good for the greatest number" *(Triage and Justice*, p. 21).

15. Neglect of this distinction is one of the few weaknesses in Winslow's fine book, *Triage and Justice;* see, for example, pp. 21–23. For a discussion of the

distinction, see Tom L. Beauchamp and James F. Childress, *Principles of Biomedical Ethics*, 3rd ed. (New York: Oxford University Press, 1989), chap. 2.

16. Having distinguished "medical utility" from "social utility," I will later distinguish two types of "social utility": (a) specific functions of individuals in social institutions, such as the military, and (b) general, overall social worth of individuals. By contrast, "medical utility" focuses on the greatest good for the greatest number of injured or wounded individuals. Medical utility involves medical judgments by medical and health care professionals about medical need and medical prognosis. This is not to suggest that it is value-neutral or objective in the sense of being value-free, for values clearly enter in determining need, benefit, etc.

17. Hinds, "On the Relations of Medical Triage to World Famine," p. 46.

18. Henry K. Beecher discusses this case: "Which group would get the penicillin? By all that is just, it would go to the heroes who had risked their lives, who were still in jeopardy, and some of whom were dying. They did not receive it, nor should they have; it was given to those infected in brothels. Before indignation takes over, let us examine the situation. First, there were desperate shortages of manpower at the front. Second, those with broken bodies and broken bones would not be swiftly restored to the battle line even with penicillin, whereas those with venereal disease, on being treated with penicillin, would in a matter of days free the beds they were occupying and return to the front. Third, no one will catch osteomyelitis from his neighbor; the man with venereal disease remains, until he is cured, a reservoir of infection and a constant threat. In view of customary morality, a great injustice was done; in view of the circumstances, I believe that the course chosen was the proper one." See Beecher, *Research and the Individual: Human Studies* (Boston: Little, Brown, and Company, 1970), pp. 209–10.

19. For an analysis of San Francisco's triage plans, see Winslow, *Triage and Justice*, esp. chap. 2.

20. Thomas O'Donnell, "The Morality of Triage," *Georgetown Medical Bulletin* 14 (August 1960).

21. Gilbert W. Beebe and Michael E. DeBakey, *Battle Casualties: Incidence, Mortality, and Logistic Considerations* (Springfield: Charles C. Thomas, 1952), p. 216; quoted in Winslow, *Triage and Justice*, p. 11 (italics added).

22. Paul Ramsey, *Ethics at the Edges of Life: Medical and Legal Intersections* (New Haven: Yale University Press, 1978), p. 245.

23. See Ramsey, *Patient as Person*, chap. 7, and *Ethics at the Edges of Life*. In the latter, he writes: "Triage has been justified heretofore only when *one* social value is at stake—survival, for example" (p. 232, n. 5). It is not clear why only one goal, whatever it might be, is permitted; it is not clear why two or more goals could not function in this way.

24. John M. Taurek, "Should the Numbers Count?" *Philosophy and Public Affairs* 6 (Summer 1977): 293–316.

25. Derek Parfit, "Innumerate Ethics," *Philosophy and Public Affairs* 7 (Summer 1978): 301. Parfit offers a thorough and devastating critique of Taurek's argument. See also the exchange between Parfit and Charles Fried in "Correspondence," *Philosophy and Public Affairs* 8 (Summer 1979): 393–97.

26. See Nora K. Bell, "Triage in Medical Practices: An Unacceptable Model?" *Social Science and Medicine* 15F (1981): 151–56.

27. See Beauchamp and Childress, *Principles of Biomedical Ethics*, chap. 5.

28. Taurek, "Should the Numbers Count?" p. 311. Taurek contends that people deciding on a policy of distributing a scarce resource, such as an evacuation ship (or an ICU), to which they have contributed, would want the policy to reflect "equal concern" and their "equal claim on the resource." Thus they would reject priority for "medical researchers, high-powered managerial types, and people with IQs over 120." He also argues that "under certain conditions" they would reject a policy of saving the

larger number over the smaller number (pp. 314–15). I agree that there are such conditions, but he seems to concede that there is no necessary incompatibility between "equal concern" and saving the large number over the smaller number (what I have called medical utility). Taurek ought also to concede that his general principle of expressing "equal concern" when deciding to give his lifesaving drug to the five or the one does not require a lottery. There are, of course, situations in which flipping a coin would be morally preferable to express equal concern by providing an equal chance. Derek Parfit proposes three examples: "(1) We can save X or Y. Nothing would be lost by flipping coins. Something would be gained. (2) We can save X's life or Y's arm. Something would again be gained. We would give Y a chance. But if Y wins X would die. The case for flipping coins seems here to be outweighed. (3) We can save David or the five. There is again a case for flipping coins. But I believe it is again outweighed." See Parfit, "Innumerate Ethics," p. 300, n. 17.

29. Jonsen and Garland, eds., *Ethics of Newborn Intensive Care*, pp. 142–55. Hereafter page numbers in the text refer to this volume. See also Albert R. Jonsen et al., "Critical Issues in Newborn Intensive Care: A Conference Report and Policy Proposal," *Pediatrics* 55 (June 1975): 756–68, which reports on the conference that produced the volume *Ethics of Newborn Intensive Care.* Jonsen subsequently identified analogical reasoning as a major ingredient in moral casuistry, which he has defended over against an emphasis on principles and rules in applied or practical ethics, including biomedical ethics. See, for instance, Albert R. Jonsen, "Can an Ethicist Be a Consultant?" in *Frontiers in Medical Ethics,* ed. Virginia Abernathy (Cambridge: Ballinger, 1980), pp. 157–71; "On Being a Casuist," in *Clinical Medical Ethics: Exploration and Assessment,* ed. Terrence F. Ackerman et al. (Lanham, MD: University Press of America, 1987), pp. 117–27; and with Stephen Toulmin, *The Abuse of Casuistry: A History of Moral Reasoning* (Berkeley: University of California Press, 1988).

30. They might avoid the charge of equating relevance and weight by insisting that when they discuss the former, they indicate that "considerations of the common good *become* relevant." When they discuss the latter, they indicate that "a common good . . . *may be* the predominant consideration." If so, they are *not* making identical claims about the common good in both settings (military and civil disasters on the one hand, and NICUs on the other), and the analogy suffers.

31. Much of the debate depends on what one defines as the resource to be distributed. While I have concentrated on lifesaving medical treatment, particularly in the form of intensive care, others sometimes concentrate on familial resources (both emotional and financial) and argue that decisions about treatment of seriously ill babies should include the impact on the "family commons." Thus, Raymond Duff argues that "the family may . . . be said to have a commons, those limited, private resources on which individuals and their families depend and must protect in order to function or even to survive." He concludes that "there are times, for example, when one must decide between the well-being of the unfortunate and that of others in competition for limited resources. By direct or proxy decision, it may be reasonable at times to make a tragic choice of neglect, even death, for one in order to protect others" (Duff, "On Deciding the Use of the Family Commons," pp. 73, 79). But Duff fails to consider the burden of justification that would have to be met in order to justify letting a particular newborn die in order to protect the family. Rarely, if ever, would it be justifiable to let a newborn die in order to protect familial interests. See David H. Smith, "On Letting Some Babies Die," *Hastings Center Studies* 2 (May 1974): 37–46. Elsewhere, Duff invokes a military metaphor, insisting that we should avoid "a medical Vietnam" in caring for defective newborns. See Duff, "Counseling Families and Deciding Care of Severely Defective Children: A Way of Coping with 'Medical Vietnam,'" *Pediatrics* 67 (March 1981).

32. Anthony Shaw, "Defining the Quality of Life," *Hastings Center Report* 7

(October 1977): 11. Jonsen and Garland rightly emphasize the hazards of such an approach: "the hope of survival with maximal function is predicated not only on the infant's physical potential, but also on the nature of the socioeconomic world it enters. Thus, estimates of the quality of future care may affect its selection for continued intensive care" (pp. 152–53). For another approach to the allocation of intensive care, mainly for adults, see H. Tristram Engelhardt, Jr., and Michael A. Rie, "Intensive Care Units, Scarce Resources, and Conflicting Principles of Justice," *Journal of the American Medical Association* 255 (March 7, 1986): 1159–64. Stressing that moral arguments can only establish a range of acceptable models and that societies must choose their own model through common agreement, they argue that decision making in such cases should be directed by the metaphor of choice of insurance policies and by general criteria regarding the probability of successful outcome, including both length of life and quality of life, as well as costs, all expressed in an explicit ICU treatment entitlement index (ICU-EI). For an approach to the allocation of critical care that concentrates on the neediest patients, see Robert M. Veatch, "The Ethics of Resources Allocation in Critical Care," *Critical Care Clinics* 2 (January 1986): 73–89. While generally holding that nonconsequentialist principles should not be balanced against consequentialist principles, Veatch does indicate that such a balancing might be necessary in order to avoid the conclusion that "a medically hopeless patient has a claim that exhausts all resources." In practice, he contends, it would probably be better for physicians to avoid any role in resource allocation in order to protect their role as patient advocate.

33. In earlier writings, I argued, as I do now, for randomization or queuing rather than social worth to determine recipients of scarce lifesaving medical care from the pool of medically acceptable candidates. I did not then take into account relative chances of success as I now do. Relative chances of success are made relevant by medical utility and also by equal concern for life under conditions of scarcity. Contrast Childress, "Who Shall Live When Not All Can Live?" (the previous chapter in this volume).

34. See my discussion of "medical utility" earlier. Winslow adopts a principle of "conservation" as an important proviso for disaster triage and the totally implantable artificial heart—the two cases he considers—"so long as it could be reasonably demonstrated that at least one less life would probably be lost." Such a principle thus would override in some circumstances his egalitarian principle of "medical neediness." See Winslow, *Triage and Justice*, pp. 164, 74–76. John F. Kilner offers "disproportionate resources" as an acceptable exception to an equality and need-based conception of justice in the distribution of scarce lifesaving medical resources: "The most basic consideration is the value of life. We are morally bound to save as many lives as possible. Not only is this not a violation of the equal right to life of all involved—it is demanded by it.... The real claim of each person, then, is that his life be valued equally with all others—which in turn necessitates that two rather than one be selected." See Kilner, "A Moral Allocation of Scarce Lifesaving Medical Resources," *Journal of Religious Ethics* 9 (Fall 1981): 264. See also Kilner, *Who Lives? Who Dies? Ethical Criteria in Patient Selection* (New Haven: Yale University Press, 1990).

35. See Jonsen and Garland, eds., *Ethics of Newborn Intensive Care*, pp. 191–94, and Jonsen et al., "Critical Issues in Newborn Intensive Care," pp. 767–68.

36. See Beauchamp and Childress, *Principles of Biomedical Ethics*, chap. 4.

37. It might be argued that I have emphasized the limitations of the metaphor of triage because I focused on it in relation to microallocation—who would receive the particular, scarce good of intensive care?—rather than on macroallocation—how much of a particular good such as intensive care should be made available? Paul Ramsey appears to make this argument: "To introduce the notion of triage *into* intensive care policy—distinguishing the worse from the next better prognosis—is an error *ab initio*. Triage starts with those least in need, in a crisis disposition of sparse

medical resources. To bring the notion of triage, where justifiable, to bear at all upon neonatal intensive care would impel us to conclude that, in allocating a society's normal but still limited medical resources, intensive and costly care is the *last* thing we should undertake" (*Ethics at the Edges of Life*, p. 245).

Ramsey's interpretation of triage is debatable. As Hinds emphasizes, medical triage has classified and sorted casualties "into priority groups according to their individual needs, without reference to rank, birth, or favor. When for any reason shortages of essential facilities complicate the picture, the sorting will be influenced further by the *prognosis rating of each casualty*" (Hinds, "On the Relations of Medical Triage to World Famine," p. 33 [italics added]). Ramsey emphasizes broad categories rather than the comparison of individuals. However, it is not clear that triage can really illuminate policies of macroallocation as Ramsey suggests. Such policies would include whether to concentrate resources on reduction of neonatal morbidity and mortality rather than on other medical needs, and, if so, whether to concentrate resources on NIC rather than on prevention of neonatal morbidity and mortality through, for example, maternal education and prenatal care. By contrast, Bell argues that "the *less immediate* the crisis the *more anonymous* the persons who will benefit—the more nameless, and faceless the 'unsalvageable' are—the more disanalogous the triage model to the allocation context, hence the more unacceptable it becomes to make appeal to the triage model" (Bell, "Triage in Medical Practices," p. 156 [italics added]).

38. Charles Fried, "Rights and Health Care—Beyond Equity and Efficiency," *New England Journal of Medicine* 293 (July 31, 1975): 241–45.

39. I am grateful for the helpful comments and suggestions offered by the respondents at the symposium "The Beginnings of Life" at the University of Virginia Law School (November 5–6, 1982), where this paper was presented in its original form: Haavi Morreim, James Canfield, Lynn Cook, Donald Jones, and Paul Lombardo. I regret that I have not been able to deal adequately with all of their criticisms and suggestions.

40. See Robert Baker and Martin Strosberg, "Triage and Equality: An Historical Reassessment of Utilitarian Analyses of Triage," *Kennedy Institute of Ethics Journal* 2 (June 1992): 103–23.

41. It is, of course, difficult to specify "very substantial differences," which the Council on Ethical and Judicial Affairs of the American Medical Association considers sufficient to warrant selecting one patient over another for such medical resources as IC and transplantable organs. The Council recognizes the potential relevance of several criteria for scarce IC (urgency of need, likelihood of benefit, change in quality of life, duration of benefit, and in some cases the amount of resources required for successful treatment), but "only when the differences among patients are very substantial. . . . Where no relevant ethical distinctions can be made among patients, priority should be given according to the order in which patients present to the ICU." See the Council's "Ethical Considerations in the Allocation of Organs and Other Scarce Medical Resources among Patients," *Archives of Internal Medicine* 155 (January 9, 1995): 36.

42. See n. 18 above.

43. See David J. Rothman, *Strangers at the Bedside* (New York: Basic Books, 1991), p. 40; and Baker and Strosberg, "Triage and Equality," p. 121.

44. Hinds, "On the Relations of Medical Triage to World Famine," p. 37.

45. Baker and Strosberg, "Triage and Equality," p. 121.

46. Ibid., p. 115.

47. Ibid., pp. 107–108.

48. Ibid., p. 113.

49. The language of "medical utility" does not appear in Winslow's comprehensive 1982 book, *Triage and Justice*. When I served on the federal Task Force on Organ Transplantation a couple of years later, I pressed this distinction, which was featured

in its recommendations about policies to allocate organs for transplantation. See Task Force on Organ Transplantation, *Organ Transplantation* (Washington, D.C.: Department of Health and Human Services, 1986) and the next chapter in this volume.

50. Robert Zussman, *Intensive Care: Medical Ethics and the Medical Profession* (Chicago: University of Chicago Press, 1992). Even though I concentrated on NIC in my discussion of triage, most of the points also extend to adult IC, with the exception of determination of function since the older patients have a past, which provides a different indication of function than is possible for newborns.

51. Ibid., p. 210.

52. Ibid., p. 211.

53. Ibid., pp. 211–16.

54. Robert D. Truog, "Triage in the ICU," *Hastings Center Report* 22 (May/June 1992): 13–17.

55. William Knaus, "Criteria for Admission to Intensive Care Units," in *Rationing of Medical Care for the Critically Ill* (Washington, D.C.: Brookings Institution, 1989), pp. 44–51. By partial contrast, Zussman contends that physicians do not invite family participation in triage decisions, even though they usually disclose such decisions as a matter of courtesy and honesty (*Intensive Care*, p. 187). Truog argues for the importance of prospective notification regarding triage policies. The military does have clearly formulated policies of triage, as some cities do for disasters and hospitals for some units. However, it is not clear how often hospitals with such policies for ICUs actually notify patients and or surrogates in advance about the conditions under which discharge might be made when it is not in the patient's best interests but is rather dictated by (substantial) medical utility.

56. For a strong defense of publicity, see Gerald R. Winslow, "Rationing and Publicity," in *The Price of Health,* ed. George J. Agich and Charles E. Begley (Boston: D. Reidel, 1986).

12. FAIRNESS IN THE ALLOCATION AND DELIVERY OF HEALTH CARE

1. This case study is largely drawn from Case 33 in Tom L. Beauchamp and James F. Childress, *Principles of Biomedical Ethics,* 3rd ed. (New York: Oxford University Press, 1989), pp. 446–49 (where complete references are given). References of particular importance include Roger W. Evans et al., *The National Heart Transplantation Study: Final Report* (Seattle: Battelle Human Affairs Research Center, 1984), and H. Gilbert Welch and Eric B. Larson, "Dealing with Limited Resources: The Oregon Decision to Curtail Funding for Organ Transplantation," *New England Journal of Medicine* 319 (1988): 171–73. See also Richard Rettig, "The Politics of Organ Transplantation: A Parable of Our Time," *Journal of Health Politics, Policy and Law* 14 (Spring 1989): 191–227; and United States General Accounting Office, *Heart Transplants: Concerns about Cost, Access, and Availability of Donor Organs,* Report to the Chairman, Subcommittee on Health, Committee on Ways and Means, House of Representatives (Washington, D.C.: U.S. General Accounting Office, May 1989).

2. See Martin Benjamin, Carl Cohen, and Eugene Grochowski for the Ethics and Social Impact Committee of the Transplant and Health Policy Center, "Sounding Board: What Transplantation Can Teach Us about Health Care Reform," *New England Journal of Medicine* 330 (March 24, 1994): 858–60.

3. See Task Force on Organ Transplantation, *Organ Transplantation: Issues and Recommendations* (Rockville, MD: Office of Organ Transplantation, Health Resource and Services Administration, U.S. Department of Health and Human Services, April 1986).

4. Public Law No. 98–507, "The National Organ Transplant Act."

5. See Task Force, *Organ Transplantation,* pp. 8–9, 85–86 and passim.

6. On the "moral connections" between procurement and distribution, see James F. Childress, "Some Moral Connections between Organ Procurement and Organ Distribution," *Journal of Contemporary Health Law and Policy* 3 (1987): 85–110. See also chapter 14 below.

7. Jeffrey M. Prottas, "Nonresident Aliens and Access to Organ Transplant," *Transplantation Proceedings* 21 (June 1989): 3428.

8. Jeffrey M. Prottas, *The Most Useful Gift: Altruism and the Public Policy of Organ Transplants,* A Twentieth Century Fund Book (San Francisco: Jossey-Bass, 1994).

9. Task Force, *Organ Transplantation.*

10. Michigan Transplant Policy Center, *The Case for Federalism in Organ Sharing* (Ann Arbor: Michigan Transplant Policy Center, 1987).

11. For a discussion of criteria of justice, see Beauchamp and Childress, *Principles of Biomedical Ethics,* 4th ed. (New York: Oxford University Press, 1994), chap. 6.

12. Task Force, *Organ Transplantation,* chaps. 4 and 5.

13. For the distinction between "medical utility" and "social utility," see James F. Childress, "Triage in Neonatal Intensive Care: The Limitations of a Metaphor," *Virginia Law Review* 69 (April 1983): 547–61, which appears in revised form as the preceding chapter of this volume.

14. See P. W. Eggers, "Effect of Transplantation on the Medicare End Stage Renal Disease Program," *New England Journal of Medicine* 318 (1989): 223–29; C. M. Kjellstrand, "Age, Sex, and Race Inequality in Renal Transplantation," *Archives of Internal Medicine* 148 (1988): 1305–1309; P. J. Held et al., "Access to Kidney Transplantation: Has the United States Eliminated Income and Racial Differences?" *Archives of Internal Medicine* 148 (December 1988): 2594–2600; and, for hearts, U.S. General Accounting Office, *Heart Transplants* (see n. 1 above).

15. Held et al., "Access to Kidney Transplantation." A report from the Office of the Inspector General indicates that blacks on kidney transplant waiting lists have to wait almost twice as long as whites for a first transplant (13.9 months compared with 7.6 months) and insists that this disparity cannot be explained solely by differences in "blood type, age, immunological and locational factors" because it remains even when these factors are controlled. See Office of the Inspector General, U.S. Department of Health and Human Services, *The Distribution of Organs for Transplantation: Expectations and Practices,* OEI-01-89-00550 (1991). See also F. P. Sanfilippo, W. K. Vaughn, T. G. Peters et al., "Factors Affecting the Waiting Time of Cadaveric Transplant Candidates in the United States," *Journal of the American Medical Association* 267 (1992): 247–52.

16. *1994 Annual Report of the U.S. Scientific Registry for Transplant Recipients and the Organ Procurement and Transplantation Network—Transplant Data: 1988–1993* (Richmond, VA: UNOS, and Bethesda, MD: Division of Organ Transplantation, Bureau of Health Resources Development, Health Resources and Services Administration, U.S. Department of Health and Human Services, 1994), V-3.

17. The UNOS board in 1987 and 1988 voted to permit but then to prohibit and, finally, following public comment and public hearings, again to permit multiple listing, and in 1994 and 1995 recommended publication of a proposed policy to prohibit multiple listing but later, after public comment, tabled it.

18. V. Smirnow, "National Affairs," *Dialysis and Transplantation* (1988): 272.

19. UNOS, *Policy Proposal Statement: UNOS Policy Regarding the Listing of Patients on Multiple Transplant Waiting Lists* (1988).

20. *1994 Annual Report.*

21. New York state law restricts kidney patients from being listed at more than one transplant center in that state, but they can obviously choose to be listed in other states. See Tracy E. Miller, "Multiple Listing for Organ Transplantation: Autonomy Unbounded," *Kennedy Institute of Ethics Journal* 2 (March 1992): 43–59.

22. See T. E. Starzl, T. R. Hakala, A. Tzakis et al., "A Multifactorial System for Equitable Selection of Cadaver Kidney Recipients," *Journal of the American Medical Association* 257 (1987): 3073–75.

23. UNOS, *Final Statement of Policy: UNOS Policy Regarding Utilization of the Point System for Cadaveric Kidney Allocation* (Richmond, VA: UNOS, April 4, 1989), which gives the history of the policy development to that point, as as well as an overview of the different stages and arguments involved in the policy formation.

24. See "New Kidney Point System Goes into Effect," *UNOS Update* 11 (March 1995): 12. UNOS point systems for hearts and livers have also generated some controversy, as will be evident in the discussion that follows. See "Heart Allocation Policy," *UNOS Update* 5 (January 1989): 1–2.

25. T. E. Starzl, R. Shaprio, and L. Teperman, "The Point System for Organ Distribution," *Transplantation Proceedings* 21 (June 1989): 3434.

26. Daniel Wikler, "Equity, Efficacy, and the Point System for Transplant Recipient Selection," *Transplantation Proceedings* 21 (June 1989): 3437.

27. Dan W. Brock, "Ethical Issues in Recipient Selection for Organ Transplantation," in *Organ Substitution Technology: Ethical, Legal, and Public Policy Issues*, ed. Deborah Mathieu (Boulder: Westview Press, 1988), pp. 86–99.

28. See Daniel Wikler's argument in "Equity, Efficacy, and the Point System for Transplant Recipient Selection."

29. See ibid., and Martin Benjamin, "Value Conflicts in Organ Allocation," *Transplantation Proceedings* 21 (June 1989): 3378–80. For further discussion, see the same issue, p. 3413ff., particularly the comments by Ruth Macklin.

30. See Olga Jonasson, "Waiting in Line: Should Selected Patients Ever Be Moved Up?" *Transplantation Proceedings* 21 (June 1989): 3390–94, and her comments in the discussion on p. 3413.

31. *Report of the Massachusetts Task Force on Organ Transplantation* (Boston: Department of Public Health, October 1984).

32. See Prottas, *Most Useful Gift*.

33. Task Force, *Organ Transplantation*, chaps. 4 and 5; Jonasson, "Waiting in Line," p. 3393.

34. Robert M. Veatch, "Allocating Organs by Utilitarianism Is Seen as Favoring Whites over Blacks," *Kennedy Institute of Ethics Newsletter* 3 (July 1989): 1 and 3.

35. See, for example, *1994 Annual Report*, V6–7 passim.

36. Robert M. Veatch, *Death, Dying and the Biological Revolution*, rev. ed. (New Haven: Yale University Press, 1989), p. 210.

37. C. R. Stiller, F. N. McKenzie, and W. J. Jostuk, "Cardiac Transplantation: Ethical and Economic Issues," *Transplantation Today* 2 (February 1985): 24.

38. George J. Annas, "No Cheers for Temporary Artificial Hearts," *Hastings Center Report* 15 (October 1985), and "Death and the Magic Machine: Informed Consent to the Artificial Heart," in *Organ Substitution Technology*, ed. Mathieu, pp. 257–76.

39. See "Heart Allocation Policy."

40. Jonasson, "Waiting in Line," p. 3391.

41. Ibid.

42. Task Force, *Organ Transplantation*.

43. Comments by A. P. Monaco in roundtable discussion, *Transplantation Proceedings* 21 (June 1989): 3418.

44. Dennis H. Novack et al., "Physicians' Attitudes toward Using Deception to Resolve Difficult Ethical Problems," *Journal of the American Medical Association* 261 (May 26, 1989): 2980–85.

45. Task Force, *Organ Transplantation*, pp. 88–89.

46. James F. Childress, "Who Shall Live When Not All Can Live?" *Soundings* 53 (1970): 339–55 (chap. 10). Contrast Childress, "Triage in Neonatal Intensive Care" (chap. 11, in revised form).

47. Jonasson, "Waiting in Line," p. 3392; Ruth Macklin, "Comment: Should Selected Patients Ever Be Moved Up?" *Transplantation Proceedings* 21 (June 1989): 3397. At least Jonasson's statement (and perhaps also Macklin's statement) can be construed as an objection to the *dominance* but not necessarily to the *relevance* of queuing, ceteris paribus.

48. See Task Force, *Organ Transplantation,* chap. 5. Contrast *Report of the Massachusetts Task Force on Organ Transplantation.*

49. Jonasson, "Waiting in Line," p. 3393.

50. For an argument for an equal-opportunity mechanism, such as first come, first served, when the differences in "medical suitability" are not "very substantial," see Council on Ethical and Judicial Affairs, American Medical Association, "Ethical Considerations in the Allocation of Organs and Other Scarce Medical Resources among Patients," *Archives of Internal Medicine* 155 (January 9, 1995): 38.

51. Council on Ethical and Judicial Affairs, "Ethical Considerations in the Allocation of Organs," p. 38; and Paul Menzel, "Rescuing Lives: Can't We Count?" *Hastings Center Report* 24 (January/February 1994): 22–23.

52. Task Force, *Organ Transplantation,* p. 90, drawn from Department of Health and Human Services, Office of Organ Transplantation, *Organ Transplantation Background Information,* February 1985.

53. See John Robertson, "Patient Selection for Organ Transplantation: Age, Incarceration, Family Support, and Other Social Factors," *Transplantation Proceedings* 21 (June 1989): 3401; A. P. Monaco, "Comment: A Transplant Surgeon's Views on Social Factors in Organ Transplantation," *Transplantation Proceedings* 21 (June 1989): 3406.

54. Council on Ethical and Judicial Affairs, "Ethical Considerations in the Allocation of Organs," pp. 37–38.

55. For a fuller discussion of some of these criteria, see Childress, "Some Moral Connections."

56. U.S. Department of Health and Human Services, Office of Analysis and Inspections, Office of the Inspector General, *The Access of Foreign Nationals to U.S. Cadaver Organs,* pub. no. OAI-01-86-00107 (Washington, D.C.: U.S. Department of Health and Human Services, 1986).

57. Task Force, *Organ Transplantation.*

58. U.S. DHHS, OAI, Office of the Inspector General, *The Access of Foreign Nationals to U.S. Cadaver Organs.*

59. Brock, "Ethical Issues in Recipient Selection for Organ Transplantation."

60. U.S. DHHS, OAI, Office of the Inspector General, *The Access of Foreign Nationals to U.S. Cadaver Organs,* p. 10.

61. United Network for Organ Sharing, *Policy Proposal Statement: UNOS Policies Regarding Transplantation of Foreign Nationals and Exportation and Importation of Organs* (Richmond, VA: UNOS, 1988).

62. Ibid.; see also UNOS, *Articles of Incorporation, By Laws, and Policies* (1988).

63. See Task Force on Organ Transplantation, *Report to the Secretary on Immuno-suppressive Therapies* (Washington, D.C.: U.S. Department of Health and Human Services, October 1985), pp. 1–3.

64. See Task Force, *Organ Transplantation,* chap. 5.

65. Ibid.; see also Childress, "Ethical Criteria for Procuring and Distributing Organs for Transplantation."

66. Contrast Norman Daniels, "Comment: Ability to Pay and Access to Transplantation," *Transplantation Proceedings* 21 (June 1989): 3434. For a sharp criticism see F. M. Kamm, "The Report of the U.S. Task Force on Organ Transplantation: Criticisms and Alternatives," *Mount Sinai Journal of Medicine* 56 (May 1989): 207–20. Kamm has a number of important insights but in the final analysis mistakenly imports a rigid philosophical model of argumentation into public justification, using such language as "proof" and downplaying the importance of various arguments that fall short of her standard of "proof."

67. See PL 98–507; see also Susan Denise, "Regulating the Sale of Human Organs," *Virginia Law Review* 71 (September 1985): 1015–38.

68. See Childress, "Some Moral Connections."

69. H. G. Welch and E. B. Larson, "Dealing with Limited Resources," pp. 171–73.

70. President's Commission for the Study of Ethical Problems in Medicine and Biomedical and Behavioral Research, *Securing Access to Health Care,* vol. 1 (Washington, D.C.: U.S. Government Printing Office, 1983), introduction and chap. 1.

71. H. Tristram Engelhardt, Jr., *Foundations of Bioethics* (New York: Oxford University Press, 1986), p. 369, n. 7.

72. Roger W. Evans, "Money Matters: Should Ability to Pay Ever Be a Consideration in Gaining Access to Transplantation?" *Transplantation Proceedings* 21 (June 1989): 3423; see Roger W. Evans, Christopher R. Blagg, and Fred A. Bryan, Jr., "Implications for Health Policy: A Social and Demographic Profile of Hemodialysis Patients in the United States," *Journal of the American Medical Association* 245 (1981): 478–91.

73. Beauchamp and Childress, *Principles of Biomedical Ethics.*

74. For a sketch of a model of balancing, which is, however, not fully consistent, see "The UNOS Statement of Principles and Objectives of Equitable Organ Allocation," *UNOS Update* (August 1994), pp. 20–38. For a strong argument against balancing medical utility against justice, with justice interpreted as requiring "opportunities for equality of health," see Robert M. Veatch's response to an earlier version of this chapter in "Equality, Justice, and Rightness in Allocating Health Care: A Response to James Childress," in *A Time to Be Born and a Time to Die: The Ethics of Choice,* ed. Barry S. Kogan (New York: Aldine de Gruyter, 1991), pp. 205–16. My reasons for rejecting much of Veatch's position appear in the argument for medical utility in the previous chapter as well as in the overall argument of this chapter, even though I do not directly address his arguments.

75. Norman Daniels, "Comment: Ability to Pay and Access to Transplantation," p. 3424. For a fuller presentation of his position, see Daniels, *Just Health Care* (Cambridge: Cambridge University Press, 1985), which has greatly influenced these concluding remarks.

76. Guido Calabresi and Philip Bobbitt, *Tragic Choices* (New York: Norton, 1977).

13. RIGHTS TO HEALTH CARE IN A DEMOCRATIC SOCIETY

1. Daniel Yankelovich and John Immerwahr, "A Perception Gap," *Health Management Quarterly* 13 (Third Quarter 1991): 11–14.

2. For a similar distinction between two uses of the term "rationing," see Henry J. Aaron and William B. Schwartz, "Rationing Health Care: The Choice before Us," *Science* 247 (January 26, 1990): 418. For a helpful analysis of the on-going debate about "rationing," see Robert Baker, "Rationing, Rhetoric, and Rationality: A Review of the Health Care Rationing Debate in America and Europe," in *Allocating Health Care Resources, Biomedical Ethics Reviews 1994,* ed. James M. Humber and Robert F. Almeder (Totowa, NJ: Humana Press, 1995), pp. 55–84.

3. Henry J. Aaron and William B. Schwartz, *The Painful Prescription: Rationing Hospital Care* (Washington, D.C.: Brookings Institution, 1984).

4. In this chapter I use the phrase "health care" mainly to refer to services provided by physicians and other health care professionals. Medicine and nursing are primary components of health care.

5. Another major issue is the weight or stringency of the obligations to provide or the right to receive health care, but I will not address this issue, assuming against

libertarians that it is sometimes justified for a democratic society to take individuals' property through taxation in order to benefit other individuals in the society. Thus, I will not examine the weight of various liberties that many claim as rights.

6. See Charles Fried, "Rights and Medical Care—Beyond Equity and Efficiency," *New England Journal of Medicine* 293 (July 31, 1975): 242.

7. Report of the President's Commission for the Study of Ethical Problems in Medicine and Biomedical and Behavioral Research, *Securing Access to Health Care*, vol. 1 (Washington, D.C.: U.S. Government Printing Office, March 1983), p. 34. Former Commission staff philosopher Dan Brock notes that the first draft, for which he had responsibility, endorsed a right to health care, but that the Commission changed this language to societal obligation, in part because "in the real world of politics and policy in the United States in the 1980s and 90s, the strongest moral claims, and in particular the strongest claims grounded in justice, were and are typically expressed in terms of rights, not obligations." For Brock, these strongest moral claims better recognize "the moral status of the victim of a failure to fulfill the obligation or respect the right." See Brock, "The President's Commission on the Right to Health Care," in *Health Care Reform: A Human Rights Approach,* ed. Audrey R. Chapman (Washington, D.C.: Georgetown University Press, 1994), pp. 69, 79.

8. For an analysis of rights as justified claims, see Tom L. Beauchamp and James F. Childress, *Principles of Biomedical Ethics*, 3rd ed. (New York: Oxford University Press, 1989), chap. 2. Joel Feinberg prefers the narrower term "validity" over the broader term "justification." Validity is justification within a system of rules. See Feinberg, *Social Philosophy* (Englewood Cliffs: Prentice-Hall, 1973), p. 67.

9. David Braybrooke, "The Firm but Untidy Correlativity of Rights and Obligations," *Canadian Journal of Philosophy* 1 (March 1972): 351–63. I will use the terms "obligation" and "duty" interchangeably in this chapter.

10. Dan W. Brock and Norman Daniels, "Ethical Foundations of the Clinton Administration's Proposed Health Care System," *Journal of the American Medical Association* 271 (April 20, 1994): 1189. See also Charles J. Dougherty, "Ethical Values at Stake in Health Care Reform," *Journal of the American Medical Association* 268 (November 4, 1992): 84–107; and Reinhard Preister, "A Values Framework for Health System Reform," *Health Affairs* 11 (Spring 1992): 84–107.

11. See the discussions in Audrey R. Chapman, ed., *Health Care Reform: A Human Rights Approach.*

12. Humphrey Taylor and Uwe E. Reinhardt, "Does the System Fit?" *Health Management Quarterly* 13 (Third Quarter 1991): 2–10.

13. William Temple, *Citizen and Churchmen* (1941), as quoted by Ronald Preston, "Welfare State," in *The Westminster Dictionary of Christian Ethics,* 2nd ed., ed. James F. Childress and John Macquarrie (Philadelphia: Westminster Press, 1986), pp. 657–58.

14. H. Tristram Engelhardt, Jr., "Health Care Allocations: Responses to the Unjust, the Unfortunate, and the Undesirable," in *Justice and Health Care, Philosophy and Medicine* 8, ed. Earl E. Shelp (Dordrecht: D. Reidel, 1981), pp. 121–37. His most recent statement of this view appears in *The Foundations of Bioethics,* 2nd ed. (New York: Oxford University Press, 1995), chap. 8.

15. I do not suppose that these examples are all uncontroversial. They are rather intended to indicate how specific rights to health care can be established within the society, apart from a general right. For a couple of these, see Allen E. Buchanan, "The Right to a Decent Minimum of Health Care," *Philosophy and Public Affairs* 13 (Winter 1984): 66–67.

16. Gene Outka, Letter to the Editor, *Perspectives in Biology and Medicine* 19 (Spring 1976): 452.

17. Gene Outka, "Social Justice and Equal Access to Health Care," *Journal of Religious Ethics* 2, no. 1 (1974): 22.

18. See President's Commission, *Securing Access to Health Care,* vol. 1, pp. 23–24, and David Gauthier, "Unequal Need: A Problem of Equity in Access to Health Care," *Securing Access to Health Care,* vol. 2, appendix H (Washington, D.C.: U.S. Government Printing Office, March 1983).

19. Norman Daniels, "Equity of Access to Health Care: Some Conceptual and Ethical Issues," *Milbank Memorial Fund Quarterly/Health and Society* 60, no. 1 (1982), and *Securing Access to Health Care,* vol. 2, appendix B, as well as several of his other essays, some of which appear in *Just Health Care* (New York: Cambridge University Press, 1985).

20. L. H. LaRue, "A Comment on Fried, Summers, and the Value of Life," *Cornell Law Review* 57 (1972): 630.

21. *Securing Access to Health Care,* vol. 1, p. 17.

22. See Henk Ten Have and Helen Keasberry, "Equity and Solidarity: The Context of Health Care in the Netherlands," *Journal of Medicine and Philosophy* 17 (August 1992): 463–77.

23. Larry R. Churchill, *Rationing Health Care in America: Perceptions and Principles of Justice* (Notre Dame: Notre Dame University Press, 1987).

24. For possible connections between justice and care in feminist ethics, see Virginia Held, ed., *Justice and Care: Essential Readings in Feminist Ethics* (Boulder: Westview Press, 1995).

25. Buchanan, "The Right to a Decent Minimum of Health Care" (1984), pp. 69 and 72.

26. Buchanan, "The Right to a Decent Minimum of Health Care" (1983), *Securing Access to Health Care,* vol. 2, appendix I, p. 234.

27. Another argument for a political-legal right to health care is utilitarian. Although utilitarianism may not recognize moral rights at all, it may support a legal right to health care in some contexts. In determining whether such a right would probably produce a net balance of aggregate or average welfare, utilitarianism could consider the value of health and the instrumental value of health care, as measured either by objective values or by subjective preferences.

28. Gary E. Jones, "The Right to Health Care and the State," *The Philosophical Quarterly* 33 (July 1983): 278–87.

29. See Loren E. Lomasky, "Medical Progress and National Health Care," *Philosophy and Public Affairs* 10 (1980): 72–73.

30. Michael Walzer, *Spheres of Justice: A Defense of Pluralism and Equality* (New York: Basic Books, 1983), pp. 86–94.

31. See Ronald Dworkin's review in *New York Review of Books,* April 14, 1983, and the exchange with Walzer in *New York Review of Books,* July 21, 1983.

32. See Daniels, *Just Health Care,* and Alan Gibbard, "The Prospective Pareto Principle and Equity of Access to Health Care," in *Securing Access to Health Care,* vol. 2, appendix G.

33. Ronald Dworkin, "Will Clinton's Plan Be Fair?" *New York Review of Books,* January 13, 1994, pp. 20–25. See also Dworkin, "Justice in the Distribution of Health Care," *McGill Law Journal* 38 (1993). According to Dworkin, the "prudent insurance test" offers an alternative to the "rescue principle," which treats health care as different from all other goods because life and health are chief among all goods, and it must be distributed equally even if no other goods are so distributed.

34. For a careful critique of several hypothetical choice models, see Madison Powers, "Hypothetical Choice Approaches to Health Care Allocation," in *Allocating Health Care Resources,* pp. 145–77. Powers contends that hypothetical choice models suffer from several flaws, including the "indeterminacy of prudential rationality, added to the lack of decisive reasons in favor of one account of the circumstances of hypothetical choice and its decision rule. . . ."

35. Larry R. Churchill, *Self-interest and Universal Health Care: Why Well-Insured*

Americans Should Support Coverage for Everyone (Cambridge: Harvard University Press, 1994). There is clearly a difference of emphasis, if not fundamental tension, between his approach in this work and his earlier, more communitarian argument in *Rationing Health Care in America.*

36. *Securing Access to Health Care,* vol. 1, p. 4.

37. Daniel Wikler, "Philosophical Perspective on Access to Health Care: An Introduction," *Securing Access to Health Care,* vol. 2, appendix F, p. 114.

38. For a fuller discussion of the implications of various theories of justice (libertarian, egalitarian, communitarian, and utilitarian) for debates about health care, see Beauchamp and Childress, *Principles of Biomedical Ethics,* 4th ed. (New York: Oxford University Press, 1994), chap. 6.

39. Paul Starr, "Medical Care and the Pursuit of Equality in America," *Securing Access to Health Care,* vol. 2, appendix A, p. 7.

40. Amy Gutmann, "For and against Equal Access to Health Care," *Securing Access to Health Care,* vol. 2, appendix C, p. 52.

41. Robert Baker, "Rationing, Rhetoric, and Rationality: A Review of the Health Care Rationing Debate in American and Europe," in *Allocating Health Care Resources,* pp. 72–73.

42. See, e.g., Outka, "Social Justice and Equal Access to Health Care." Buchanan notes that the slogan, "'To each according to his need,' looks more like a principle describing distribution that would be possible if there were no problem of scarcity than a prescriptive principle of distributive justice designed to cope with the problem of scarcity" (Buchanan, "The Right to a Decent Minimum," pp. 230–31).

43. On "moral traces," see Robert Nozick, "Moral Complications and Moral Structures," *Natural Law Forum* 13 (1968): 1–50.

44. Ronald Dworkin, *Taking Rights Seriously* (Cambridge: Harvard University Press, 1977).

45. See Charles Fried, *Right and Wrong* (Cambridge: Harvard University Press, 1978). Fried also recognizes that there are nonpaternalistic reasons for health care or vouchers, instead of money.

46. President's Commission, *Securing Access to Health Care,* vol. 1.

47. Wikler, "Philosophical Perspective on Access to Health Care: An Introduction," p. 118.

48. Daniels, "Equity of Access to Health Care," *Securing Access to Health Care,* vol. 2, appendix B.

49. Tom L. Beauchamp and Ruth R. Faden appear to use cost-benefit analysis for both purposes; see "The Right to Health and the Right to Health Care," *Journal of Medicine and Philosophy* 4 (June 1979): 118–31. For a fuller assessment of these analytic techniques, see Beauchamp and Childress, *Principles of Biomedical Ethics,* 4th ed., chap. 5.

50. This case is presented in Marc D. Basson, ed., *Rights and Responsibilities in Modern Medicine: The Second Volume in a Series on Ethics, Humanism, and Medicine* (New York: Alan R. Liss, 1980), where it is discussed by Norman Daniels and others. For a thorough and helpful analysis of "costworthy" care, see Paul T. Menzel, *Medical Costs, Moral Choices: A Philosophy of Health Care Economics in America* (New Haven: Yale University Press, 1983).

51. See the essays in Steven E. Rhoads, ed., *Valuing Life: Public Policy Dilemmas* (Boulder: Westview Press, 1980), and the discussion in Beauchamp and Childress, *Principles of Biomedical Ethics.*

52. *Securing Access to Health Care,* vol. 1, p. 42.

53. For reasons of space I have not considered some important variables of process that influence what people view as adequate health care, for example, the humanistic qualities of medicine. I do not dismiss these variables as mere amenities or "extras," for they constitute an important part of health care and frequently provide important

benefits. Furthermore, since they often involve more time with patients, they will be significantly affected by allocation policies.

54. See, for instance, Norman Daniels, "Is the Oregon Rationing Plan Fair?" *Journal of the American Medical Association* 265 (May 1, 1991): 2232–35, and the discussion in Beauchamp and Childress, *Principles of Biomedical Ethics*, 4th ed., pp. 366–69.

55. Eli Ginsberg, "Cost Containment—Imaginary and Real," *New England Journal of Medicine* 308 (May 19, 1983): 1223.

56. See Howard M. Leichter, "Public Policy and the British Experience," *Hastings Center Report* 11 (October 1981): 32–39, which was incorporated into his *Free to Be Foolish: Politics and Health Promotion in the United States and Great Britain* (Princeton: Princeton University Press, 1991). See also Robert Crawford, "You Are Dangerous to Your Health: The Ideology and Politics of Victim Blaming," *International Journal of Health Services* 7 (1977): 663–80.

57. Ginsberg, "Cost Containment—Imaginary and Real," p. 1223.

58. Leichter, "Public Policy and the British Experience," p. 38.

59. See Robert M. Veatch, "Voluntary Risks to Health: The Ethical Issues," *Journal of the American Medical Association* 243 (January 4, 1980): 54.

60. For a fuller discussion, see James F. Childress, "Rights to Health Care in a Democratic Society," in *Biomedical Ethics Reviews,* vol. 2, ed. Robert Almeder and James Humber (Totowa, NJ: Humana Press, 1984), on which this whole chapter is based. See also Childress, *Who Should Decide? Paternalism in Health Care* (New York: Oxford University Press, 1982).

61. There has been a tendency, especially among proponents of equal access, to consider access separately from the other questions and to treat the other questions as "nonaccess" questions. The interrelation of the questions can be seen when we consider some of the major issues in macroallocation (in contrast to the microallocation issues raised when I discussed rationing above). First, how much of the society's budget should go to health care and how much to other social goods? Equal access, as we have seen, may accept too little or too much, as long as it is equally distributed. If, however, there is a right to a decent minimum or an adequate level of health care, then there should be at least enough in the health care budget to provide this level. Second, how much of the society's budget should go into prevention and how much into critical care (and also chronic care)? For analysis of these priority questions, see James F. Childress, "Priorities in the Allocation of Health Care Resources," *Soundings* 62 (Fall 1979): 256–74.

62. Paul F. Camenisch, "The Right to Health Care: A Contractual Approach," *Soundings* 62 (Fall 1979): 293–310.

63. Victor Sidel, "The Right to Health Care: An International Perspective," in *Bioethics and Human Rights: A Reader for Health Professionals,* ed. Elsie L. Bandman and Bertram Bandman (Boston: Little, Brown and Company, 1978), p. 347.

64. See A. I. Melden, *Rights and Right Conduct* (Oxford: Basil Blackwell, 1959).

65. Although this distinction between obligation-meeting and obligatory overlaps the traditional distinction between imperfect and perfect obligations, it also covers an area not covered by the traditional distinction.

66. Mark Siegler, "A Physician's Perspective on a Right to Health Care," *Journal of the American Medical Association* 244 (October 1980): 1591–96. An excellent discussion of physicians' obligations appears in Norman Daniels, "What Is the Obligation of the Medical Profession in the Distribution of Health Care?" *Social Science and Medicine,* vol. 15F (1981): 129–33.

67. See the arguments offered by William F. May, "Code and Covenant of Philanthropy and Contract," in *Ethics in Medicine: Historical Perspectives and Contemporary Concerns,* ed. Stanley Joel Reiser, Arthur J. Dyck, and William J. Curran (Cambridge: MIT Press, 1977), pp. 65–76.

68. *Securing Access to Health Care,* vol. 1, pp. 4, 22 and passim.

69. Ibid., p. 5.

70. See Paul Starr, "Medical Care and the Pursuit of Equality in America," *Securing Access to Health Care,* vol. 2, p. 22, for the metaphor of "Medicaid-private insurance corridor." For a debate about physicians' special duties, see Mark Siegler and Harry Schwartz, "Treating the Jobless for Free: Do Doctors Have a Special Duty?" *Hastings Center Report* 13 (August 1983): 12–13.

71. This section of this chapter draws liberally from Carolyn Engelhard and James F. Childress, "Caveat Emptor: The Cost of Managed Care," *Trends in Health Care, Law and Ethics* 10 (Winter/Spring 1995): 11–14. I am grateful to Carolyn Engelhard for her collaboration and her permission to draw from our coauthored essay.

72. See, for example, K. M. Langwell, V. S. Staines, and N. Gordon, *The Effects of Managed Care on Use and Costs of Health Services and the Potential Impact of Certain Forms of Managed Care on Health Care Expenditures* (Washington, D.C.: Congressional Budget Office, 1992); A. C. Enthoven, "Why Managed Care Has Failed to Contain Costs," *Health Affairs* 12, no. 3 (1993): 27–43; R. H. Miller and H. S. Luft, "Managed Care Plan Performance since 1980: A Literature Analysis," *Journal of the American Medical Association* 271 (1994): 1512–19.

73. John K. Iglehart, "The American Health Care System: Managed Care," *New England Journal of Medicine* 327 (1992): 742–47.

74. J. Gabel et al., "The Health Insurance Picture in 1993: Some Rare Good News," *Health Affairs* 13 (1994): 327–35.

75. See George J. Annas, "Reframing the Debate on Health Care Reform by Replacing Our Metaphors," *New England Journal of Medicine* 332 (March 16, 1995): 744–47, which is discussed in chapter 1.

76. See, for example, D. P. Sulmasy, "Physicians, Cost Control, and Ethics," *Annals of Internal Medicine* 116 (1992): 920–26; Ezekiel J. Emanuel and Nancy N. Dubler, "Preserving the Physician-Patient Relationship in the Era of Managed Care," *Journal of the American Medical Association* 273 (1995): 323–29; Council on Ethical and Judicial Affairs of the American Medical Association, "Ethical Issues in Managed Care," *Journal of the American Medical Association* 273 (1995): 330–35.

77. E. Haavi Morreim, *Balancing Act: The New Medical Ethics of Medicine's New Economics* (Dordrecht: Kluwer, 1991).

78. Robert M. Veatch, "Healthcare Rationing through Global Budgeting: The Ethical Choices," *Journal of Clinical Ethics* 5 (Winter 1994): 276.

79. Norman Daniels, "Why Saying No to Patients in the United States Is So Hard: Cost Containment, Justice, and Provider Autonomy," *New England Journal of Medicine* 314 (1986): 1381–83.

80. David Mechanic, *From Advocacy to Allocation: The Evolving American Health Care System* (New York: Free Press, 1986), and "Managed Care: Rhetoric and Realities," *Inquiry* 31 (1994): 124–28.

81. Henry J. Kaiser Family Foundation, Commonwealth Fund, Louis Harris and Associates Poll (Storrs, CT: Roper Center for Public Opinion Research, August 6, 1993).

82. R. J. Blendon, M. Brodie, and J. Benson, "What Should Be Done Now That National Health System Reform Is Dead?" *Journal of the American Medical Association* 273 (1995): 243–44.

83. E. Haavi Morreim, "Ethical Issues in Managed Care: Economic Roots, Economic Resolutions," *Managed Care Medicine* 1, no. 6 (1994): 52–55.

84. S. Butler, "What to Do Now on Health Care System Reform," *Journal of the American Medical Association* 273 (1995): 253–54.

85. Henry J. Kaiser Family Foundation, Commonwealth Fund, Louis Harris and Associates Poll, August 6, 1993.

14. ETHICAL CRITERIA FOR POLICIES TO OBTAIN ORGANS FOR TRANSPLANTATION

1. National Conference of Commissioners on Uniform State Laws, Uniform Anatomical Gift Act, 1987 (Chicago: NCCUSL, 1987). Stephen R. Munzer defines "body parts" to include "any organs, tissues, fluids, cells, or genetic material on the contours of or within the human body, or removed from it, except for waste products such as urine and feces." See Munzer, "An Uneasy Case against Property Rights in Body Parts," in *Property Rights,* ed. Ellen Frankel Paul, Fred D. Miller, Jr., and Jeffrey Paul (New York: Cambridge University Press, 1994), p. 260.

2. Stuart J. Youngner, "Psychological Impediments to Procurement," *Transplantation Proceedings* 24, no. 5 (1992): 2159–61.

3. For these points about language, see James F. Childress, "Ethical Criteria for Procuring and Distributing Organs for Transplantation," in *Organ Transplantation Policy: Issues and Prospects,* ed. James F. Blumstein and Frank A. Sloan (Durham: Duke University Press, 1989), pp. 87–113.

4. Paul Ramsey, *The Patient as Person* (New Haven: Yale University Press, 1970), chap. 5.

5. See Tom L. Beauchamp and James F. Childress, *Principles of Biomedical Ethics,* 3rd ed. (New York: Oxford University Press, 1989). See also National Commission for the Protection of Human Subjects of Biomedical and Behavioral Research, *The Belmont Report: Ethical Guidelines for the Protection of Human Subjects of Research,* Washington, D.C.: U.S. Department of Health, Education and Welfare, pub. no. (OS) 78–0012.

6. James F. Childress, "Some Moral Connections between Organ Procurement and Organ Distribution," *Journal of Contemporary Health Law and Policy* 3 (1987): 85–110.

7. U.S. Congress, Office of Technology Assessment, *New Developments in Biotechnology: Ownership of Human Tissue and Cells—Special Report,* OTA-BA-337 (Washington, D.C.: U.S. Government Printing Office, March 1987); P. Matthews, "Whose Body? People as Property," *Current Legal Problems 1983* (1983): 193–239.

8. National Conference of Commissioners on Uniform State Laws, Uniform Anatomical Gift Act, 1968 (Chicago: NCCUSL, 1968).

9. Ramsey, *Patient as Person,* chap. 5.

10. "Most in U.S. Found Willing to Donate Organs," *New York Times,* January 17, 1968, p. 18.

11. Jeffrey M. Prottas, "Organ Procurement in Europe and the United States," *Milbank Memorial Fund Quarterly/Health and Society* 63 (1985): 94–126.

12. Gallup Organization, *The U.S. Public's Attitudes toward Organ Transplants/ Organ Donation* (Princeton: Gallup Organization, 1985, 1986, 1987).

13. Gallup Organization, *U.S. Public's Attitudes toward Organ Transplants/ Organ Donation,* 1985.

14. Ibid.; Clive Callender et al., "Attitudes among Blacks toward Donating Kidneys for Transplantation: A Pilot Project," *Journal of the National Medical Association* 74 (1982): 807–809; Task Force on Organ Transplantation, *Organ Transplantation: Issues and Recommendations* (Washington, D.C.: U.S. Department of Health and Human Services, 1986).

15. Task Force on Organ Transplantation, *Organ Transplantation*; Prottas, "Organ Procurement in Europe and the United States."

16. National Conference of Commissioners on Uniform State Laws, Uniform Anatomical Gift Act, 1987.

17. Judith Areen, "A Scarcity of Organs," *Journal of Legal Education* 38 (December 1988): 555–65.

18. American Medical Association (AMA) Council on Ethical and Judicial Affairs, "Strategies for Cadaveric Organ Procurement: Mandated Choice and Presumed Consent," *Journal of the American Medical Association* 272 (1994): 809. For another proposal of mandated choice, see Robert M. Veatch, *Death, Dying and the Biological Revolution,* rev. ed. (New Haven: Yale University Press, 1989), pp. 211–33.

19. Laura Siminoff et al., "Public Policy Governing Organ and Tissue Procurement in the United States," *Annals of Internal Medicine* 123 (1995): 10–17.

20. I develop this line of argument in the 1995 Andre Hellegers Memorial Lecture at the Kennedy Institute of Ethics at Georgetown University (forthcoming).

21. AMA Council, "Strategies for Cadaveric Organ Procurement," p. 810.

22. Ibid.

23. Ibid.

24. P. M. Quay, "Utilizing the Bodies of the Dead," *St. Louis University Law Journal* 28 (1984): 890–927.

25. Gallup Organization, *U.S. Public's Attitudes toward Organ Transplants/Organ Donation,* 1985, 1986, 1987.

26. Gallup Organization, *U.S. Public's Attitudes toward Organ Transplants/Organ Donation,* 1985.

27. Arthur L. Caplan, "Ethical and Policy Issues in the Procurement of Cadaver Organs for Transplantation," *New England Journal of Medicine* 314 (1984): 981–83.

28. Jeffrey M. Prottas and Helen Levine Batten, "Health Professionals and Hospital Administrators in Organ Procurement: Attitudes, Reservations, and Their Resolutions," *American Journal of Public Health* 78 (1988): 642–45.

29. Arthur L. Caplan, "Professional Arrogance and Public Misunderstanding," *Hastings Center Report* 18 (April/May 1988): 34–37.

30. George J. Annas, "The Paradoxes of Organ Transplantation," *American Journal of Public Health* 78 (1988): 621–22.

31. S. Martyn, R. Wright, and L. Clark, "Required Request for Organ Donation: Moral, Clinical, and Legal Problems," *Hastings Center Report* 18 (April/May 1988): 27–34.

32. Caplan, "Professional Arrogance and Public Misunderstanding."

33. Martyn, Wright, and Clark, "Required Request for Organ Donation."

34. Beth A. Virnig and Arthur L. Caplan, "Required Request: What Difference Has It Made?" *Transplantation Proceedings* 24 (October 1992): 2156.

35. Jeffrey M. Prottas and Helen Levine Batten, "The Willingness to Give: The Public and the Supply of Transplantable Organs," *Journal of Health Politics, Policy and Law* 16 (1991): 121–34.

36. Arthur L. Caplan, Laura Siminoff, Robert Arnold, and Beth A. Virnig, "Increasing Organ Donation and Tissue Donation: What Are the Obstacles, What Are Our Options?" in *The Surgeon General's Workshop on Increasing Organ Donation: Background Papers, Washington, D.C., July 8–10, 1991* (Washington, D.C.: U.S. Department of Health and Human Services, Public Health Service, 1992).

37. Ramsey, *Patient as Person.*

38. Alexander Capron, "Anencephalic Donors: Separate the Dead from the Dying," *Hastings Center Report* 17 (February 1987): 5–9; and "Ethical and Legal Issues in Organ Replacement," in *Organ Transplants: Hearings before the Subcommittee on Investigations and Oversight of the Committee on Science and Technology, U.S. House of Representatives,* 98th Cong., 1st sess., Washington, D.C.: U.S. Government Printing Office, 1983.

39. Walter Land and B. Cohen, "Postmortem and Living Organ Donation in Europe: Transplant Laws and Activities," *Transplantation Proceedings* 24 (1992): 2165–67.

40. F. P. Stuart, F. J. Veith, and R. E. Cranford, "Brain Death Laws and Patterns of Consent to Remove Organs for Transplantation from Cadavers in the United States and 28 Other Countries," *Transplantation* 31 (1981): 238–44.

41. A. John Simmons, "Tacit Consent and Political Obligation," *Philosophy and Public Affairs* 5 (1976): 274–95.

42. Prottas, "Organ Procurement in Europe and the United States."

43. Arthur Caplan, "Organ Procurement: It's Not in the Cards," *Hastings Center Report* 14 (October 1984): 9–12.

44. James L. Muyskens, "An Alternative Policy for Obtaining Cadaver Organs for Transplantation," *Philosophy and Public Affairs* 8 (1978): 88–99.

45. Ramsey, *Patient as Person.*

46. Gallup Organization, *U.S. Public's Attitudes toward Organ Transplants/Organ Donation,* 1985.

47. National Conference of Commissioners on Uniform State Laws, *Uniform Anatomical Gift Act,* 1987.

48. James Lindemann Nelson, "The Rights and Responsibilities of Potential Organ Donors: A Communitarian Approach," Washington, D.C.: Communitarian Network, 1992.

49. William F. May, "Attitudes toward the Newly Dead," *Hastings Center Studies* 1 (1973): 3–13.

50. Prottas, "Organ Procurement in Europe and the United States," and Caplan, "Ethical and Policy Issues."

51. Land and Cohen, "Postmortem and Living Organ Donation in Europe."

52. Theodore Silver, "The Case for a Post-Mortem Organ Draft and a Proposed Model Organ Act," *Boston University Law Review* 68 (1988): 681–728.

53. Albert R. Jonsen, "Transplantation of Fetal Tissue: An Ethicist's Viewpoint," *Clinical Research* 36 (1988): 215–19.

54. See Michael Walzer, *Just and Unjust Wars* (New York: Basic Books, 1977).

55. See *Moore v. Regents of the University of California,* Cal. App. 2d (1988) (88 Daily Journal D.A.R. 9520).

56. "Man's Heart Is Transplanted without Permission," *New York Times,* May 24, 1988, sec. 1, p. 22.

15. HUMAN BODY PARTS AS PROPERTY: AN ASSESSMENT OF OWNERSHIP, SALES, AND FINANCIAL INCENTIVES

1. National Conference of Commissioners on Uniform State Laws, Uniform Anatomical Gift Act (1968).

2. E. Stason, "The Uniform Anatomical Gift Act," *Business Lawyer* 23 (1968): 919–29, quoted in Henry Hansmann, "The Economics and Ethics of Markets for Human Organs," in *Organ Transplantation Policy: Issues and Prospects,* ed. James F. Blumstein and Frank A. Sloan (Durham: Duke University Press ,1989), pp. 57–85.

3. Hansmann, "The Economics and Ethics of Markets for Human Organs," p. 59.

4. U.S. Congress, House of Representatives, Committee on Science and Technology, Subcommittee on Investigations and Oversight, *Hearings on Organ Transplants,* 98th Cong., 1st sess., April 13, 14, 27, 1983 (Washington, D.C.: U.S. Government Printing Office, 1983), p. 373.

5. See U.S. Congress, House of Representatives, Committee on Energy and Commerce, Subcommittee on Health and the Environment, *Hearings on H.R. 4080, National Organ Transplant Act,* 98th Cong., 1st sess., July 29, October 17 and 31, 1983 (Washington, D.C.: U.S. Government Printing Office, 1984), and U.S. Congress, House of Representatives, Committee on Science and Technology, Subcommittee on Investigations and Oversight, *Hearings on Procurement and Allocation of Human Organs for Transplantation,* 98th Congress, 1st Sess., November 7 and 9, 1983 (Washington, D.C.: U.S. Government Printing Office, 1984).

6. Susan Denise, "Regulating the Sale of Human Organs," *Virginia Law Review* 71 (September 1985): 1015–38.

7. See James F. Childress, "Ethical Criteria for Procuring and Distributing Organs for Transplantation," in *Organ Transplantation Policy: Issues and Prospects,* ed. James F. Blumstein and Frank A. Sloan (Durham: Duke University Press, 1989), pp. 87–113.

8. These figures were obtained from the United Network for Organ Sharing (UNOS), the National Organ Procurement and Transplantation Network in Richmond, VA. The most recent data appear in *Annual Report of the U.S. Scientific Registry for Transplant Recipients and the Organ Procurement and Transplantation Network— Transplant Data: 1988–1993* (Richmond, VA: UNOS, and Bethesda, MD: Division of Organ Transplantation, Bureau of Health Resources Development, Health Resources and Services Administration, U.S. Department of Health and Human Services, 1994).

9. Hansmann, "Economics and Ethics of Markets for Human Organs."

10. Jeremy Waldron, *The Right to Private Property* (Oxford: Clarendon Press, 1988).

11. Lawrence C. Becker, *Property Rights: Philosophic Foundations* (London: Routledge and Kegan Paul, 1977).

12. Judith Jarvis Thomson, *The Realm of Rights* (Cambridge: Harvard University Press, 1990).

13. D. W. Meyers, *The Human Body and the Law,* 2nd ed. (Stanford: Stanford University Press, 1990).

14. Becker, *Property Rights: Philosophic Foundations.*

15. Michael Walzer, *Spheres of Justice* (New York: Basic Books, 1983).

16. George Mavrodes, "The Morality of Selling Organs," in *Ethics, Humanism, and Medicine,* ed. Marc D. Basson (New York: Alan R. Liss, 1980).

17. Paul Ramsey, *The Patient as Person* (New Haven: Yale University Press, 1970).

18. Lori Andrews, "The Body as Property: Some Philosophical Reflections—A Response to J. F. Childress," *Transplantation Proceedings* 24 (October 1992): 2149–51. For a fuller statement of Andrews's position, see her "My Body, My Property," *Hastings Center Report* 16 (October 1986): 28–38.

19. Andrews, "Body as Property," p. 2151.

20. H. Tristram Engelhardt, Jr., and Kevin W. Wildes, "Postmodernity and Limits on the Human Body: Libertarianism by Default," in *Medicine Unbound: The Human Body and the Limits of Medical Intervention,* ed. Robert H. Blank and Andrea L. Bonnicksen (New York: Columbia University Press, 1994), pp. 61–71.

21. See Thomas Murray, "On the Human Body as Property: The Meaning of Embodiment, Markets, and the Meaning of Strangers," *Journal of Law Reform* 20 (Summer 1987): 1055–88; William F. May, "Attitudes toward the Newly Dead," *Hastings Center Studies* 1 (1973): 3–13; May, "Religious Justifications for Donating Body Parts," *Hastings Center Report* 15 (February 1985): 38–42; and U.S. Congress, Office of Technology Assessment, *New Developments in Biotechnology: Ownership of Human Tissues and Cells—Special Report,* OTA-BA-337 (Washington, D.C.: U.S. Government Printing Office, March 1987).

22. Margaret Jane Radin, "Justice and the Market Domain," in *Markets and Justice,* Nomos Series XXXI, ed. John W. Chapman and J. Roland Pennock (New York: New York University Press, 1989), pp. 165–97.

23. Ibid., p. 187.

24. See R. A. Sells, "The Case against Buying Organs and a Futures Market in Transplants," *Transplantation Proceedings* 24 (October 1992): 2198–2202.

25. Hansmann, "The Economics and Ethics of Markets for Human Organs," p. 74.

26. Becker, *Property Rights: Philosophic Foundations,* p. 114, and see p. 129. See also Leon R. Kass, "Organs for Sale? Property, Propriety and the Price of Progress," *The Public Interest* (April 1992): 65–86.

27. Stephen Munzer, "An Uneasy Case against Property Rights in Body Parts," in *Property Rights*, ed. Ellen Frankel Paul, Fred D. Miller, Jr., and Jeffrey Paul (Cambridge: Cambridge University Press, 1994), p. 279.

28. Andrews, "Body as Property," p. 2150.

29. Joel Feinberg, "The Mistreatment of Dead Bodies," *Hastings Center Report* 15 (February 1985): 31–37.

30. National Conference of Commissioners on Uniform State Laws, Uniform Anatomical Gift Act (1987).

31. U.S. Congress, Office of Technology Assessment, *New Developments in Biotechnology: Ownership of Human Tissues and Cells.*

32. Munzer, "An Uneasy Case against Property Rights in Body Parts," pp. 259–89. All of the references to Munzer's arguments are to this essay, which, because of its intricate and subtle argument, deserves more attention than I am able to give it.

33. Lloyd R. Cohen, "Increasing the Supply of Transplant Organs: The Virtues of a Futures Market," *George Washington Law Review* 58 (1989): 1–51.

34. Hansmann, "Economics and Ethics of Markets for Human Organs," p. 67.

35. Gallup Organization, *The U.S. Public's Attitudes toward Organ Transplants/ Organ Donation* (Princeton: Gallup Organization, 1985), and *The U.S. Public's Attitudes toward Organ Transplants/Organ Donation* (Princeton: Gallup Organization, 1987). See also the discussion in the previous chapter of this volume.

36. Cohen, "Increasing the Supply of Transplant Organs: The Virtues of a Futures Market." Cohen's proposal is, as I note in the text, the most promising one for a (procurement) market, and I will later discuss variations that attempt to convert it into the provision of financial incentives for donation, in contrast to sales. Hansmann's proposal would provide a modest annual payment (perhaps around $30) to the seller, who in turn promises to provide body parts upon his or her death. The amount is heavily discounted because upon death the seller's parts may not be available (e.g., because of the nature of the death) or may not be usable even if available.

37. Radin, "Justice and the Market Domain."

38. Jeffrey M. Prottas, *The Most Useful Gift: Altruism and the Public Policy of Organ Transplants* (San Francisco: Jossey-Bass, 1994), p. 50. It is notoriously difficult to determine actual motivations in human action, for instance, to determine which motives were necessary and/or sufficient in particular human conduct. Nevertheless, many interpretations of our system of organ procurement assume that donations necessarily presuppose altruism, compassion, and the like. Along these lines Roger Evans writes: "Organ procurement activities in the United States are based on altruism. Out of their compassion for others, people agree to donate the organs of a deceased relative" ("Organ Procurement Expenditures and the Role of Financial Incentives," *Journal of the American Medical Association* 269 [June 23/30, 1993]: 3113). Evans's otherwise splendid analysis simply assumes what has not been established; for we do not actually know (and may be unable to determine) what motivates people to donate. At any rate, allowing only organ donations largely eliminates the possibility of financial motivations, but we cannot conclude that the only remaining motivation is altruism or compassion or that, even if present, altruism or compassion is sufficient in most cases.

39. See Evans, "Organ Procurement Expenditures and the Role of Financial Incentives," which argues that financial incentives would threaten the cost-effectiveness of transplantation by adding per-procedure expenditures while benefits remain unchanged.

40. The National Kidney Foundation, Inc., "Financial Incentives," Consensus Conference, Controversies in Organ Donation, February 25–26, 1991.

41. Ibid.

42. Ibid., and the Surgeon General's Workshop on Increasing Organ Donation, July 8–10, 1991, *Proceedings* (U.S. Department of Health and Human Services, Public Health Service, February 1982).

43. William F. May, *The Patient's Ordeal* (Bloomington: Indiana University Press, 1991), p. 181.

44. American Medical Association Council on Judicial and Ethical Affairs, "Financial Incentives for Organ Procurement," *Archives of Internal Medicine* 155 (March 27, 1995): 581–89. All references in this paragraph are to this publication.

45. Thomas G. Peters, "Life or Death: The Issue of Payment in Cadaveric Organ Donation," *Journal of the American Medical Association* 265 (1991): 1302–1304.

46. Dilip S. Kittur et al., "Incentives for Organ Donation?" *UNOS Update* (January 1992): 8–10 (first published in *The Lancet* in December 1991).

47. See Task Force on Organ Transplantation, *Organ Transplantation: Issues and Recommendations* (Washington, D.C.: U.S. Department of Health and Human Services, 1986).

48. National Commissioners on Uniform State Laws, Uniform Anatomical Gift Act (1987), section 6.

49. Task Force on Organ Transplantation, *Organ Transplantation: Issues and Recommendations*, p. 86.

50. This paper is a revised and expanded version of a paper originally delivered at a conference on organ transplantation sponsored by the Michigan Transplant Center in September 1991 and subsequently discussed at the Park Ridge Center in April 1992. The author is grateful for helpful criticisms and suggestions in both contexts, particularly from Lori Andrews and David Smith.

16. ETHICS, PUBLIC POLICY, AND HUMAN FETAL TISSUE TRANSPLANTATION RESEARCH

1. *Report of the Human Fetal Tissue Transplantation Research Panel,* Consultants to the Advisory Committee to the Director, National Institutes of Health, December 1988, 2 vols. References to this report, including concurring and dissenting statements, will appear in the text in parentheses; subsequent references to this report in the notes will appear as *Report.* This essay draws on my experiences as a member of that Panel, as a member of an Office of Technology Assessment Advisory Panel on New Developments in Neuroscience, and on some materials I developed in preparing a case study for the Institute of Medicine, from which I have drawn some ideas and formulations. That case study contains more information about the context, process and procedures of the HFTTR Panel. See "Deliberations of the Human Fetal Tissue Transplantation Research Panel," in *Biomedical Politics,* ed. Kathi Hanna (Washington, D.C.: National Academy Press, 1991), pp. 215–48. The basic structure and content of the current essay are drawn from my "Ethics, Public Policy, and Human Fetal Tissue Transplantation Research," *Kennedy Institute of Ethics Journal* 1 (June 1991): 93–121, with the addition of materials, particularly in the section on consensus, from my "Consensus in Ethics and Public Policy: The Deliberations of the U.S. Human Fetal Tissue Transplantation Research Panel," in *The Concept of Consensus: The Case of Technological Interventions into Human Reproduction,* ed. Kurt Bayertz (Dordrecht: Kluwer, 1995), and, very briefly, from my "Disassociation from Evil: The Case of Human Fetal Tissue Transplantation Research," *Social Responsibility: Business, Journalism, Law, Medicine,* vol. 16, ed. Louis W. Hodges (Lexington, VA: Washington and Lee University, 1990), pp. 32–49.

2. See Robert J. Levine, *Ethics and Regulation of Clinical Research,* 2nd ed. (Baltimore: Urban and Schwarzenberg, 1986), chap. 13.

3. For an excellent discussion of the use of fetal tissue in various contexts, see Dorothy E. Vawter et al., *The Use of Human Fetal Tissue: Scientific, Ethical, and Policy Concerns,* (Center for Biomedical Ethics, University of Minnesota, January 1990). See also Dorothy Lehrman, *Summary: Fetal Research and Fetal Tissue Research* (Washington, D.C.: Association of American Medical Colleges, 1988).

4. "Fact Sheet: NIH Fetal Tissue Research," prepared by OSPL-OD-NIH, July 22, 1988.

5. See Kathleen Nolan, "Genug ist Genug: A Fetus Is Not a Kidney," *Hastings Center Report* 18 (December 1988): 13–19.

6. See Rebecca H. Buckley, "Fetal Thymus Transplantation for the Correction of Congenital Absence of the Thymus (DiGeorge's Syndrome)," HFTTR Panel, *Report,* vol. 2, pp. D50–57. See also Vawter et al., *Use of Human Fetal Tissue,* pp. 21–56.

7. See Vawter et al., *Use of Human Fetal Tissue,* pp. 57–82, and Dr. Barry J. Hoffer, "Summary of Current Literature," HFTTR Panel, *Report,* pp. 19–22.

8. Ibid.

9. James Wyngaarden, M.D., Memorandum to Robert E. Windom, "Approval to Perform Experimental Surgical Procedure at NIH Clinical Center—ACTION," October 23, 1987.

10. Advisory Committee to the Director, National Institutes of Health, *Human Fetal Tissue Transplantation Research,* December 14, 1988, Bethesda, MD.

11. Louis W. Sullivan, M.D., Secretary, DHHS, letter to Dr. William F. Raub, Acting Director, NIH, November 2, 1989.

12. Vawter et al., *Use of Human Fetal Tissue,* pp. 135–36.

13. See Daniel C. Maguire, "Cooperation with Evil," *The Westminster Dictionary of Christian Ethics,* 2nd ed., ed. James F. Childress and John Maquarrie (Philadelphia: Westminister Press, 1986).

14. See James Tunstead Burtchaell, "University Policy of Experimental Use of Aborted Fetal Tissue," *IRB: A Review of Human Subjects Research* 10 (July/August 1988): 7–11; Burtchaell, "The Use of Aborted Fetal Tissue in Research: A Rebuttal," *IRB: A Review of Human Subjects Research* 11 (March/April 1989): 9–12; James Bopp and James T. Burtchaell, "Human Fetal Tissue Transplantation Research Panel: Statement of Dissent," HFTTR Panel, *Report,* vol. 1, pp. 45–71. For responses on complicity, see John A. Robertson, "Rights, Symbolism, and Public Policy in Fetal Tissue Transplants," *Hastings Center Report* 18 (December 1988): 5–12; Robertson, "Fetal Tissue Transplant Research Is Ethical," *IRB: A Review of Human Subjects Research* 10 (November/December 1988): 5–8; Benjamin Freedman, "The Ethics of Using Human Fetal Tissue," *IRB: A Review of Human Subjects Research* 10 (November/December 1988): 1–4. For excellent analyses of issues surrounding complicity, see Vawter et al., *Use of Human Fetal Tissue,* esp. pp. 251–67; and Richard C. Sparks, C.S.P., "Ethical Issues of Fetal Tissue Transplantation Research, Procurement, and Complicity with Abortion," pp. 199–221 in *The Annual of the Society of Christian Ethics,* ed. Diane Yeager (Knoxville: Society of Christian Ethics, 1990). See also Childress, "Disassociation from Evil: The Case of Human Fetal Tissue Transplantation Research."

15. Vawter et al., *The Use of Human Fetal Tissue,* pp. 251–67.

16. Burtchaell, "University Policy on Experimental Use of Aborted Fetal Tissue," p. 9.

17. Burtchaell, "The Use of Aborted Fetal Tissue in Research: A Rebuttal," p. 10. See also Bopp and Burtchaell, "Human Fetal Tissue Transplantation Research Panel: Statement of Dissent," p. 68, and the discussion of this analogy and others in *Transcript of the Meeting of the Human Fetal Tissue Transplantation Research Panel,* September 14–16, 1988 (Bethesda, MD: National Institutes of Health, 1988), pp. 374–403.

18. HFTTR Panel, *Report,* vol. 2, p. E14 (emphasis added).

19. John Robertson, "Concurring Statement," *Report* (signed by ten other panelists), p. 31.

20. For a discussion of "association" and "disassociation," see Childress, "Disassociation from Evil." See also Maguire, "Cooperation with Evil"; Leslie Griffin, "The Problem of Dirty Hands," *Journal of Religious Ethics* 17 (Fall 1989): 31–61; and Thomas E. Hill, Jr., "Symbolic Protest and Calculated Silence," *Philosophy and Public Affairs* 9 (1979).

21. See Bopp and Burtchaell, "Human Fetal Tissue Transplantation Research Panel: Statement of Dissent," pp. 63–71.

22. Robertson, "Concurring Statement," pp. 32–33. See also Aaron A. Moscona, "Concurring Statement," *Report*, pp. 27–28.

23. For the debate about the use of Nazi data, see, for example, Michael Weisskoph, "EPA Bars Use of Nazis' Human Test Data after Scientists Object," *Washington Post*, March 24, 1988, p. A10; and Mark Weitzman, "The Ethics of Using Nazi Medical Data: A Jewish Perspective," *Second Opinion: Health, Faith, and Ethics* 14 (July 1990): 26–38.

24. Burtchaell, "University Policy on Experimental Use of Aborted Fetal Tissue," p. 9.

25. *Transcript of the Meeting of the Human Fetal Tissue Transplantation Research Panel*, September 14–16, 1988, pp. 688–704 and 723.

26. Burtchaell, "University Policy on Experimental Use of Aborted Fetal Tissue," p. 9.

27. Vawter et al., *Use of Human Fetal Tissue*, p. 259, which also provides a good analysis of legitimation charges.

28. Louis W. Sullivan, M.D., Secretary, DHHS, letter to Dr. William F. Raub, Acting Director, NIH, November 2, 1989.

29. John Fletcher, "Moratorium on Federal Support for Fetal Tissue Transplant Research," testimony before the Subcommittee on Health and the Environment, Committee on Energy and Commerce, U.S. House of Representatives, April 2, 1990.

30. Judge Arlin B. Adams, "Concurring Statement," *Report*, pp. 25–28.

31. Daniel N. Robinson, letter to Dr. Jay Moskowitz, *Report*, p. 73.

32. Nolan, "Genug ist Genug."

33. 42 USC 274e.

34. Vawter et al., *Use of Human Fetal Tissue*, p. 175.

35. Larry Thompson, "Fetal Cells Hold the Most Promise—and Raise Ethical Questions," *Washington Post Health*, July 14, 1987.

36. Louis W. Sullivan, M.D., Secretary, DHHS, letter to Dr. William F. Raub, Acting Director, NIH, November 2, 1989. See also Rabbi J. David Bleich's dissenting statement, "Fetal Tissue Research and Public Policy," *Report*, p. 40.

37. See Vawter et al., *Use of Human Fetal Tissue*, p. 261.

38. C. D. Broad, "Conscience and Conscientious Action," in *Moral Concepts*, ed. Joel Feinberg (New York: Oxford University Press, 1970), pp. 74–79.

39. See *Transcript of the Meeting of the Human Fetal Tissue Transplantation Research Panel*, September 14–16, 1988, pp. 420, 478, 780–81, 791–95, 800–802 and passim. See also Robin Duke's presentation, *Report*, vol. 2, pp. D113–21. For an analysis see Vawter et al., *Use of Human Fetal Tissue*, esp. pp. 261–67.

40. James O. Mason, "Should the Fetal Tissue Research Ban Be Lifted?" *Journal of NIH Research* 2 (January/February 1990): 17–18.

41. Ibid.

42. For a fuller analysis, see James F. Childress, "Ethical Criteria for Procuring and Distributing Organs for Transplantation," *Journal of Health Politics, Policy and Law* 14 (Spring 1989): 87–113, much of which appears in an earlier chapter in this volume. In that chapter, I also now draw a distinction between presumed donation and routine salvaging with the possiblity of opting out. The former rests on a presuppostion of individual or familial dispositional authority over cadaveric organs and tissues, while the latter rests on a presupposition of society's dispositional authority over those organs and tissues.

43. Vawter et al., *Use of Human Fetal Tissue*, p. 199.

44. See James F. Childress, "Statement to the Advisory Committee to the Director, NIH," *Human Fetal Tissue Transplantation Research*, Report of the Advisory Committee to the Director, National Institutes of Health, pp. C7–8. Contrast Kathleen Nolan's warning in "Genug ist Genug" about the dangers of viewing fetal

material in terms of gift and donation and her argument for the language of contribution.

45. See Burtchaell, "University Policy on Experimental Use of Aborted Fetal Tissue," pp. 8–9.

46. Vawter et al., *Use of Human Fetal Tissue,* p. 217.

47. For summaries of state laws and federal regulations, see Lori Andrews, "State Regulation of Human Fetal Tissue Transplantation"; Judith Areen, "Statement on Legal Regulation of Fetal Tissue Transplantation"; and David H. Smith et al., "Using Human Fetal Tissue for Transplantation and Research: Selected Issues," HFTTR Panel, *Report,* vol. 2. pp. D1–20, D21–26, F1–43. See also Patricia King and Judith Areen, "Legal Regulation of Fetal Tissue Transplantation," *Clinical Research* 36 (1988): 205–208; and Vawter et al., *Use of Human Fetal Tissue,* pp. 169–87.

48. Steven Maynard-Moody, *The Dilemma of the Fetus: Fetal Research, Medical Progress, and Moral Politics* (New York: St. Martin's Press, 1995), p. 151.

49. Committee to Review the Guidance on the Research Use of Fetuses and Fetal Material, *Review of the Guidance on the Research Use of Fetuses and Fetal Material,* July 1989 (London: Her Majesty's Stationery Office, 1989) (the Polkinghorne Report).

50. Vawter et al., *Use of Human Fetal Tissue,* p. 239f.

51. For other arguments, see ibid., p. 244f.

52. See J. Fox, "Overview of Panel Meetings," *Report,* vol. 2, appendix A1.

53. This was evident in subsequent congressional hearings. See U.S. Congress, House of Representatives, Committee on Energy and Commerce, Subcommittee on Health and the Environment, *Fetal Tissue Transplantation Research,* Hearing, 101st Cong., 2nd sess., April 2, 1990, serial no. 101–35, pp. 66, 81.

54. Advisory Committee, *Human Fetal Tissue Transplantation Research,* p. 4.

55. Ibid., p. C5, statement by LeRoy Walters; and Maynard-Moody, *Dilemma of the Fetus,* p. 156.

56. Peter Caws, "Committees and Consensus: How Many Heads Are Better than One?" *Journal of Medicine and Philosophy* 16 (1991): 375–91.

57. See Jonathan Moreno, "Consensus by Committee: Philosophical and Social Aspects of Ethics Committees," in *Concept of Consensus,* ed. Bayertz. See also Moreno, *Deciding Together: Bioethics and Moral Consensus* (New York: Oxford University Press, 1995).

58. U.S. Congress, *Fetal Tissue Transplantation Research,* p. 67.

59. Ibid., pp. 68–69; see also Advisory Committee, *Human Fetal Tissue Transplantation Research,* p. C5.

60. King, "Commentary," in *Biomedical Politics,* ed. Hanna, p. 250.

61. See *Transcript of the Meeting of the Human Fetal Tissue Transplantation Research Panel,* October 20–21, 1988, National Institutes of Health, Bethesda, MD, p. 140 and passim. See also Judge Adams's testimony before the Subcommittee on Health and the Environment, Committees on Energy and Commerce, U.S. House of Representatives, *Fetal Tissue Transplantation Research,* Hearing, 101st Cong., 2nd sess., April 2, 1990, serial no. 101–35.

62. Mason, "Should the Fetal Tissue Research Ban Be Lifted?" p. 17.

63. When I note deficiencies in the work of the Panel, I do not blame any particular individuals, and I certainly do not exempt myself from these criticisms. In addition, constraints on the Panel's collective deliberations included the limited time allowed, tight schedules, and limited staff, all of which may have reduced the Panel's effectiveness in contrast to that of some other national bodies.

64. See King, "Commentary," in *Biomedical Politics,* ed. Hanna, pp. 251–52.

65. *Transcript of the Meeting of the Human Fetal Tissue Transplantation Research Panel,* September 14–16, 1988, pp. 660–63.

66. Daniel Callahan, lecture and discussion at the Hastings Center, June 14, 1991.

67. Michael Walzer, *Interpretation and Social Criticism* (Cambridge: Harvard University Press, 1987), p. 89.

68. Walter Harrelson, "Commentary," in *Biomedical Politics,* ed. Hanna, p. 256.

69. See Bopp and Burtchaell's dissent, *Report,* p. 70; Mason, "Should the Fetal Tissue Research Ban Be Lifted?" p. 17.

70. These questions were developed by a member of the Assistant Secretary's staff through an analysis of the existing literature and consultation with some academic bioethicists.

71. King, "Commentary," in *Biomedical Politics,* ed. Hanna, p. 252.

72. This is another area of overlap between principlists and casuists, as argued in other chapters in this volume. See Albert R. Jonsen's arguments in "Transplantation of Fetal Tissue: An Ethicist's Viewpoint," *Clinical Research* 36 (1988): 215–19. On the importance of framing and reframing problems, see Deborah A. Stone, *Policy Paradox and Political Reason* (San Francisco: HarperCollins, 1988).

73. Subcommittee on Health and the Environment, Committees on Energy and Commerce, U.S. House of Representatives, *Human Fetal Tissue Transplantation Research,* Hearings, April 2, 1990.

74. See LeRoy Walters, "Statement to the Advisory Committee to the Director, NIH," *Human Fetal Tissue Transplantation Research,* Report of the Advisory Committee to the Director, National Institutes of Health, December 14, 1988 (Bethesda, MD), p. C5. See also Walters, "Testimony on Human Fetal Tissue Research," Subcommittee on Health and the Environment, Committees on Energy and Commerce, U.S. House of Representatives, *Human Fetal Tissue Transplantation Research,* Hearings, April 2, 1990. These two statements are the sources for Walters's comments in subsequent paragraphs.

75. Walters's testimony in *Human Fetal Tissue Transplantation Research,* p. 15.

76. Mary Ann Glendon, "A World without Roe: How Different Would It Be?" *Hastings Center Report* 19 (July/August 1989): 30–31.

77. George J. Annas and Sherman Elias, "The Politics of Transplantation of Human Fetal Tissue," *New England Journal of Medicine* 320 (1989): 1079–82.

78. The use of the label "fetal research" to include HFTTR mars the otherwise helpful book by Maynard–Moody, *Dilemma of the Fetus.* See also the testimony of John Fletcher, U.S. Congress, *Fetal Tissue Transplantation Research,* pp. 130–47.

79. Maynard-Moody, *Dilemma of the Fetus,* chap. 12, "The Turning Point."

80. Statement of Guy Walden, U.S. Senate, Committee on Labor and Human Resources, Hearing, *Finding Medical Cures: The Promise of Fetal Tissue Transplantation Research,* 102nd Cong., 1st sess., November 21, 1991 (Washington, D.C.: U.S. Government Printing Office, 1992).

81. Ibid., p. 179.

82. Testimony of Assistant Secretary for Health James Mason, U.S. Congress, *Finding Medical Cures: The Promise of Fetal Tissue Transplantation Research,* pp. 4–8.

INDEX

Abortion, 304–305; analogical reasoning about, 20–22; culture wars over, 11, 12; international consensus on, 324–25; motivation for, 311–13; societal legitimation of, 309–11. *See also* Human fetal tissue in transplantation research

Abram, Dr. Harry S., 178, 349*n*25

Adams, Judge Arlin, 305, 325

AIDS: caring metaphor for, 117, 342*n*77; patient selection for experimental drugs, 191; punishment metaphor of, 116; warfare metaphor of, 7, 8, 10–11, 96, 116–18. *See also* HIV antibodies, screening for

Akiba, Rabbi, 182

Alexander, Shana, 180

Allocation of biomedical resources, 123, 170; analogies for, 19; for HIV-positive persons, 102; lifesaving drugs, 191, 200, 353*n*28; penicillin in the military, 170, 198, 202–203, 208–209, 352*n*18; and the "playing God" metaphor, 14; symbolic rationality in, 23, 332*n*67; triage as, 197; and the warfare metaphor, 8. *See also* Scarce Lifesaving Medical Resources, allocation of

Altruism: in abortion for fetal tissue donation, 312–13; in organ donation, 294–95, 296, 370*n*38

Ambrogi, D. M., 71

Ambrose, Saint, 181–82

Amman, Arthur, 110

Analogical reasoning, 3–4, 17–18; and casuistry, 19–20, 353*n*29; about fetal tissue research, 307; and HIV screening, 22, 96, 100, 108, 110, 111, 113; for ill newborns, 195; about maternal-fetal relations, 20–22; Nazi analogies, 18–19, 308, 322, 323; and patient selection, 181–83; for triage in neonatal ICUs, 201–202, 353*n*30, 353*n*31

Andrews, Lori, 288, 289, 291

Annas, George J., 10, 114, 226, 350*n*39

Anscombe, G. E. M., 142

Aquinas, Thomas, 19

Aristotle, 18, 74

Arras, John D., 39, 40

Artificial organs, 174, 226

Assisted suicide, 7, 15, 86, 124, 127

Attribution, analogy of, 19–20

Austin, J. L., 138

Authenticity, 63, 72, 128–29

Autonomy, 50; and authenticity, 63, 72, 128–29; as constraint and limit, 52–53; as effective deliberation, 126–27; as an end-state, 127–28; as freedom of action, 126, 127; of health care professionals, 14, 124; limits to, 123, 126–28; in negotiation between patient and physician, 51, 52–53; in the partnership model, 46; as a side-constraint, 127

Autonomy, principle of respect for personal (PRA), 25, 30, 33, 35, 59–60, 89–90; as the avoidance of controlling influences, 61; of cancer patients, 77–78; conditions for, 60–61; and consent, 62–64; cultural factors in, 91–94; and first-order autonomy, 60–61, 62; vs. the ideal of autonomy, 60; infringement of, 66–67; limits to, 60, 64–67, 72, 123, 126–28; and mandatory screening for HIV antibodies, 66–67, 101, 104; in nursing home residents, 68–75; and organ procurement, 271–72, 288–91; overconcentration on, 67–68; paternalism vs., 122–23, 125; "respect" element of, 61; and rules of liberal society, 97–98; and second-order autonomy, 61, 62

Autopsy, 34–35

Avendano, Dr. Alberto, 191

AZT (Zidovudine), 97, 109, 110, 111

and allowance for time and community, 123; and beneficence, 125, 132, 134, 153, 154–55, 160; and cardiopulmonary resuscitation, 84; and irony in case narration, 88; in a liberal society, 97–98; and obtaining organs for transplant, 267; social context of, 132–33. *See also* Autonomy, principle of respect for personal

Peters, Thomas G., 297

Physician-patient relations: in managed care programs, 259–60; shared decision making in, 78; trust in, 176–77

—metaphors and models of, 5, 13–14; friendship, 47–48; negotiation, 50–55, 336*n*23; partnership, 46–47, 335*n*4, 335*n*5; paternalism, 44–46; rational contracts, 47; technician, 48

"Playing God" metaphor, 14–15, 331*n*38

Please Let Me Die (videotape), 121, 127, 132, 138–40, 343*n*1, 344*n*32

Polkinghorne report, 325

Pray, Lawrence, 6

Pregnancy: analogical reasoning about, 20–22; HIV screening during, 107–12

President's Commission for the Study of Ethical Problems in Medicine and Biomedical and Behavioral Research, 234, 240, 241, 244–45, 249–50, 254, 258, 361*n*7

Price, Monroe, 116

Principles of Biomedical Ethics (PBE), 25–26; cases in, 39; casuistry-based critique of, 34–40; specification in, 38, 336*n*12; theory-based critique of, 30–34, 333*n*18

Principlism, 16, 25–26, 332*n*2; casuistry-based critique of, 26, 34–40; "ethics mantra" criticized in, 40–41; feminist/feminine critique of, 41–43; justificatory conditions for overriding principles of, 98–100; metaphor and analogy in, 3–4; and obtaining organs for transplant, 266–68; and patient selection, 181–83; and public health protection, 96–98; specified, 38; theory-based critique of, 26, 30–34, 333*n*16; and withholding artificial nutrition and hydration, 153–55

Prisoners: mandatory HIV testing of, 112–13

Privacy, rule of, 25, 65, 66–67, 70, 97; and HIV antibody screening, 101, 104; infringement of, 71–72, 73, 99–100

Property rights: and organ procurement, 285–95, 299

Proportion, analogy of, 20

Prostitutes: mandatory HIV testing of, 112–13

Prottas, Jeffrey M., 219, 294–95

Prudence: and a right to health care, 247–49, 362*n*33

Public health, protection of, 95–98; and the justificatory conditions for overriding principles, 98–100

Quay, P. M., 273

Quinlan, Karen, 145, 345*n*14

Radin, Margaret Jane, 290

Ramsey, Paul: on donation of organs, 266, 269, 276; on the focused community, 189, 190; on nutrition and hydration for the dying, 345*n*14; on patient selection, 19, 176, 180, 181, 183, 184, 185, 191; on playing God, 14, 331*n*38; on triage, 199–200, 354*n*37; on the universalizability principle, 15

Random selection of patients for scarce resources, 175–80, 181, 183–90, 348*n*22, 349*n*23, 354*n*33; in newborn ICUs, 204–206, 365*n*33; for organ transplants, 227–28, 359*n*47

Rationing of health care resources, 8, 238–39, 259–62, 360*n*2

Rawls, John, 40, 189, 225, 241, 245, 348*n*21

Realism, 133

Rebirth metaphor, 6, 7, 136

Reinhardt, Uwe E., 242

Refugees: mandatory HIV screening of, 113–14

Refusal of treatment by patients, 62–64, 121–22; discerning patient preferences over time, 128–32; and internal limits to patient autonomy, 125–28; and narrative approach to case studies, 133–39; and paternalism, 122–25; and the social context of patients, 132–33

Rescher, Nicholas, 178, 181, 184, 188, 349*n*23

Rettig, Richard, 8

Richardson, Henry S., 37–38

Right to health care, 239–41, 243–49, 360*n*5; costs of, 254–56; cost-benefit analysis of, 253–54, 363*n*49; and decent minimum of care, 252–54; and equal access, 250–52, 364*n*61; moral vs. political-legal, 242–43; optional vs. mandatory, 93; and risk-taking behavior, 255–56; utilitarian argument for, 362*n*27

Robertson, John, 314, 316, 322, 372*n*14

Robinson, Daniel N., 310

Ross, W. D., 89

Rothman, David J., 209

Rule utilitarianism, 157, 197, 351*n*15

Rules, 25, 37; justificatory conditions for overriding, 98–100; theory-based, 30, 33

Sale of organs for transplant, 234, 268, 282–84, 299–300; and autonomy, 288–91; and fail-

United Network for Organ Sharing (UNOS), 214, 217, 219, 220; heart allocation criteria, 226; and multi-organ transplants, 229–30; point system for kidneys, 222–23, 226, 228; waiting list policies, 221, 357n17
Universalizability, principle of, 15, 18, 37, 100
U.S. v. Holmes, 170–71, 172, 175, 176, 182–83, 190
Utilitarianism: act-utilitarianism, 197, 209, 351n15; and patient selection, 174–80, 183–84; and a right to health care, 362n27; rule-utilitarianism, 157, 197, 351n15; and the sale of organs, 288–89, 293; symbol utilitarianism, 24, 158, 160; triage justified in, 197–201, 210, 351n14, 352n23, 352n28
Utility: failed, 165; medical vs. social, 184–85, 198–203, 205, 207–12, 220, 224, 352n16

Veatch, Robert M., 12, 47, 48, 225, 226, 333n19, 335n5, 360n74, 367n18
Virtues, 42; virtuous case narration, 90–91

Waiting lists of transplantation candidates, 220–22, 227–28, 359n47
Walden, Guy and Terri, 326–27
Walters, LeRoy, 324, 375n74
Walzer, Michael, 247, 249, 286, 322
Wanglie, Helga, 163–64

Waxman, Henry, 324
Weber, Max, 23
White, Dr. Robert, 121, 124–27, 138
Wikler, Daniel, 224, 249
Wildes, Kevin W., 289
William Brown tragedy, 170–71, 172, 175, 176, 182–83, 190
Willrich, Mason, 179, 349n25
Windom, Robert, 302
Winslade, William, 35
Winslow, Gerald R., 184, 351n14, 351n15, 352n19, 354n34
Withholding/withdrawal of artificial nutrition and hydration, 141–43, 151–53, 345n14, 347n20; arguments against, 155–59; distancing devices used in, 161–62; inappropriate moral constraints on, 146–50, 345n14, 346n17; Nazi analogy, 18; obligatory conditions for, 159–61; principlist analysis of, 153–55; when a disproportionate burden, 145–46, 163; when no possibility of benefit, 145, 163; when treatment is futile, 144, 163–66
Wolchok, Carol, 113
Wyngaarden, James, 302

Zawacki, Bruce, 126
Zidovudine. *See* AZT
Zussman, Robert, 210–11, 356n55

JAMES F. CHILDRESS is Edwin B. Kyle Professor of Religious Studies and Professor of Medical Education at the University of Virginia, where he also codirects the Virginia Health Policy Center. He served as chair of the Department of Religious Studies for over a decade. He is the author of over a hundred articles, many of them in biomedical ethics, and several books, including *Principles of Biomedical Ethics* (with Tom L. Beauchamp), now in its fourth edition; *Who Should Decide? Paternalism in Health Care*; and *Priorities in Biomedical Ethics.* He was Joseph P. Kennedy, Sr., Professor at the Kennedy Institute of Ethics from 1975 to 1979, and he has also been a visiting professor at Princeton University and the University of Chicago Divinity School.

Dr. Childress was vice-chair of the federal Task Force on Organ Transplantation, and he has served on the Board of Directors of the United Network for Organ Sharing (UNOS) and its Ethics Committee, on the Human Fetal Tissue Transplantation Research Panel, on the Recombinant DNA Advisory Committee of the National Institutes of Health, the Human Gene Therapy Subcommittee, the Data and Safety Monitoring Board for AIDS Drug Trials, as well as with several other governmental committees and advisory groups. In July 1996 President Clinton appointed Dr. Childress to the new National Bioethics Advisory Commission.